Computational Fluid Dynamics - Analysis, Simulations, and Applications

Edited by Mahboub Baccouch

Published in London, United Kingdom

Computational Fluid Dynamics - Analysis, Simulations, and Applications
http://dx.doi.org/10.5772/intechopen.1003390
Edited by Mahboub Baccouch

Contributors
Abdelrahman K. Hamed, Ahmed A. Abuhatira, Alessandro Masci, Alexandre Bonatto, Camilla Cortesi, Corrado Tomasi, Cristiana Corsi, Cristian Bontoiu, Gengxin Zhang, Guoxing Xia, Haytham F. Isleem, Javier Resta-López, Jing Sun, Keagan Church-Forbes, Mahboub Baccouch, Matteo Falanga, Mohamed K. Elshaarawy, Mohammad Zamir A. Koshuriyan, Nadia D'Alessandro, Neelakantan Padmanabhan, Pablo Martín-Luna, Penghua Shi, Roel Verstappen, Rukiye Kara, Sachal Hussain, Yingjie Tang, Yu Jin

First published in London, United Kingdom, 2025 by IntechOpen
IntechOpen is the global imprint of INTECHOPEN LIMITED, registered in England and Wales, registration number: 11086078, 167-169 Great Portland Street, London, W1W 5PF, United Kingdom

For EU product safety concerns: IN TECH d.o.o., Prolaz Marije Krucifikse Kozulić 3, 51000 Rijeka, Croatia, info@intechopen.com or visit our website at intechopen.com.

British Library Cataloguing-in-Publication Data
A catalogue record for this book is available from the British Library

Computational Fluid Dynamics - Analysis, Simulations, and Applications
Edited by Mahboub Baccouch
p. cm.
Print ISBN 978-0-85014-850-3
Online ISBN 978-0-85014-849-7
eBook (PDF) ISBN 978-0-85014-851-0

Meet the editor

Mahboub Baccouch is a Professor in the Department of Mathematical and Statistical Sciences at the University of Nebraska at Omaha (UNO) in the USA. He earned his Ph.D. in Mathematics from Virginia Tech in 2008 and joined UNO the same year as a tenure-track assistant professor. He was promoted to associate professor with tenure in 2014 and to full professor in 2019. Dr. Baccouch specializes in computational and applied mathematics, focusing on developing advanced numerical methods for solving deterministic and stochastic differential equations. His research emphasizes higher-order methods, such as finite element and discontinuous Galerkin methods, applied to problems in engineering, biology, and financial mathematics. He has an extensive publication record, with more than 75 papers in leading journals of numerical analysis, numerical methods, and scientific computing. Additionally, he has authored four book chapters and edited three textbooks. Dr. Baccouch has been recognized for his contributions to research and teaching. He received awards such as the College of Arts and Sciences Excellence in Research Award in 2017 and the Kerrigan Award for Excellence in Teaching Mathematics in 2014. Over his career, he has taught a wide range of undergraduate and graduate courses and continues to inspire the next generation of mathematicians.

Contents

Preface

The field of Computational Fluid Dynamics (CFD) stands as a cornerstone in modern science and engineering, enabling the analysis, simulation, and application of fluid flow phenomena with unprecedented accuracy. By combining mathematical modeling, numerical techniques, and high-performance computing, CFD empowers engineers and researchers to tackle complex fluid systems that were once beyond reach. From optimizing industrial designs to advancing biomedical applications and environmental sustainability, CFD continues to reshape our understanding of fluid behavior and its role across various domains.

This textbook, "Computational Fluid Dynamics – Analysis, Simulations, and Applications", aims to provide a comprehensive and accessible resource for students, researchers, and practitioners. It balances theoretical foundations, advanced numerical techniques, and practical applications, making it suitable for both academic study and professional development. Each chapter addresses cutting-edge developments and challenges in CFD, delivering insights into its transformative capabilities across diverse fields.

Below is a summary of the ten chapters featured in this book, highlighting the depth and breadth of topics covered:

1. Numerical Methods for the Viscid and Inviscid Burgers Equations
By Mahboub Baccouch

This chapter explores the viscid and inviscid Burgers' equations, fundamental models in fluid dynamics and shock wave theory. It introduces key mathematical concepts such as weak solutions, entropy conditions, and numerical approximations. Methods like finite difference, finite volume, and discontinuous Galerkin techniques are detailed, offering readers a solid foundation for solving nonlinear conservation laws.

2. Fundamentals and Theory of Large Eddy Simulation
By Rukiye Kara

This chapter provides a detailed comparison of RANS, LES, and Direct Numerical Simulation (DNS) in turbulence modeling. It explains the theoretical distinctions between these methods, emphasizing their strengths and limitations in capturing turbulent flows. Practical applications, computational complexities, and the future potential of LES in engineering disciplines are thoroughly explored, offering readers a solid understanding of turbulence modeling.

3. A Numerical Study of the Minimum-Dissipation Model for Large-Eddy Simulation in OpenFOAM
By Jing Sun and Roel Verstappen

The chapter presents a numerical study of the minimum-dissipation (QR) model applied to turbulent channel flows and other complex scenarios using OpenFOAM. The QR model's performance is evaluated against DNS and experimental data, demonstrating its computational efficiency and accuracy. Special attention is given to the role of discretization methods, with symmetry-preserving schemes outperforming traditional approaches in predicting turbulence dynamics.

4. Recent Advancements in Large Eddy Simulations of Compressible Real Gas Flows
By Neelakantan Padmanabhan

This chapter delves into the recent progress in Large Eddy Simulations (LES) for compressible real gas flows. It begins by outlining the scales of turbulence and the advantages of LES over alternative methods like Reynolds-Averaged Navier-Stokes (RANS). The chapter then explores the governing equations for LES, focusing on modifications for real gases. Advanced filtering techniques and subgrid-scale modeling are discussed, including stochastic methods and machine-learning approaches. These innovations push the boundaries of LES, making it more robust for complex real-gas applications.

5. Heat Transfer and Thermal Analysis with Computational Fluid Dynamics
By Yingjie Tang

Focusing on heat transfer and thermal management, this chapter illustrates how CFD tools evaluate conduction, convection, and radiation mechanisms. Practical applications, including heat exchangers and power systems, demonstrate how CFD aids in optimizing designs, preventing thermal hotspots, and reducing prototype testing costs. The chapter provides essential guidance for integrating CFD into thermal analysis workflows.

6. Plasmonic Excitations in Carbon Nanotubes: PIC Simulations vs Hydrodynamic Model
By Pablo Martín-Luna, Alexandre Bonatto, Cristian Bontoiu, Guoxing Xia and Javier Resta-López

This chapter examines the electromagnetic plasmonic modes in carbon nanotubes (CNTs), which hold promise for particle acceleration technologies. The research highlights their strengths, limitations, and compatibility by comparing particle-in-cell (PIC) simulations and hydrodynamic models. This work opens new avenues for high-energy physics and nanotechnology applications.

7. Influence of Working Fluid on the Mean Flow Vortex Structure in Industrial Flow Pipelines of Arbitrary Bend Angle via RANS Simulation
By Keagan Church-Forbes, Mohammad Zamir A. Koshuriyan and Ahmed A. Abuhatira

This chapter investigates the effects of working fluids and bend angles on vortex structures in industrial pipelines using Reynolds-Averaged Navier-Stokes (RANS) simulations. The findings provide insights into flow properties, turbulence behavior, and wall shear stress for various gases, offering practical guidance for optimizing pipeline designs in the oil and gas industry.

8. Analysis of Flow Dynamics and Energy Dissipation in Piano Key and Labyrinth Weirs Using Computational Fluid Dynamics
By Haytham F. Isleem, Mohamed K. Elshaarawy and Abdelrahman K. Hamed

This chapter investigates the flow dynamics and energy dissipation of Piano Key Weirs (PKWs) and Labyrinth Weirs (LWs). Using FLOW-3D and ANSYS models, it evaluates the impact of geometric parameters on discharge efficiency and energy dissipation. Design optimizations, such as trapezoidal PKWs and adjusted overhang placements, are proposed to enhance performance, making this chapter a valuable guide for engineers in flood control and reservoir management.

9. Computational Fluid Dynamics Simulations to Deepen Understanding of the Hemodynamic Underlying Atrial Fibrillation and Improve Therapeutic Approaches
By Camilla Cortesi, Matteo Falanga, Sachal Hussain, Nadia D'Alessandro, Corrado Tomasi, Alessandro Masci and Cristiana Corsi

This chapter explores patient-specific CFD simulations to study atrial fibrillation (AF). It identifies key fluid dynamic parameters, such as wall shear stress and endothelial cell activation, influencing disease progression and treatment outcomes. The research underscores CFD's potential in precision medicine for improving stroke risk prediction and therapeutic strategies.

10. Applications of Particle Image Velocimetry in Internal Combustion Engines Research
By Gengxin Zhang, Penghua Shi and Yu Jin

Particle image velocimetry (PIV) is a critical internal combustion engine research tool. This chapter reviews its applications in fuel injection, combustion processes, and in-cylinder flow studies. Challenges like tracer particle selection and flow field complexity are addressed with innovative solutions like AI-enhanced data processing. The chapter emphasizes the role of PIV in advancing carbon-neutral engine technologies.

Through these chapters, this textbook aspires to bridge theoretical principles with practical applications, providing readers with the tools to navigate the complexities of computational fluid dynamics. Whether you are a student, researcher, or practitioner, this book offers a comprehensive resource to deepen your knowledge and advance your expertise in CFD.

Mahboub Baccouch
Department of Mathematical and Statistical Sciences,
University of Nebraska at Omaha,
Omaha, Nebraska, USA

Chapter 1

Numerical Methods for the Viscid and Inviscid Burgers Equations

Mahboub Baccouch

Abstract

The Burgers equation is the simplest equation combining both nonlinear propagation effects and diffusive effects. It describes a weak shock phenomena in gas dynamics. This equation has been a center of interest for researchers studying various physical phenomena such as theory of shock waves, fluid dynamics, turbulent flow and gas dynamics. In this chapter, the viscid and inviscid Burgers equations are studied. Two important mathematical models described by the Burgers equations are presented. Then a very brief review of theoretical results for conservation laws are provided. In particular, the characteristic method is used to demonstrate the development of shocks. Furthermore, the notion of weak solutions, state the Rankine-Hugoniot jump condition, and introduce the notion of an entropy condition are introduced. Finally, several important numerical methods to approximate the solution to the viscid and inviscid Burgers equations are developed. For the inviscid Burgers equation, the classical finite difference and finite volume methods are presented in details including Up-wind nonconservative, Up-wind conservative, Lax-Friedrichs, Lax-Wendroff, MacCormack, Godunov methods. In addition, the discontinuous Galerkin method is described in details. For the viscid Burgers equation, a finite difference method, a Chebyshev collocation method, an ultra-weak discontinuous Galerkin method, and a local discontinuous Galerkin method are presented.

Keywords: viscid and inviscid Burgers equation, finite difference method, finite volume method, discontinuous Galerkin method, computational fluid dynamics, fluid mechanics, gas dynamics, traffic flow

1. Introduction

Burgers' equation or Bateman-Burgers equation is a fundamental partial differential equation (PDE) occurring in various areas of applied mathematics, such as fluid mechanics, nonlinear acoustics, gas dynamics, and traffic flow. The typical solutions to Burgers equations may exhibit boundary or interior layers. The design and development of accurate, efficient, and robust numerical schemes to address these difficulties has survived as a classical challenge since the earliest days of digital computation. For convection problems, the main reasons for the limitations involve the development of discontinuous solution (called shocks) in finite time even from smooth initial data. For convection-diffusion problems, the main reasons for the difficulties involve the development of sharp transition layers. Another difficulty involves the design of numerical approximations that satisfy physical principles such as the conservation of energy. An active research direction now is the search for more robust and high order numerical methods.

In this chapter, the viscid and inviscid Burgers equations are analyzed. The numerical methods presented here are applicable for other PDEs. The one-dimensional viscid Burgers equation is a nonlinear PDE

$$u_t + uu_x = \nu u_{xx}, \tag{1}$$

where $\nu > 0$ is called the viscosity. When $\nu = 0$, (1) reduces to the inviscid Burgers equation

$$u_t + uu_x = 0, \tag{2}$$

which is a prototype for conservation equations that can develop discontinuities (or shock waves). Notice that (2) can be rewritten in the following scalar hyperbolic conservation law form

$$u_t + (f(u))_x = 0, \tag{3}$$

where $f(u) = \frac{u^2}{2}$. The nonconservative form of (3) is

$$u_t + f'(u)u_x = 0. \tag{4}$$

The classical Burgers Eq. (3) models wave motion, where $u(x,t)$ is the height of the wave at point x and time t. It has been a center of interest for researchers studying various physical phenomena such as theory of shock waves, fluid dynamics, turbulent flow and gas dynamics [1, 2]. In the next section, we show that the Navier-Stokes equations reduce to the viscid Burgers equation.

2. Derivation of Burgers equations

We may obtain Burgers's equation as a simplified version of the Navier-Stokes equations. The basic equations that govern the viscid flow of simple substances like air and water under normal conditions are called the Navier-Stokes equations (nonlinear system of PDEs):

$$\rho(\mathbf{u}_t + \mathbf{u} \cdot \nabla \mathbf{u}) = -\nabla p + \mu \Delta \mathbf{u} + \mathbf{F}, \tag{5}$$

where $\mathbf{u}$ is the velocity of the fluid, p is the fluid pressure, ρ is the fluid density, μ is the fluid dynamic viscosity, and $\mathbf{F}$ is an external force. The Kinematic viscosity is $\nu = \frac{\mu}{\rho}$. The term $\rho(\mathbf{u}_t + \mathbf{u} \cdot \nabla \mathbf{u})$ corresponds to the inertial forces, $-\nabla p$ corresponds to pressure forces, $\mu \Delta \mathbf{u}$ corresponds to viscid forces, and $\mathbf{F}$ corresponds to the external forces applied to the fluid. The Navier-Stokes equations were derived by Navier, Poisson, Saint-Venant, and Stokes between 1827 and 1845. These equations are always solved together with the continuity equation $\rho_t + \nabla \cdot (\rho \mathbf{u}) = 0$. For incompressible flows (density of the fluid is constant), the continuity equation yields $\nabla \cdot \mathbf{u} = 0$. The

Navier-Stokes equations represent the conservation of momentum, while the continuity equation represents the conservation of mass.

If we drop the pressure term we get the Burgers' equation

$$\rho(\mathbf{u}_t + \mathbf{u} \cdot \nabla \mathbf{u}) = \mu \Delta \mathbf{u} + \mathbf{F}. \tag{6}$$

When the external force **F** is equal to zero and when ρ is a constant (for an incompressible fluid), the above Eq. (6) simplifies to

$$\mathbf{u}_t + \mathbf{u} \cdot \nabla \mathbf{u} = \nu \Delta \mathbf{u}, \tag{7}$$

where the new constant $\nu = \mu/\rho$ is called the kinematic viscosity. An even further simplification arises when we assume the viscosity is zero. Then we obtain the inviscid Burgers' equation

$$\mathbf{u}_t + \mathbf{u} \cdot \nabla \mathbf{u} = \mathbf{0}. \tag{8}$$

If one consider a one-dimensional problem with no pressure gradient, the Eq. (7) reduces to the viscid Burgers Eq. (1). When the viscosity ν of the fluid is zero, the Eq. (8) reduces to the inviscid Burgers Eq. (2).

Burgers equations arise in the modeling of many physical phenomena, such as the mathematical model of turbulence [1], the traffic flow, the approximate theory of flow through a shock wave traveling in viscid fluid [2], and many others. For instance, traffic flow along a long road can be described by the Burgers equation. It was developed in [3, 4] and it was discussed more extensively in [5].

3. Inviscid Burgers' equation and the development of shocks

In this section, we consider the following Cauchy problem

$$\begin{cases} u_t + (f(u))_x = 0, & x \in \mathbb{R},\ t > 0, \\ u(x, 0) = u_0(x), & x \in \mathbb{R}, \end{cases} \tag{9}$$

where $f(u) = u^2/2$ and u_0 is the initial data.

3.1 Characteristic method

The Cauchy problem (9) can be solved using the so-called method of characteristics. Let the initial curve Γ be parameterized by s, i.e., $x(s, 0) = s$, $t(s, 0) = 0$, $u(s, 0) = u_0(s)$. The characteristic equations for the curves $(x(s, r), t(s, r), u(s, r))$ are

$$\frac{\partial x}{\partial r} = u,\ \frac{\partial t}{\partial r} = 1,\ \frac{\partial u}{\partial r} = 0, \text{ subject to } x(s, 0) = s,\ \ t(s, 0) = 0,\ \ u(s, 0) = u_0(s).$$

These yield $t = r$, $u(s, r) = u_0(s)$, $x = u_0(s)r + s$. We note that (s, r) can be expressed in terms of (x, t) when the Jacobian of this transformation is nonsingular i.e.,

$$J = \begin{vmatrix} x_s & x_r \\ t_s & t_r \end{vmatrix} = \begin{vmatrix} u_0'(s)r + 1 & u_0(s) \\ 0 & 1 \end{vmatrix} = u_0'(s)r + 1 \neq 0.$$

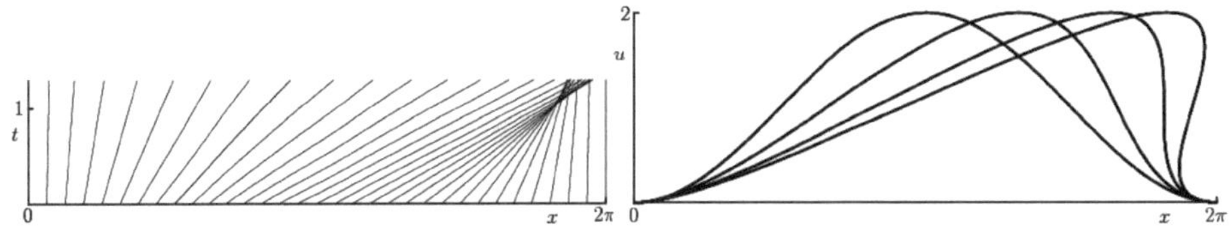

Figure 1.
Characteristics (left) and profiles at times t = 0, .5, 1, and 1.3 (right) for Burgers' equation with $u_0(x) = 1 - \cos(x)$.

Provided $t = r \neq -1/u_0'(s)$ we can find x, t in terms of s, r. Since at $r = 0$ we have $J = 1 \neq 0$, which shows that the initial curve Γ is noncharacteristic. Thus, at least for small t, we can solve for x, t in terms of s, r. Since $r = t$ and $s = x - ut$, we obtain an implicit solution $u(x,t) = u_0(x - tu(x,t))$. The curves parametrized by $(x = u_0(s)r + s, t = r)$ in the xt-plane are called the *projected characteristic curves*. For fixed s, the equations of these curves are $x = s + u_0(s)t$, which are lines with slopes $\frac{dt}{dx} = \frac{1}{u_0(s)}$. Moreover $u = u_0(s)$ has the constant value $u_0(s)$ along the line $x = s + u_0(s)t$.

3.2 The development of shocks

Considering two characteristics lines $x = u_0(s_1)r + s_1$ and $x = u_0(s_2)r + s_2$ passing through $(s_1, 0)$ and $(s_2, 0)$ with $s_2 > s_1$. They intersect at some point P at positive time obtained by solving $u_0(s_1)r + s_1 = u_0(s_2)r + s_2$ for r to get $t_B = \frac{s_2 - s_1}{u_0(s_1) - u_0(s_2)}$. Depending on u_0, the characteristic lines may or may not cross. There are 3 cases:

Case 1: If $u_0(s)$ =constant, then the characteristic lines are parallel. In this case the solution will exist globally for $t > 0$.

Case 2: Suppose there is an $s_1 < s_2$ such that $u_0(s_1) < u_0(s_2)$. Then $t_B < 0$ and the projected characteristic lines fan out. In this case the solution will exist globally for $t > 0$. The solution is like a fan and it is called a *rarefaction wave*.

Case 3: Suppose there is an $s_1 < s_2$ such that $u_0(s_1) > u_0(s_2)$. In this case, $t_B > 0$ and the projected characteristic lines intersect at $(x = u_0(s_1)r + s_1, r)$, where $r = \frac{s_2 - s_1}{u_0(s_1) - u_0(s_2)} > 0$. This will result in a difficulty in defining the solution beyond the point of intersection. This phenomenon occurs as a result of wave breaking. Geometrically, even if $u(x,0) = u_0(x)$ is sufficiently smooth, the surface $u = u(x,t)$ is folded over on itself (into a backward S-shape) at some point (x_1, t_1) and the solution $u(x,t)$ becomes multi-valued for $t > t_1$ for some critical t_1 that depends on u_0. This phenomenon is illustrated in **Figure 1** for the initial condition $u_0(x) = 1 - \cos(x)$. The implicit solution $u(x,t) = u_0(x - tu(x,t))$ is no longer valid. The solution then continues with a shock wave. The solution is not even continuous at the shock, but the solution still makes sense, because the PDE expresses a conservation law and the shock preserves conservation. We note that u_x and u_t approach infinity at the positive time as $t \to -\frac{1}{u_0'(s)}$ and the minimum time for which the solution becomes multi-valued is $t_B = -\frac{1}{\min u_0'(s)}$. The solution suffers a gradient catastrophe type of singularity (u_x and u_t become discontinuous). Such a jump discontinuities are called *shocks*. The solution fails to exist globally.

3.3 Weak solution

As we showed in the previous subsection, smooth solutions of conservation laws can blow up in finite time. After the shock, the notion of classical solution fails and a new concept is needed. The only way to establish a solution after the breaking time is to allow discontinuities of u. This type of solution is known as a weak solution which means we have to consider solutions in the sense of distributions.

Definition 3.1 We say that u is a weak solution of (9) if

$$\int_0^\infty \int_{-\infty}^\infty (uv_t + f(u)v_x)dxdt + \int_{-\infty}^\infty u_0(x)v(x,0)dx = 0, \tag{10}$$

for any smooth function $v \in C^\infty(\mathbb{R} \times [0,\infty))$ with compact support.

The idea of weak solution allows for solutions which need not be differentiable or even continuous. We would lie to mention that weak solutions have some restrictions on types of discontinuities. To illustrate this, we assume u is a weak solution of (9) and is discontinuous across a curve defined by $x = s(t)$, but u is smooth on both sides of the curve $s(t)$. We use $u^-(x,t)$ to denote the limit of $u(x,t)$ from the left. We also use $u^+(x,t)$ to denote the limit of $u(x,t)$ from the right. It turns out that the curve $x = s(t)$ cannot be any curve. We present the relation between $x = s(t)$, u^- and u^+ in the next theorem.

Theorem 1.1 If u is a weak solution of (9) such that u is discontinuous across the curve $x = s(t)$, but u is smooth on either side of $x = s(t)$, then u must satisfy the following condition, called the *Rankine-Hugoniot jump condition*, across the curve of discontinuity

$$s'(t) = \frac{f(u^+) - f(u^-)}{u^+ - u^-} = \frac{[f(u)]}{[u]}, \tag{11}$$

where $[f(u)]$ and $[u]$ are, respectively, used to denote the jumps of $f(u)$ and u across the discontinuity curve and $s'(t)$ is the speed of the curve of discontinuity.

Consequently, if we assume that u is a weak solution with discontinuity along a curve $x = s(t)$, then the solution must satisfy the condition (11). We note that the weak solutions defined in (10) are not unique. A necessary condition to the jump discontinuity is given by the Rankine-Hugoniot condition (11). This condition can be used to give a discontinuous solution u. However, it turns out that there are still some situations in which this concept is not enough to guarantee the uniqueness of the solution after the shock. It turns out that solutions of quasilinear equations of the form (10) which satisfy an *entropy condition* are considered more physically realistic [6]. It is one of the most widely used. The entropy condition is introduced to eliminate nonphysical weak solutions. There is a number of variations in which this condition can be presented, we will mention only the simplest one.

For the Eq. (3), to select solutions which do have a physical sense, we only allow for a curve of discontinuity in our solution $u = u(x,t)$ if the wave to the left is moving faster than the wave to the right. In other words, we only allow for a curve of discontinuity between u^- and u^+ if $f'(u^+) < s' < f'(u^-)$. This condition is known as the entropy condition. A curve of discontinuity is said to be a shock curve for a solution $u(x,t)$ if the curve satisfies the Rankine-Hugoniot jump condition and the entropy condition for that solution $u(x,t)$. Thus, to select solutions which do have a physical sense and eliminate the physically less realistic solutions, we consider solutions u for which curves of discontinuity in the solution are shock curves. For the classical Burgers equation, the entropy condition reduces to the requirement that if a discontinuity is propagating with speed $s' = \frac{ds}{dt}$ then $u^+ < u^-$. As its name suggests, this entropy condition is the mathematical translation of the condition that indicates that in every physical process the entropy of the system is nondecreasing. It is well-known this is a fundamental assumption in thermodynamics.

4. Viscid Burgers' equation

In this section, we consider the following initial-value problem

$$\begin{cases} u_t + uu_x = \nu u_{xx}, & x \in \mathbb{R},\ t > 0,\ \nu > 0, \\ u(x,0) = u_0(x), & x \in \mathbb{R}. \end{cases} \tag{12}$$

This problem ban be solved using the Cole-Hopf transformation. This transformation defined by $u = -2\nu\frac{\phi_x}{\phi}$, named after Eberhard Hopf, [7], and Julian D. Cole, [2], is to transform the strongly nonlinear Burgers Eq. (12) to the linear initial-value problem for the heat equation

$$\begin{cases} \phi_t = \nu\phi_{xx}, & x \in \mathbb{R},\ t > 0,\ \nu > 0, \\ \phi(x,0) = \phi_0(x) = e^{-\int_0^x \frac{u_0(y)}{2\nu}dy}, & x \in \mathbb{R}. \end{cases} \tag{13}$$

The unique solution of (13) is $\phi(x,t) = \frac{1}{\sqrt{4\pi\nu t}}\int_{-\infty}^{\infty}\phi_0(y)e^{-\frac{(x-y)^2}{4\nu t}}dy$. Consequently, we obtain the analytic solution for the problem (12)

$$u(x,t) = \frac{\displaystyle\int_{-\infty}^{\infty}\frac{x-y}{t}\phi_0(y)e^{-\frac{(x-y)^2}{4\nu t}}dy}{\displaystyle\int_{-\infty}^{\infty}\phi_0(y)e^{-\frac{(x-y)^2}{4\nu t}}dy}.$$

5. Numerical methods for the inviscid Burgers equation

Consider the one-dimensional hyperbolic partial differential equation for $u(x,t)$ of the form (3) or (4) on the domain $x \in [a,b]$, $t \in [0,T]$ with initial condition $u(x,0) = u_0(x)$ and the boundary conditions $u(a,t) = u_1(t)$, $u(b,t) = u_2(t)$. For Burgers equation we have $f(u) = \frac{u^2}{2}$. We discretize the domain $(a,b) \times (0,T)$ to a grid with equally spaced points with a spacing of h in the x-direction and k in the t-direction, we define $u_i^n = u(x_i,t_n)$ with $x_i = a + ih$, $t_n = nk$ for $i = 0, 1, \ldots, N$ and $n = 0, 1, \ldots, M$, where $N = \frac{b-a}{h}$ and $M = \frac{T}{k}$ are integers representing the number of grid intervals.

Next, we define the notion of conservative numerical methods.

Definition 5.1 A numerical scheme is called conservative if it has the form

$$u_i^{n+1} = u_i^n - \frac{k}{h}\left(\hat{f}\left(u_i^n, u_{i+1}^n\right) - \hat{f}\left(u_{i-1}^n, u_i^n\right)\right), \tag{14}$$

where $\hat{f}$ is Lipschitz continuous and consistent *i.e.*, $\hat{f}(u,u) = f(u)$.

Consistency is a property required for a numerical method. A numerical method is said to be consistent if the local errors introduced at each step tend to zero when the discretization steps tend to zero. A numerical method is said to be convergent when the method is stable and consistent, which means that the discrete solution tends to the exact continuous one when the time step tends to zero. However, consistency is a necessary but not sufficient condition for solution convergence.

For time and space dependent problems, the estimation of error in time and space is often independent and has the form $|u(x_i,t_n) - \hat{u}_i^n| \le C_1h^p + C_2k^q$, where C_1 and C_2

are constants independent of the step sizes h and k. The numbers p and q are called the orders of convergence in space and time, respectively. We say that the method is p-th order accurate in space and q-th order accurate in time.

For some time-dependent PDEs and numerical schemes, we have a constraint in the choice of h and k (called the CFL condition) in order to obtain a stable method.

Definition 5.2 Any numerical scheme of the form (14) is stable if the so-called Courant-Friedrichs-Levy (CFL) condition $\sup_{x\in[a,b],t\in[0,T]}|f'(u(x,t))|\frac{k}{h}\leq 1$ is satisfied.

We note that the CFL condition is necessary but not sufficient in order to have a convergent numerical scheme.

5.1 Upwind nonconservative method

We consider the inviscid Burgers equation in the quasilinear form $u_t + uu_x = 0$. A natural finite difference method obtained by a forward discretization in time $u_t(x_i,t_n)\approx\frac{u_i^{n+1}-u_i^n}{k}$ and backward discretization in space $u_x(x_i,t_n)\approx\frac{u_i^n-u_{i-1}^n}{h}$ is

$$\frac{u_i^{n+1}-u_i^n}{k}+u_i^n\frac{u_i^n-u_{i-1}^n}{h}=0 \quad \text{or} \quad u_i^{n+1}=u_i^n-\frac{k}{h}u_i^n\left(u_i^n-u_{i-1}^n\right), \tag{15}$$

where the initial values and boundary nodes are taken from $u_i^0 = u_0(x_i)$, $u_0^n = u_1(t_n)$, and $u_N^n = u_2(t_n)$. This method is called the upwind nonconservative scheme. Although this method is consistent with (9) and is adequate for smooth solutions, it will not converge in general to a discontinuous weak solution as the grid is refine.

5.2 Upwind conservative method

Here, we consider the inviscid Burgers equation in conservation form $u_t + (f(u))_x = 0$. A finite difference method obtained by a forward in time $u_t(x_i,t_n)\approx\frac{u_i^{n+1}-u_i^n}{k}$ and backward in space $(f(u(x_i,t_n)))_x\approx\frac{f(u_i^n)-f(u_{i-1}^n)}{h}$ discretization of the derivatives is

$$\frac{u_i^{n+1}-u_i^n}{k}+\frac{f(u_i^n)-f(u_{i-1}^n)}{h}=0 \quad \text{or} \quad u_i^{n+1}=u_i^n-\frac{k}{h}\left(f(u_i^n)-f(u_{i-1}^n)\right), \tag{16}$$

where the initial values and boundary nodes are taken from $u_i^0 = u_0(x_i)$, $u_0^n = u_1(t^n)$, and $u_N^n = u_2(t^n)$. This method is called the upwind conservative scheme and is consistent and convergent.

5.3 Lax-Friedrichs method

The Lax-Friedrichs method for (3) takes the form

$$u_i^{n+1}=\frac{1}{2}\left(u_{i-1}^n+u_{i+1}^n\right)-\frac{k}{2h}\left(f(u_{i+1}^n)-f(u_{i-1}^n)\right), \tag{17}$$

where the initial values and boundary nodes are taken from $u_i^0 = u_0(x_i)$, $u_0^n = u_1(t_n)$, and $u_N^n = u_2(t_n)$. This scheme can be derived by using a finite difference method in time $u_t(x_i, t_n) \approx \frac{u_{i+1}^{n+1} - \frac{1}{2}(u_{i+1}^n + u_{i-1}^n)}{k}$ and a centred difference method in space $(f(u(x_i, t_n)))_x \approx \frac{f(u_{i+1}^n) - f(u_{i-1}^n)}{2h}$. Under the CFL condition, this method is explicit and first-order accurate (first order accurate in time and first order accurate in space). In addition, this method is conservative and hence quite dissipative. We note that this method can be written in conservation form:

$$u_i^{n+1} = u_i^n - \frac{k}{h}\left(\hat{f}_{i+1/2}^n - \hat{f}_{i-1/2}^n\right),$$

where $\hat{f}_{i-1/2}^n = \frac{1}{2}(f(u_{i-1}^n) + f(u_i^n)) - \frac{h}{2k}(u_i^n - u_{i-1}^n)$. Without the extra terms u_i^n and u_{i-1}^n in the discrete flux, $\hat{f}_{i-1/2}^n$, one ends up with the following Forward Time Centered Space (FTFS) scheme, which is well-known to be unconditionally unstable for hyperbolic problems

$$u_i^{n+1} = u_i^n - \frac{k}{2h}(f(u_{i+1}^n) - f(u_{i-1}^n)). \tag{18}$$

5.4 Godunov method

Godunov's scheme is a conservative numerical scheme, proposed by S. K. Godunov in 1959 [8], for solving PDEs

$$u_i^{n+1} = u_i^n - \frac{k}{h}(F(u_i^n, u_{i+1}^n) - F(u_{i-1}^n, u_i^n)), \tag{19}$$

where the numerical flux F is defined by

$$F(u_l, u_r) = \begin{cases} \min\limits_{u_l \le u \le u_r} f(u) & \text{if } u_l \le u_r, \\ \max\limits_{u_r \le u \le u_l} f(u) & \text{if } u_l > u_r. \end{cases} \tag{20}$$

This method can be viewed as a conservative finite volume method, which solves exact, or approximate Riemann problems at each inter-cell boundary. We would like to mention that Godunov's method is first-order accurate scheme in both space and time. Furthermore, it can be used as a base scheme for developing higher-order methods. We skip the details to save space.

5.5 Lax-Wendroff method

The Lax-Wendroff method to nonlinear conservation laws takes the form

$$u_i^{n+1} = u_{i+1}^n - \frac{k}{2h}(f(u_{i+1}^n) - f(u_{i-1}^n)) + \frac{k^2}{2h^2}\left(A_{i+\frac{1}{2}}(f(u_{i+1}^n) - f(u_i^n)) - A_{i-\frac{1}{2}}(f(u_i^n) - f(u_{i-1}^n))\right), \tag{21}$$

where $A_{i\pm\frac{1}{2}} = f'\left(\frac{u_i^n + u_{i\pm1}^n}{2}\right)$. The Lax-Wendroff method is second order accurate in space and time.

5.6 MacCormack method

MacCormack method is a second-order finite difference method. It is a widely used discretization scheme for the numerical solution of hyperbolic PDEs arizing in computational fluid dynamics. The method was introduced by Robert W. MacCormack in 1969 [9]. The MacCormack method is a variation of the Lax-Wendroff scheme (21). It proceeds in two steps: a predictor step which is followed by a corrector step

$$\overline{u}_i^n = u_i^n - \frac{k}{h}\left(f\left(u_{i+1}^n\right) - f\left(u_i^n\right)\right) \tag{22}$$

$$u_i^{n+1} = \frac{1}{2}\left(u_i^n + \overline{u}_i^n\right) - \frac{k}{2h}\left(f\left(\overline{u}_i^n\right) - f\left(\overline{u}_{i-1}^n\right)\right). \tag{23}$$

This method uses first forward difference method and then backward difference method to achieve second-order accuracy. We remark that, for linear equations, the MacCormack scheme is equivalent to the Lax-Wendroff method.

5.7 The DG method for hyperbolic conservation laws

Here, we present the discontinuous Galerkin (DG) method for hyperbolic conservation laws. DG methods are a class of finite element methods. They use discontinuous basis functions (usually chosen as piecewise polynomials). Unlike the basis functions for finite element methods, the basis functions for DG methods can be discontinuous. Therefore, these methods have the flexibility which is not shared by typical finite element methods including high order accuracy, geometric flexibility, suitability for $h-$ and $p-$ adaptivity, extremely local data structure, high parallel efficiency and a good theoretical foundation for stability and error estimates. These methods have found their way into the main stream of computational fluid dynamics and other areas of applications. The first DG method was introduced in 1973 by Reed and Hill [10] for neutron transport equations, and since that time there has been an active development of DG methods for many PDEs. A major development of the DG method is the Runge-Kutta DG (RKDG) framework introduced for solving nonlinear hyperbolic conservation laws containing first-order spatial derivatives in a series of papers by Cockburn et al. [11–15]. The proceeding review article of Shu [16–18] contain a more complete and current survey of the DG methods and their applications.

Multiplying (3) by a test function v, integrating over an arbitrary element $I_i = (x_{i-1}, x_i)$, and using integration by parts, we obtain the week form

$$\int_{I_i} u_t v dx - \int_{I_i} f(u) v_x dx + f(u_i) v_i - f(u_{i-1}) v_{i-1} = 0. \tag{24}$$

where $v_i = v(x_i)$ and $u_i = u(x_i, t)$. We introduce the following discontinuous finite element approximation space $V_h^p = \{v : v\big|_{I_i} \in P^p(I_i),\ \ i = 1, 2, \ldots, N\}$. Here, $P^p(I_i)$ is the space of polynomials of degree at most p on I_i. Let u_i^+ and u_i^-, respectively, denote the values of u at $x = x_i$ from the right cell I_{i+1} and from the left cell I_i.

Next, we replace the exact solution $u(\cdot, t)$ by $u_h(\cdot, t) \in V_h^p$. We also replace the boundary terms $f(u_{i-1})$ and $f(u_i)$ by single-valued numerical fluxes $\hat{f}_{i-1} = \hat{f}\left(u_{i-1}^-, u_{i-1}^+\right)$

and $\hat{f}_i = \hat{f}(u_i^-, u_i^+)$, respectively. These numerical fluxes in general depend both on the left limit and on the right limit. The idea is to treat these terms by an upwinding mechanism (information from characteristics), borrowed from successful high resolution finite volume schemes. For the Eq. (3), the flux $\hat{f}_i$ is taken as a monotone numerical flux, *i.e.*, it is Lipschitz continuous in both arguments u^- and u^+, consistent *i.e.*, $\hat{f}(u,u) = f(u)$, non-decreasing in the first argument and non-increasing in the second argument. Some examples of monotone fluxes which are suitable for DG methods can be found in Ref. [13]. For instance, we could use the simple Lax-Friedrichs flux

$$\hat{f}(u^-, u^+) = \frac{1}{2}(f(u^-) + f(u^+) - \alpha(u^+ - u^-)), \quad \text{where } \alpha = \max|f'(u)|.$$

Here the maximum is taken over a relevant range of u. Finally, we replace the test function v at the boundaries x_i and x_{i-1} by v_i^- and v_{i-1}^+. The resulting DG scheme consists of finding $u_h \in V_h^p$ such that $\forall\ i = 1, 2, \dots, N$,

$$\int_{I_i} (u_h)_t v\,dx - \int_{I_i} f(u_h) v_x\,dx + \hat{f}_i v_i^- - \hat{f}_{i-1} v_{i-1}^+ = 0. \tag{25}$$

The resulting system of ODEs (26), (27), (28) is then discretized by the nonlinearly stable high order Runge-Kutta time discretizations [19]. The most popular scheme in this classis the third-order Runge-Kutta method for solving $\mathbf{u}_t = \mathbf{F}(t, \mathbf{u})$

$$\mathbf{u}^{(1)} = \mathbf{u}^n + k\mathbf{F}(t_n, \mathbf{u}^n), \tag{26}$$

$$\mathbf{u}^{(2)} = \frac{3}{4}\mathbf{u}^n + \frac{1}{4}\mathbf{u}^{(1)} + \frac{k}{4}\mathbf{F}\left(t_n + k, \mathbf{u}^{(1)}\right), \tag{27}$$

$$\mathbf{u}^{n+1} = \frac{1}{3}\mathbf{u}^n + \frac{2}{3}\mathbf{u}^{(2)} + \frac{2k}{3}\mathbf{F}\left(t_n + \frac{k}{2}, \mathbf{u}^{(2)}\right), \tag{28}$$

To complete the definition of the DG scheme, we still need to define the discrete initial condition $u_h(x, 0) \in V_h^p$. One could use a special projection of the exact initial condition $u_0(x)$ as

$$u_h(x, 0) = P_h u_0(x), \quad x \in I_i, \quad i = 1, 2, \dots, N, \tag{29}$$

where P_h is the standard L^2-projection.

5.8 A numerical example

Consider the one-dimensional Burgers Eq. (3) on the domain $x \in [-2, 5]$, $t \in [0, 10]$ with initial condition $u(x, 0) = e^{-(x-1)^2}$ and the boundary conditions $u(-2, t) = u(5, t) = 0$. We use $N = 20$ and $k = 0.01$ and display the numerical solutions using the Upwind non conservative, Upwind conservative, Lax-Friedrichs, and Godunov schemes in **Figure 2**.

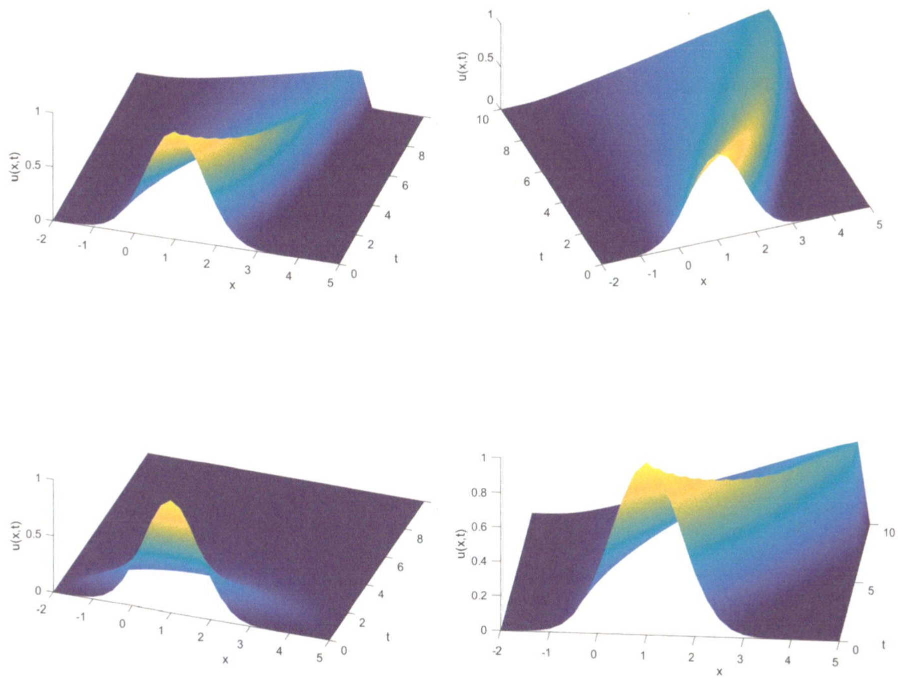

Figure 2.
Upwind non conservative, Upwind conservative, Lax-Friedrichs, and Godunov respectively.

6. Numerical methods for the viscid Burgers equation

We consider the following problem with $f(u) = u^2/2$.

$$\begin{cases} u_t + (f(u))_x = \nu u_{xx}, \quad x \in [a,b], \quad t \in [0,T], \ \nu > 0, \\ u(x,0) = u_0(x), \quad x \in [a,b], \\ u(a,t) = u_a(t), \quad u(c,t) = u_b(t), \quad t \in [0,T]. \end{cases} \tag{30}$$

6.1 A finite difference method

We approximate $u_t(x_i, t_n)$ with the forward in time $u_t(x_i, t_n) \approx \frac{u_i^{n+1} - u_i^n}{k}$, we approximate $(f(u(x_i, t_n)))_x$ with the central difference method $(f(u(x_i, t_n)))_x \approx \frac{f(u_{i+1}^n) - f(u_{i-1}^n)}{2h}$, and we approximate $u_{xx}(x_i, t_n)$ with the finite difference method $u_{xx}(x_i, t_n) \approx \frac{u_{i+1}^n - 2u_i^n + u_{i-1}^n}{h^2}$. Thus, we get the explicit method

$$u_i^{n+1} = u_i^n + k\left(\nu \frac{u_{i+1}^n - 2u_i^n + u_{i-1}^n}{h^2} + \frac{f(u_{i+1}^n) - f(u_{i-1}^n)}{2h}\right), \tag{31}$$

where the initial values and boundary nodes are taken from $u_i^0 = u_0(x_i)$, $u_0^n = u_1(t_n)$, and $u_N^n = u_2(t_n)$.

6.2 The Chebyshev collocation method

Spectral methods are one of the most powerful solution techniques for ordinary and partial differential equations [20]. Here, we present an efficient and accurate numerical method for solving (30). The method is based on the Chebyshev collocation technique in space and the fourth-order Runge-Kutta in time. This proposed scheme is robust, fast, flexible, and easy to implement using modern mathematical software such as MATLAB.

We first consider the viscid Burgers' equation on the reference interval $[-1, 1]$

$$u_t + uu_\xi = \nu u_{\xi\xi}, \quad \xi \in [-1, 1], \quad t > 0, \tag{32}$$

$$u(\xi, 0) = u_o(\xi), \quad u(-1, t) = u_a(t), \quad u(1, t) = u_a(t). \tag{33}$$

A numerical solution of the problem (32) and (33) using the Chebyshev pseudospectral method consists of the following steps:

1. Let $N \geq 1$ be any integer. We introduce the Chebyshev extreme points (also called the Gauss-Lobatto-Chebyshev points)

$$\xi_i = \cos\left(\frac{i\pi}{N}\right), \quad i = 0, 1, \dots, N. \tag{34}$$

These collocation points are the extreme points in $[-1, 1]$ of the Chebyshev polynomial $T_N(\xi) = \cos(N\cos^{-1}(\xi)), \ \xi \in [-1, 1]$. For simplicity we will call them the Chebyshev points.

1. Let $P^N(-1, 1)$ be the set of polynomials of degree $\leq N$ on the interval $[-1, 1]$. We choose $L_i(\xi), \ i = 0, \dots, N$, as a basis of $P^N(-1, 1)$, where $L_i(\xi) = \prod_{j=0, j\neq i}^{N} \frac{\xi - \xi_j}{\xi_i - \xi_j}$ is the polynomial of degree N satisfying $L_i\left(\xi_j\right) = \delta_{ij}$. Here δ_{ij} is the Kronecker symbol.

2. We assume that the numerical solution at fixed time t is a linear combination of the basis functions $\{L_i\}_{i=0}^{i=N}$. In other words, we approximate the exact solution $u(\xi, t)$ by $u_N(\xi, t) = \sum_{i=0}^{N} c_i(t) L_i(\xi)$ and require that u_N satisfies the PDE at the collocation points ξ_i, *i.e.*,

$$\frac{\partial u_N(\xi_i, t)}{\partial t} + u_N(\xi_i, t)\frac{\partial u_N(\xi_i, t)}{\partial \xi} = \nu \frac{\partial^2 u_N(\xi_i, t)}{\partial \xi^2}, \quad i = 1, \dots, N - 1,$$

and $u_N(\xi_i, 0) = u_0(\xi_i)$, $u_N(-1, t) = u_a(t)$, $u_N(1, t) = u_b(t)$. This yields a system of ODEs for the coefficients $c_i(t)$.

1. We use an ODE solver to approximate the unknown coefficients $c_i(t)$.

A simple way to implement the Chebyshev pseudospectral method is via the so-called differentiation matrices. It turns out that when the polynomials $L_i(\xi)$ are used as basis functions and the collocation points are chosen to be the Chebyshev points, the entries of the differentiation matrix are explicitly known.

Given the approximate values $u_N(\xi_i,t),\ \ i=0,\ldots,N$ to $u(\xi_i,t)$ at a fixed time t. Let $u_N(\xi,t)=\sum_{j=0}^{N}u_N\left(\xi_j,t\right)L_j(\xi)$. Then, we obtain highly accurate approximations to $u_\xi(\xi_i,t)$ and $u_{\xi\xi}(\xi_i,t)$ at fixed time t using [21].

$$\frac{\partial u_N(\xi_i,t)}{\partial\xi}=\sum_{j=0}^{N}u_N\left(\xi_j,t\right)L_j'(\xi_i)=\sum_{j=0}^{N}d_{ij}u_N\left(\xi_j,t\right),\quad \text{where } d_{ij}=L_j'(\xi_i), \tag{35}$$

$$\frac{\partial^2 u_N(\xi_i,t)}{\partial^2\xi}=\sum_{j=0}^{N}u_N\left(\xi_j,t\right)L_j''(\xi_i)=\sum_{j=0}^{N}b_{ij}u_N\left(\xi_j,t\right),\quad \text{where } b_{ij}=L_j''(\xi_i). \tag{36}$$

Letting $\mathbf{u}=[u_N(\xi_0,t),\ldots,u_N(\xi_N,t)]^t$. It turns out that a highly accurate approximation to $\mathbf{u}_\xi=\left[\frac{\partial u_N(\xi_0,t)}{\partial\xi},\ldots,\frac{\partial u_N(\xi_N,t)}{\partial\xi}\right]^t$ is given by $\mathbf{u}_\xi=D_N\mathbf{u}$, where D_N is the first-order spectral differentiation matrix $D_N=\left(d_{ij}\right)_{0\le i,j\le N}$. The entries of D_N are given by [28, Theorem 7]

$$\begin{aligned}
d_{00}&=-d_{NN}=\frac{2N^2+1}{6},\\
d_{ii}&=\frac{\xi_i}{2(\xi_i^2-1)},\quad i=1,\ldots,N-1,\\
d_{ij}&=\frac{c_i(-1)^{i+j}}{c_j\left(\xi_i-\xi_j\right)},\quad i\neq j,\ \ i,j=0,\ldots,N,
\end{aligned}$$

where $c_0=c_N=2$ and $c_i=1,\ \ i=1,\ldots,N-1$.

The construction of the differentiation matrix D_N does not explicitly include the boundary conditions. In order to satisfy the boundary conditions, we must manually set $u_N(\xi_0,t)=u_N(1,t)=u_b(t)$ and $u_N(\xi_N,t)=u_N(-1,t)=u_a(t)$.

To calculate higher derivatives, simply raise the matrix to the appropriate power. For instance, to approximate the m-th derivative, simply raise the matrix D_N to the mth power. In particular, a highly accurate approximation to $\mathbf{u}_{\xi\xi}=\left[\frac{\partial^2 u_N(\xi_0,t)}{\partial^2\xi},\ldots,\frac{\partial^2 u_N(\xi_N,t)}{\partial^2\xi}\right]^t$ is given by $\mathbf{u}_{\xi\xi}=D_N^2\mathbf{u}$. Now, let us discretize problem (32) and (33) using the Chebyshev pseudospectral collocation method. If we denote by $u_i(t)=u_N(\xi_i,t)$ the approximation to $u(\xi_i,t)$, then the semi-discrete Chebyshev method is given by the following system of nonlinear ODEs

$$\frac{du_i(t)}{dt}=\sum_{j=0}^{N}\left(\nu b_{ij}-u_i(t)d_{ij}\right)u_j(t),\quad u_i(0)=f(\xi_i),\quad i=1,\ldots,N-1, \tag{37}$$

where $u_0(t)=u_b(t)$, and $u_N(t)=u_b(t)$. The system (37) can be written as

$$\frac{d\mathbf{u}(t)}{dt}=\mathbf{F}(t,\mathbf{u}),\quad \mathbf{u}(0)=\mathbf{u}_0, \tag{38}$$

where $\mathbf{u}(t) = [u_1, \dots, u_{N-1}]^t$, $\mathbf{u}_0 = [u_0(\xi_1), \dots, u_0(\xi_{N-1})]^t$, $\mathbf{F}(t, \mathbf{u}) = [F_1, \dots, F_{N-1}]^t$ with

$$F_i = \sum_{j=0}^{N} (\nu b_{ij} - u_i(t) d_{ij}) u_j(t), \quad i = 1, \dots, N-1. \tag{39}$$

This system of ODEs (38) (39) can be approximated using time-stepping schemes for ODEs. Here, we employ the classical Runge-Kutta (RK4) method of fourth-order to discretize the system as follows: First, we use $h = T/M$ to denote the time step and we set $t_n = nh$, $n = 0, \dots, M$. Then the RK4 method consists of finding approximations $\mathbf{u}^{(n)}$ to $\mathbf{u}(t_n)$ using the following algorithm: Given $\mathbf{u}^{(0)} = \mathbf{u}_0$. For $n = 0,1,2, \dots, M-1$, we compute

$$\begin{aligned} &\mathbf{K}_1 = \mathbf{F}\left(t_n, \mathbf{u}^{(n)}\right), \qquad \mathbf{K}_2 = \mathbf{F}\left(t_n + \frac{h}{2}, \mathbf{u}^{(n)} + \frac{h}{2}\mathbf{K}_1\right), \\ &\mathbf{K}_3 = \mathbf{F}\left(t_n + \frac{h}{2}, \mathbf{u}^{(n)} + \frac{h}{2}\mathbf{K}_2\right), \qquad \mathbf{K}_4 = \mathbf{F}\left(t_n + h, \mathbf{u}^{(n)} + h\mathbf{K}_3\right), \\ &\mathbf{u}^{(n+1)} = \mathbf{u}^{(n)} + \frac{h}{6}(\mathbf{K}_1 + 2\mathbf{K}_2 + 2\mathbf{K}_3 + \mathbf{K}_4). \end{aligned} \tag{40}$$

In our numerical examples, the time step Δt is chosen so that temporal errors are small relative to spatial errors. After obtaining the approximations $u_N(\xi_i, t_n)$, $i = 0, \dots, N$, $n = 0, \dots, M$ to $u(\xi_i, t_n)$, the approximation $u_N(\xi, t_n)$, $\xi \in [-1, 1]$ at fixed time t_n can be found using interpolation as $u_N(\xi, t_n) = \sum_{i=0}^{N} u_N(\xi_i, t_n) L_i(\xi)$, $\quad n = 0, 1, \dots, M$. Next, we extend the above method to approximate the Burgers' Eq. (30) on the arbitrary interval $[a, b]$. We note that the shifted Chebyshev points in the interval $[a, b]$ are defined by

$$x_i = \frac{b-a}{2}\xi_i + \frac{a+b}{2} = \frac{b-a}{2}\cos\left(\frac{i\pi}{N}\right) + \frac{a+b}{2}, \quad i = 0, 1, \dots, N. \tag{41}$$

Next, we illustrate how to approximate the solution $u(x,t)$ to the one-dimensional viscid Burgers' Eq. (30) at the locations x_i, $i = 1, 2, \dots, N-1$ and at a fixed time t. Suppose that $u_N(x_i, t)$ approximates $u(x_i, t)$. Then $u_x(x_i, t)$ and $u_{xx}(x_i, t)$ can be approximated by

$$\frac{\partial u_N(x_i, t)}{\partial x} = \frac{2}{b-a}\sum_{j=0}^{N} d_{ij} u_N(x_i, t), \qquad \frac{\partial^2 u_N(x_i, t)}{\partial^2 x} = \frac{4}{(b-a)^2}\sum_{j=0}^{N} b_{ij} u_N(x_i, t), \tag{42}$$

for $i = 0, \dots, N$. Evaluating (30) at $x = x_i$, approximating $u(x_i, t)$ by $u_i(t) = u_N(x_i, t)$, and using (42), we obtain the following system of nonlinear ODEs

$$\frac{du_i(t)}{dt} + \frac{2u_i(t)}{b-a}\sum_{j=0}^{N} d_{ij} u_j(t) = \frac{4\nu}{(b-a)^2}\sum_{j=0}^{N} b_{ij} u_j(t), \quad u_i(0) = f(x_i), \quad i = 1, \dots, N-1, \tag{43}$$

where $u_N(t) = u_a(t)$ and $u_0(t) = u_b(t)$. As before, d_{ij} and b_{ij} are, respectively, the entries of the differentiation matrices D_N and D_N^2.

The system (43) can be written as

$$\frac{d\mathbf{u}(t)}{dt} = \mathbf{F}(t, \mathbf{u}), \quad \mathbf{u}(0) = \mathbf{u}_0, \tag{44}$$

where $\mathbf{u}(t) = [u_1, \dots, u_{N-1}]^t$, $\mathbf{u}_0 = [f(x_1), \dots, f(x_{N-1})]^t$, $\mathbf{F}(t, \mathbf{u}) = [F_1, \dots, F_{N-1}]^t$ with

$$F_i = \frac{2}{b-a}\sum_{j=0}^{N}\left(\frac{2\nu b_{ij}}{b-a} - d_{ij}u_i(t)\right)u_j(t), \quad i = 1, \dots, N-1. \tag{45}$$

We would like to mention that the Chebyshev collocation method reduces any PDE to a system of ODEs, which can be approximated using RK4 (40). The proposed numerical schemes produce accurate results and are quite easy to implement.

Remark 1 The classical RK4 method to solve systems of ODEs is not the most efficient method for the temporal integration of the proposed Chebyshev collocation scheme. There are other schemes which are of higher order or are of low storage available in the literature (see for example [17] and the references therein). In practice, the time step Δt should be chosen so that temporal errors are small relative to spatial errors.

6.3 An ultra-weak DG method

For equations with higher-order spatial derivatives, such as (30), the DG method cannot be directly applied due to the discontinuous finite element space which is not regular enough to handle higher order derivatives. However, there are different types of DG methods proposed and analyzed in the literature. The most known schemes in the literature include the symmetric interior penalty DG (SIPG) methods [22, 23], non-symmetric interior penalty DG (NIPG) methods [24], direct DG (DDG) methods [25], local DG (LDG) methods [26], and ultra-weak DG (UWDG) methods [27].

Next, we apply the UWDG to solve (30). UWDG methods are based on repeated integration by parts so that all spatial derivatives are shifted from the exact solution to the test function in the weak formulation.

Multiplying the equation in (30) by a test function v, integrating over I_i, and using integration by parts twice, we get the UWDG weak formulation

$$\begin{aligned}&\int_{I_i}\left(u_t v - v' f(u) - \nu u\ v''\right)dx \\ &+\left(f(u_i) - \nu u_i'\right)v_i - \left(f(u_{i-1}) - \nu u_{i-1}'\right)v_{i-1} + \nu u_i v_i' - \nu u_{i-1} v_{i-1}' = 0.\end{aligned} \tag{46}$$

Next, we use the weak formulation (46) to define the UWDG scheme. Find $u_h \in V_h^p = \{v : v|_{I_i} \in \mathbb{P}^p(I_i),\ i = 1, 2, \dots, N\}$ such that for all test function $v \in V_h^p$,

$$\begin{aligned}&\int_{I_i}\left((u_h)_t v - v' f(u_h) - \nu u_h\ v''\right)dx \\ &+\left(\hat{f}_i - \nu\tilde{u}_{h,i}'\right)v_i^- - \left(f(u_{i-1}) - \nu\tilde{u}_{h,i-1}'\right)v_{i-1}^+ + \nu\hat{u}_{h,i}\left(v_i^-\right)' - \nu\hat{u}_{h,i-1}\left(v_{i-1}^+\right)' = 0,\end{aligned} \tag{47}$$

for all $i = 1, 2, \dots, N$. Next, we must choose the numerical fluxes, which should be designed based on different guiding principles for different equations to guarantee stability and convergence. Here, we take the numerical fluxes: The numerical flux $\hat{f}$ associated with the convection is taken as the upwind flux (Godunov flux): for $i = 0, 1, \ldots, N$,

$$\hat{f}(u_h^-, u_h^+) = \begin{cases} \min_{u_h^- \le u \le u_h^+} f(u), & \text{if } u_h^- < u_h^+, \\ \max_{u_h^+ \le u \le u_h^-} f(u), & \text{if } u_h^- \ge u_h^+. \end{cases} \tag{48}$$

The numerical fluxes $\hat{u}_h$ and $\tilde{u}_h'$ associated with the diffusion terms are taken as the alternating fluxes

$$\hat{u}_{h,i} = u_{h,i}^-, \quad \tilde{u}_{h,i}' = \left(u_{h,i}^+\right)', \quad i = 0, 1, \ldots, N. \tag{49}$$

It is crucial that we take $\hat{u}_h$ and $\tilde{u}_h'$ from the opposite directions. In practice, we could use the backward Euler time discretization to solve the resulting system of ODEs. However, in order for the time error not to dominate, we still use a small time step $\Delta t = \mathcal{O}(h^{p+1})$ for the V_h^p space.

6.4 A local DG method

The local DG (LDG) method is an extension of the DG method aimed at solving differential equations containing higher than first-order spatial derivatives. The first LDG method was introduced by Cockburn and Shu in [26] for solving convection-diffusion problems. LDG methods are extremely flexible in the mesh-design and can easily handle irregular meshes and elements of various types. They further can achieve stability without slope limiters, and are locally (element-wise) mass-conservative. The latter property is very useful in the area of computational fluid dynamics, especially in situations where there are shocks, steep gradients or boundary layers. In addition, LDG methods exhibit strong superconvergence that can be used to estimate the discretization errors. LDG schemes have been successfully applied to elliptic, parabolic, and hyperbolic PDEs [28–30].

In order to construct the LDG scheme for (30), we introduce an auxiliary variable $q = u_x$ and convert the equation in (30) into the following first-order system

$$u_t + (f(u))_x - \nu q_x = 0, \qquad q - u_x = 0. \tag{50}$$

Multiplying the equations in (50) by v and w, integrating over I_i, and using integration by parts, we obtain

$$\int_{I_i} u_t v \, dx + \int_{I_i} (\nu q - f(u)) v_x \, dx + (f(u_i) - \nu q_i) v_i - (f(u_{i-1}) - \nu q_{i-1}) v_{i-1} = 0,$$
$$\int_{I_i} q w \, dx + \int_{I_i} u w_x \, dx - u_i w_i + u_{i-1} w_{i-1} = 0.$$

Define the discontinuous finite element space $V_h^p = \{v : v|_{I_i} \in P^p(I_i), \ i = 1, \ldots, N\}$, where $P^p(I_i)$ is the space of polynomials of degree at most p on I_i. Then the semi-discrete LDG method is: Find $u_h \in V_h^p$ and $q_h \in V_h^p$ such that

$$\int_{I_i}(u_h)_t v\,dx + \int_{I_i}(\nu q_h - f(u_h))v_x\,dx + \left(\hat{f}_i - \nu\hat{q}_{h,i}\right)v_i^- - \left(\hat{f}_{i-1} - \nu\hat{q}_{h,i-1}\right)v_{i-1}^+ = 0, \tag{51}$$

$$\int_{I_i} q_h w\,dx + \int_{I_i} u_h w_x\,dx - \hat{u}_{h,i}w_i^- + \hat{u}_{h,i-1}w_{i-1}^+ = 0. \tag{52}$$

The numerical flux $\hat{f}$ associated with the convection is taken as the upwind flux (Godunov flux): for $i = 0, 1, \dots, N$,

$$\hat{f}(u_h^-, u_h^+) = \begin{cases} \min\limits_{u_h^- \le u \le u_h^+} f(u), & \text{if } u_h^- < u_h^+, \\ \max\limits_{u_h^+ \le u \le u_h^-} f(u), & \text{if } u_h^- \ge u_h^+. \end{cases} \tag{53}$$

The numerical fluxes $\hat{u}_h$ and $\hat{q}_h$ associated with the diffusion terms are taken as the alternating fluxes

$$\hat{u}_{h,i} = u_{h,i}^-, \qquad \hat{q}_{h,i} = q_{h,i}^+, \qquad i = 0, 1, \dots, N. \tag{54}$$

It is crucial that we take $\hat{u}_h$ and $\hat{q}_h$ from the opposite directions. The initial condition of the LDG scheme is

$$u_h(x, 0) = P_h u_0(x), \;\; x \in I_i, \;\; i = 1, 2, \dots, N,$$

where $P_h u$ is L^2-projection. Finally, we use the fourth-order strong stability preserving (SSP) Runge-Kutta method for the discretization in time. A time step is chosen so that temporal errors are small relative to spatial errors.

6.5 A numerical example

Consider the one-dimensional viscid Burgers Eq. (30) on the domain $x \in [-2, 5]$, $t \in [0, 10]$ with initial condition $u(x, 0) = e^{-(x-1)^2}$ and the boundary conditions $u(-2, t) = u(5, t) = 0$. We use $N = 100$ and $k = 0.001$ and display the numerical solutions using the finite difference method in **Figure 3** with $\nu = 1$ and $\nu = 0.01$.

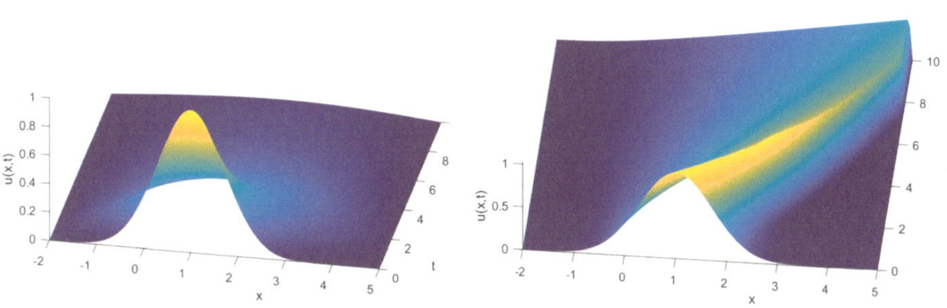

Figure 3.
The numerical solution of the viscid Burgers equation using the finite difference method (31) with $\nu = 1$ (left) and $\nu = 0.01$ (right).

Comparison and Challenges

- *Inviscid burgers equation*: The main challenge is dealing with shock formation and discontinuities. Numerical methods need to be robust enough to handle these features without introducing non-physical oscillations.

- *Viscid burgers equation*: The primary challenge is balancing accuracy and computational efficiency, especially for problems with small viscosity where solutions can exhibit sharp gradients.

The choice of numerical method for solving the Burgers equation depends on the specific form of the equation and the nature of the problem. For the inviscid form, methods that effectively capture discontinuities are crucial. For the viscid form, methods that provide a good balance between stability and accuracy are preferred. Understanding the characteristics of each method and the nature of the problem is essential for selecting the most appropriate numerical approach.

7. Numerical methods for the 2-dimensional Burgers equations

7.1 The 2-dimensional Burgers equations

The 2-dimensional Burgers equations extend the complexity of the 1-dimensional case by incorporating interactions in two spatial dimensions. These equations are used to model various physical phenomena such as fluid flow, turbulence, and gas dynamics. The equations can be written in both inviscid and viscid forms.

The 2-dimensional inviscid Burgers equations are given by:

$$u_t + uu_x + vu_y = 0,$$
$$v_t + uv_x + vv_y = 0.$$

The 2-dimensional viscid Burgers equations include viscosity terms:

$$u_t + uu_x + vu_y = \nu(u_{xx} + u_{yy}),$$
$$v_t + uv_x + vv_y = \nu(v_{xx} + v_{yy}),$$

where u and v are the velocity components in the x and y directions, respectively, and ν is the viscosity coefficient.

7.2 Numerical methods

Numerical methods for solving the 2-dimensional Burgers equations must account for the increased complexity due to interactions between the two spatial dimensions. Below are some commonly used methods:

Finite difference methods (FDM).

Explicit schemes

- *Upwind scheme*: Accounts for the direction of the wave propagation and helps handle discontinuities.

DOI: http://dx.doi.org/10.5772/intechopen.1007351

- *Lax-Friedrichs scheme*: Introduces numerical dissipation to stabilize the solution.

Implicit schemes

- *Crank–Nicolson scheme*: Provides second-order accuracy in both time and space and is unconditionally stable.

Finite volume methods (FVM).

Finite volume methods are suitable for conservation laws and ensure that fluxes across cell boundaries are accurately captured. These methods can handle complex geometries and boundary conditions more flexibly than finite difference methods.

- *Godunov's method*: Utilizes the exact or approximate Riemann solver to deal with discontinuities effectively.

- *MUSCL (monotonic upwind scheme for conservation laws)*: Enhances the accuracy of Godunov's method by using higher-order reconstruction techniques.

Finite element methods (FEM).

Finite element methods allow for flexible handling of complex geometries and boundary conditions. They are particularly useful for problems where adaptivity and local refinement are needed.

- *Standard galerkin method*: Suitable for problems with smooth solutions.

- *Discontinuous galerkin method*: Handles discontinuities and sharp gradients effectively, making it suitable for problems with shocks or strong nonlinearities.

Spectral methods.

Spectral methods use global basis functions such as Fourier series or Chebyshev polynomials to represent the solution. These methods are highly accurate for problems with smooth initial conditions and periodic domains.

- *Fourier spectral method*: Suitable for problems with periodic boundary conditions.

- *Chebyshev spectral method*: Suitable for problems defined on finite domains with non-periodic boundary conditions.

7.3 Handling boundary conditions

For both the inviscid and viscid Burgers equations, proper handling of boundary conditions is crucial. Common boundary conditions include:

- *Dirichlet boundary conditions*: Specify the values of the solution at the boundaries.

- *Neumann boundary conditions*: Specify the values of the derivative of the solution at the boundaries.

- *Periodic boundary conditions*: Assume the solution is periodic, simplifying the treatment of boundaries in spectral methods.

Challenges and considerations

- *Shock formation*: The inviscid Burgers equation can develop shocks, requiring robust numerical methods to handle discontinuities without introducing non-physical oscillations.
- *Stability and accuracy*: For the viscid Burgers equation, balancing stability and accuracy is crucial, especially for problems with small viscosity where solutions can exhibit sharp gradients.
- *Computational efficiency*: High-dimensional problems require efficient numerical methods and possibly parallel computing to handle large-scale computations effectively.

Numerical methods for solving the 2-dimensional Burgers equations must address the complexities arising from interactions in two spatial dimensions. Methods such as finite difference, finite volume, finite element, and spectral methods offer various advantages and challenges. Understanding the characteristics of each method and the specific nature of the problem is essential for selecting the most appropriate numerical approach. Proper handling of boundary conditions and addressing stability and accuracy are crucial for obtaining reliable solutions.

8. Numerical methods for the 3-dimensional Burgers equations

The 3-dimensional Burgers equations extend the complexity of the 2-dimensional case by incorporating interactions in three spatial dimensions. These equations are used to model various physical phenomena such as fluid flow, turbulence, and gas dynamics. The equations can be written in both inviscid and viscid forms.

8.1 The 3-dimensional Burgers equations

The 3-dimensional inviscid Burgers equations are given by:

$$\begin{aligned} u_t + uu_x + vu_y + wu_z &= 0, \\ v_t + uv_x + vv_y + wv_z &= 0, \\ w_t + uw_x + vw_y + ww_z &= 0. \end{aligned}$$

The 3-dimensional viscid Burgers equations include viscosity terms:

$$\begin{aligned} u_t + uu_x + vu_y + wu_z &= \nu\left(u_{xx} + u_{yy} + u_{zz}\right), \\ v_t + uv_x + vv_y + wv_z &= \nu\left(v_{xx} + v_{yy} + v_{zz}\right), \\ w_t + uw_x + vw_y + ww_z &= \nu\left(w_{xx} + w_{yy} + w_{zz}\right), \end{aligned}$$

where u, v, and w are the velocity components in the x, y, and z directions, respectively, and ν is the viscosity coefficient.

DOI: http://dx.doi.org/10.5772/intechopen.1007351

8.2 Numerical methods

Numerical methods for solving the 3-dimensional Burgers equations must account for the increased complexity due to interactions between the three spatial dimensions. Here are some commonly used methods:

Finite difference methods (FDM).

Explicit schemes

- *Upwind scheme*: Accounts for the direction of the wave propagation and helps handle discontinuities.
- *Lax-Friedrichs scheme*: Introduces numerical dissipation to stabilize the solution.

Implicit schemes

- *Crank-Nicolson scheme*: Provides second-order accuracy in both time and space and is unconditionally stable.

Finite volume methods (FVM).

Finite volume methods are suitable for conservation laws and ensure that fluxes across cell boundaries are accurately captured. These methods can handle complex geometries and boundary conditions more flexibly than finite difference methods.

- *Godunov's method*: Utilizes the exact or approximate Riemann solver to deal with discontinuities effectively.
- *MUSCL (monotonic upwind scheme for conservation laws)*: Enhances the accuracy of Godunov's method by using higher-order reconstruction techniques.

Finite element methods (FEM).

Finite element methods allow for flexible handling of complex geometries and boundary conditions. They are particularly useful for problems where adaptivity and local refinement are needed.

- *Standard galerkin method*: Suitable for problems with smooth solutions.
- *Discontinuous galerkin method*: Handles discontinuities and sharp gradients effectively, making it suitable for problems with shocks or strong nonlinearities.

Spectral methods

Spectral methods use global basis functions such as Fourier series or Chebyshev polynomials to represent the solution. These methods are highly accurate for problems with smooth initial conditions and periodic domains.

- *Fourier spectral method*: Suitable for problems with periodic boundary conditions.
- *Chebyshev spectral method*: Suitable for problems defined on finite domains with non-periodic boundary conditions.

Handling boundary conditions.

For both the inviscid and viscid Burgers equations, proper handling of boundary conditions is crucial. Common boundary conditions include:

- *Dirichlet boundary conditions*: Specify the values of the solution at the boundaries.
- *Neumann boundary conditions*: Specify the values of the derivative of the solution at the boundaries.
- Periodic boundary conditions: Assume the solution is periodic, simplifying the treatment of boundaries in spectral methods.

Challenges and considerations

- *Shock formation*: The inviscid Burgers equation can develop shocks, requiring robust numerical methods to handle discontinuities without introducing non-physical oscillations.
- *Stability and accuracy*: For the viscid Burgers equation, balancing stability and accuracy is crucial, especially for problems with small viscosity where solutions can exhibit sharp gradients.
- *Computational efficiency*: High-dimensional problems require efficient numerical methods and possibly parallel computing to handle large-scale computations effectively.

Numerical methods for solving the 3-dimensional Burgers equations must address the complexities arising from interactions in three spatial dimensions. Methods such as finite difference, finite volume, finite element, and spectral methods offer various.

9. Conclusions

In this chapter, we provided a very brief review of theoretical results for the viscid and inviscid Burgers equations. Then we presented several efficient numerical methods to approximate the solution to the viscid and inviscid Burgers equations. For the inviscid Burgers equation, we presented the classical finite difference and finite volume methods in details including Up-wind nonconservative, Up-wind conservative, Lax-Friedrichs, Lax-Wendroff, MacCormack, and Godunov methods. We also introduced the discontinuous Galerkin method. For the viscid Burgers equation, we present a finite difference method, a Chebyshev collocation method, an ultra-weak discontinuous Galerkin method, and a local discontinuous Galerkin method. We would like to mention that the procedures described above are very efficient and may be generalized without any problem to other nonlinear equations. In particular, most of these numerical methods are suitable for other two- and three-dimensional problems arising in fluid dynamics and other areas.

Acknowledgements

The author would like to thank the anonymous reviewer for the valuable comments and suggestions which improved the quality of the chapter.

Conflict of interest

The author declares no conflict of interest.

Data availability statement

Data sharing not applicable to this chapter as no datasets were generated or analyzed during the current study.

Ethical statement

The author agrees that this chapter has followed the rules of ethics presented in the publisher's Ethical Guidelines for Book Publication.

Author details

Mahboub Baccouch
University of Nebraska at Omaha, Omaha, Nebraska, USA

*Address all correspondence to: mbaccouch@unomaha.edu

References

[1] Burgers JM. A mathematical model illustrating the theory of turbulence. Advances in Applied Mechanics. 1948;**1**: 171-199

[2] Cole JD. On a quasi-linear parabolic equation occurring in aerodynamics. Quarterly of Applied Mathematics. 1951; **9**(3):225-236

[3] Lighthill MJ, Whitham GB. On kinematic waves I. Flood movement in long rivers. Proceedings of the Royal Society of London. Series A. Mathematical and Physical Sciences. 1955;**229**(1178):281-316

[4] Richards PI. Shock waves on the highway. Operations Research. 1956; **4**(1):42-51

[5] Whitham GB. Linear and Nonlinear Waves. New York, USA: John Wiley & Sons; 2011

[6] Lax PD. Hyperbolic Systems of Conservation Laws and the Mathematical Theory of Shock Waves. Philadelphia, USA: SIAM; 1973

[7] Hopf E. The partial differential equation $u_t + uu_x = \mu u_{xx}$. Communications on Pure and Applied Mathematics. 1950;**3**(3):201-230

[8] Godunov SK. A difference scheme for numerical solution of discontinuous solution of hydrodynamic equations. Math. Sbornik. 1959;**47**:271-306

[9] MacCormack R. The effect of viscosity in hypervelocity impact cratering. In: Tech. Rep., Paper AIAA-69-354. American Institute of Aeronautics and Astronautics; 1969

[10] Reed WH, Hill TR. Triangular mesh methods for the neutron transport equation. In: Tech. Rep. LA-UR-73-479. Los Alamos: Los Alamos Scientific Laboratory; 1991

[11] Cockburn B, Hou S, Shu C-W. The Runge-Kutta local projection discontinuous Galerkin finite element method for conservation laws IV: The multidimensional case. Mathematics of Computation. 1990;**54**:545-581

[12] Cockburn B, Lin S-Y, Shu C-W. TVB Runge-Kutta local projection discontinuous Galerkin finite element method for conservation laws III: One dimensional systems. Journal of Computational Physics. 1989;**84**:90-113

[13] Cockburn B, Shu CW. TVB Runge-Kutta local projection discontinuous Galerkin methods for scalar conservation laws II: General framework. Mathematics of Computation. 1989;**52**: 411-435

[14] Cockburn B, Shu C-W. The Runge-Kutta local projection P1- discontinuous-Galerkin finite element method for scalar conservation laws. Mathematical Modelling and Numerical Analysis. 1991; **25**:337-361

[15] Cockburn B, Shu C-W. The Runge-Kutta discontinuous Galerkin method for conservation laws V: Multidimensional systems. Journal of Computational Physics. 1998;**141**:199-224

[16] Cockburn B. Discontinuous Galerkin methods for computational fluid dynamics. In: Encyclopedia of Computational Mechanics. 2nd ed. New York, USA: John Wiley; 2018. pp. 1-63

[17] Shu C-W. Discontinuous Galerkin methods: General approach and stability, numerical solutions of partial differential

equations. In: Bertoluzza S, Falletta S, Russo G, Shu C-W, editors. Advanced Courses in Mathematics CRM Barcelona. Basel: Birkhäuser; 2009. pp. 149-201

[18] Shu C-W. High order WENO and DG methods for time-dependent convection-dominated PDEs: A brief survey of several recent developments. Journal of Computational Physics. 2016; **316**:598-613

[19] Shu C-W, Osher S. Efficient implementation of essentially non-oscillatory shock-capturing schemes. Journal of Computational Physics. 1988; **77**(2):439-471

[20] Baccouch M, Kaddeche S. Efficient chebyshev pseudospectral methods for viscous burgers' equations in one and two space dimensions. International Journal of Applied and Computational Mathematics. 2019;**5**(1):18

[21] Trefethen LN. Spectral Methods in MatLab. Philadelphia, PA, USA: Society for Industrial and Applied Mathematics; 2000

[22] Arnold DN. An interior penalty finite element method with discontinuous elements. SIAM Journal on Numerical Analysis. 1982;**39**:742-760

[23] Wheeler MF. An elliptic collocation-finite element method with interior penalties. SIAM Journal on Numerical Analysis. 1978;**15**:152-161

[24] Baumann CE, Oden JT. A discontinuous *hp* finite element method for convection-diffusion problems. Computer Methods in Applied Mechanics and Engineering. 1999;**175**: 311-341

[25] Liu H, Yan J. The direct discontinuous Galerkin (DDG) methods for diffusion problems. SIAM Journal on Numerical Analysis. 2009;**47**:675-698

[26] Cockburn B, Shu CW. The local discontinuous Galerkin method for time-dependent convection-diffusion systems. SIAM Journal on Numerical Analysis. 1998;**35**:2440-2463

[27] Cheng Y, Shu C-W. A discontinuous Galerkin finite element method for time dependent partial differential equations with higher order derivatives. Mathematics of Computation. 2008;**77**: 699-730

[28] Baccouch M. A local discontinuous Galerkin method for the second-order wave equation. Computer Methods in Applied Mechanics and Engineering. 2012;**209–212**:129-143

[29] Baccouch M. Asymptotically exact *a posteriori* LDG error estimates for one-dimensional transient convection-diffusion problems. Applied Mathematics and Computation. 2014; **226**:455-483

[30] Baccouch M. A superconvergent local discontinuous Galerkin method for the second-order wave equation on Cartesian grids. Computers and Mathematics with Applications. 2014;**68**: 1250-1278

Chapter 2

Fundamentals and Theory of Large Eddy Simulation

Rukiye Kara

Abstract

This chapter compares and contrasts Reynolds-Averaged Navier-Stokes (RANS), Large Eddy Simulation (LES), and Direct Numerical Simulation (DNS) in turbulence modeling within Computational Fluid Dynamics (CFD). The scope includes a thorough understanding of turbulence, subgrid-scale modeling, and the properties of RANS and LES approaches. It investigates the underlying distinctions between their methodologies and illustrates the advantages and disadvantages of each method for capturing turbulent flows. Basic definitions of length scales in fluid mechanics will be given, and in light of these, RANS, LES, and DNS methodologies will be explained. Furthermore, the chapter emphasizes RANS and LES's actual applications, computational complexities, and future prospects, providing insights into their roles in diverse engineering disciplines.

Keywords: turbulence, turbulent length scales, subgrid-scale modeling, LES RANS, DNS

1. Introduction

There is a wide range of turbulent flows, from air to blood. In fluid dynamics, turbulence refers to a flow regime characterised by stochastic and chaotic properties. The occurrence of laminar or turbulent flow depends on the Reynolds number, which represents the ratio of inertial forces to viscous forces in a fluid. Turbulent flow occurs at high Reynolds numbers, indicating the flow instabilities (**Figure 1**).

The theory of turbulence investigates the continuous variation in eddy lengths. Large eddies break down into smaller ones, and the process continues until the flow converts all the kinetic energy of the motion into heat. In the absence of an energy source, the motion's energy gradually decreases, leading to the eventual formation of motionless, laminar flow. Assuming a constant energy supply, turbulent flow persists. The motion of a fluid consists of eddies of various wavelengths. Kolmogorov's theory addresses the amount of energy contained in an eddy of a specific length.

Turbulence exhibits instability throughout an extensive range of scale motions. As a result, the extensive range that needs to be handled is the primary issue with the numerical computation of turbulence. The computational mesh must be fine enough to resolve the smallest dynamically important length scale (the Kolmogorov microscale). Direct numerical simulation (DNS), which uses a very fine mesh to

Figure 1.
A free water jet flowing through a square hole into a pool, as sketched by Leonardo da Vinci in 1508–1509 (Wikipedia).

numerically solve all the Navier-Stokes equations to capture all the scales present in a given flow, from the smallest eddies to the largest, is the most accurate method for simulating turbulent flows. DNS is therefore limited to low Reynolds number flows over simple geometries and is computationally expensive.

Significant computational time savings can be achieved when steady-state fluid flow is to be considered, that is, when detailed instantaneous flow simulation is not required. This is the basis of the Reynolds-averaged Navier-Stokes (RANS) method, which models the effect of all scales of instantaneous turbulent motion using a turbulence model and solves only for the mean magnitudes. Sometimes, this method is chosen because of its minimal computational expense. However, it is essential to understand the transient behavior of the flow and as a result, the RANS technique is inadequate and often fails to predict flow behavior such as transition.

Referred to as large eddy simulation (LES), first put forth by Smagorinsky in 1963, is an alternate technique. In contrast to the RANS method, the LES directly calculates the large-scale motions of the turbulent flow, and models only the small-scale motions. This leads to a significant reduction in computational expenses when compared to DNS. Since large eddies account for the majority of momentum transfer and turbulent mixing and contain the majority of the turbulent energy, LES is more accurate than the RANS because it directly and fully captures these eddies, while the RANS approach models them. The sizes of turbulence resolved or modeled by various simulation techniques, including RANS, LES, and DNS, are displayed in **Figure 2** [1]. While LES resolves larger scales and models smaller scales, DNS resolves all turbulence scales, from large eddies to the smallest dissipating eddies. For the most part, RANS models the entire turbulence spectrum. Moreover, a comparison of the flow quantities (Q(t)) over time resolved by DNS, LES, and RANS is shown in **Figure 3**, demonstrating how well these techniques capture flow fluctuations [2].

For large eddy simulation (LES) intended for high Reynolds number engineering applications, the near-wall resolution need to precisely resolve the boundary layer in wall-bounded flows continues to be an important item. In applications such as wall-bounded flows, where the characteristic length scale of LES is significantly larger than the characteristic turbulence length scales, essentially in the region close to the wall, LES faces computational challenges that can be quite costly. Wall models are also an important component of an LES. The wall-modeled LES can now be used as a design tool for new aircraft even in a high-lift configuration, as demonstrated, for example, by Goc et al. [3]. Numerous solutions have been proposed to address the 'near-wall problem of LES'. These approaches aim to model turbulence in the inner part of the boundary layer, eliminating the need to resolve turbulent eddies.

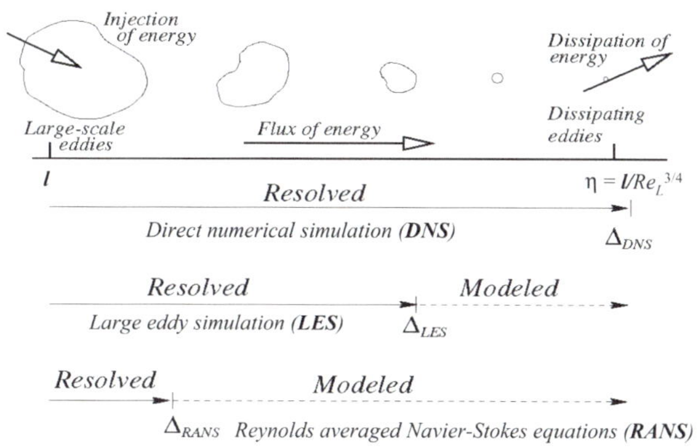

Figure 2.
Comparison of turbulence energy spectrum among direct numerical simulation (DNS), Reynolds-averaged Navier-stokes (RANS), and Large eddy simulation (LES) methods, [1].

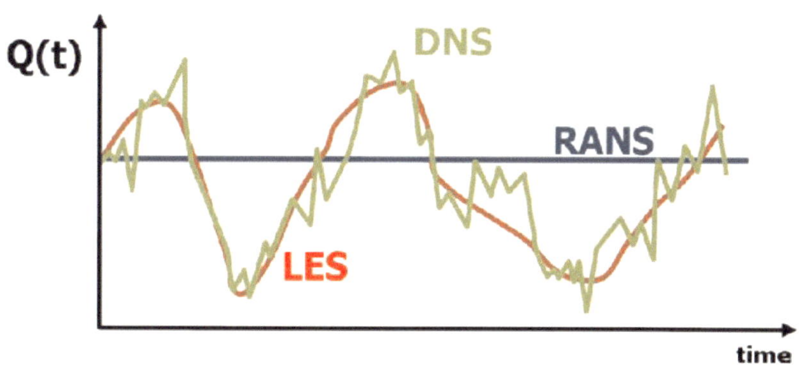

Figure 3.
Comparison between results obtained with RANS, LES simulations and DNS, [2].

Achieving a balance between physical modeling accuracy and computational costs is crucial, particularly for wall-bounded flows. Developing hybrid RANS-LES approaches is a sensible way of handling this issue. While RANS equations are not meant to deal with fluctuations, LES equations are intended to induce fluctuations (and typically require fluctuations at boundaries). Recently, it was proposed that the RANS and LES equations match consistently [4]. Various approaches to combining RANS and LES equations have been put forth, such as wall-modeled LES (WM-LES) [5], Detached Eddy Simulation (DES) coupled with multiple equations [6], Lattice Boltzmann (LB) methods [7].

One of the outstanding topics in recent years is deep learning, which has shown tremendous success in a wide variety of fields. There are recent attempts to use machine learning techniques for wall modeling in large eddy simulations. Recently, Machine Learning (ML) techniques such as neural networks and random forests have been used for subgrid-scale (SGS) modeling in Reynolds-averaged Navier-Stokes (RANS) simulations [8]. ML is also used for wall modeling in LES. There are studies using a feed-forward neural network architecture for LES wall modeling. A feed-forward neural network consists of an input layer, multiple hidden layers, and an output layer. The outputs of each layer are passed as input to the next layer, with the input layer being the starting point and the output layer being the model prediction. Each layer contains a number of neurons, which are computational units that take weighted sums of the inputs to an activation function [9, 10].

This chapter will cover the theoretical foundation of LES, including the Navier-Stokes equations, the Kolmogorov Length Scale, and the Numerical Approach to Turbulence. It primarily focuses on conventional LES while also exploring recent and classic subgrid models.

2. Navier-Stokes equations

The Navier-Stokes equations, named after Claude-Louis Navier and George Gabriel Stokes, constitute a set of equations developed in the 1840s based on conservation laws to describe the motion of fluids, including liquids and gases. Derived from the Euler equations [11–14], which were originally formulated by Leonhard Euler in 1759, these equations modeled fluid motion by applying Newton's second law, neglecting frictional forces.

Claude Navier revised Euler's equations in 1822 and adapted them for incompressible flows. In 1845, George Stokes introduced the concept of incompressibility into these equations, giving them the form known today [15, 16]. The Navier-Stokes equations state that the momentum changes acting on a unit mass in a fluid are equal to the sum of the pressure changes and the fluid forces, which represent frictional losses due to intermolecular interactions. These forces measure the resistance of the fluid to flow, making the Navier-Stokes equations a dynamic expression of the balance of forces in any region of the fluid.

The motion of incompressible Newtonian flows is described by the Navier-Stokes equations. These equations are the fundamental equations of fluid mechanics and are derived from the laws of conservation of mass and momentum. In the Navier-Stokes equations, the velocity field is defined as $\mathbf{u}(\mathbf{x}, t)$, where $\mathbf{x}$ is the position, and t is time. The Navier-Stokes equations govern the motion of incompressible flows in $\mathbb{R}^3$,

$$\frac{\partial u_i}{\partial x_i} = 0 \tag{1}$$

$$\frac{\partial u_i}{\partial t} + u_j \frac{\partial u_i}{\partial x_j} = -\frac{1}{\rho}\frac{\partial p}{\partial x_i} + \nu \frac{\partial^2 u_i}{\partial x_j^2} \tag{2}$$

where u_i is the velocity component in the direction x_i, p is the pressure, ν is the kinematic viscosity, and ρ is the density. Eq. (1) follows from the conservation of mass (incompressibility condition) and is also known as the continuity equation. It describes the divergence of the velocity at each point. Eq. (2) represents the conservation of momentum in the x, y, z directions.

The left-hand side of these momentum equations is defined as the material derivative, representing the change in velocity according to the Eulerian approach. The material derivative is divided into the temporal and spatial derivatives of the velocity. Here, the spatial derivative refers to the velocity transport due to convection. The second term on the right side of the momentum equations represents the rate of increase of the velocity due to convection [17]. Thus, the most important feature of turbulent flow is the significant variation and irregularity of the fluid's velocity field in space and time.

2.1 Kolmogorov length scale

Kolmogorov in 1941 proposed that the velocity field of incompressible turbulent flows would exhibit self-identity within a specific scale range. He utilized dimensional analysis to statistically describe turbulent motion. Faced with the mathematical challenges of solving the equations governing turbulence, Kolmogorov developed the statistical theory of turbulence, incorporating dimensional analysis and simplification approaches.

For statistically stationary turbulent flow, the energy dissipated at small scales must equal the energy supplied by large scales. Kolmogorov's first similarity hypothesis states that information about the non-homogeneous and anisotropic nature of large-scale turbulent motion at high Reynolds numbers is not transferred to the small-scale motion during the breaking-up process. The small-scale motion becomes homogeneous and isotropic, without any significant transition. Kolmogorov's second similarity hypothesis states that, for Reynolds numbers that are sufficiently large, the statistical properties of small-scale eddies, such as homogeneity and isotropy, are universal. This means that small-scale eddies exhibit these properties for all flows. The only factors influencing small-scale motion are the kinetic energy dissipation rate (or production rate) and viscosity. The unit of dissipation rate ε per unit mass, where m is the unit of length and sec is the unit of time, is denoted by m^2/sec^3, and the unit of unit of viscosity is m^2/sec. Thus, Kolmogorov's length, velocity, and time scales are determined as follows, respectively, and are referred to as Kolmogorov's scales:

$$\eta \equiv \left(\nu^3/\varepsilon\right)^{1/4} \tag{3}$$

$$u_\eta \equiv (\varepsilon\nu)^{1/4} \tag{4}$$

$$\tau_\eta \equiv (\nu/\varepsilon)^{1/2} \tag{5}$$

The rate of energy dissipation is expressed as:

$$\varepsilon = \nu\left(u_\eta/\eta\right)^2 = \nu/\tau_\eta^2. \tag{6}$$

For a wave number $k = \frac{2\pi}{l}$, regions defined by different length scales can be represented as a function of the wave number. Turbulence is visualized as energy transitioning from the low-wavenumber region to the high-wavenumber region, driven by the stretching of eddies caused by the nonlinear term in the Navier-Stokes equations. In the high-wavenumber region, dominated by viscosity, the energy transforms into heat. **Figure 4** illustrates how different simulation techniques capture distinct ranges of turbulent scales in the energy spectrum of turbulence [18].

The Fourier transform of the statistical relation between two velocities, representing the distribution of kinetic energy across various wavenumber components of turbulence, is known as the energy spectrum $\varepsilon(k)$. The kinetic energy of the turbulent flow, K, is statistically expressed by a volume average, as follows,

$$K = \frac{1}{2}\langle u_i u_i \rangle = \frac{1}{2}\frac{1}{V}\iiint u_i(x)u_i(x)dx. \tag{7}$$

The distribution of this energy among eddies of different sizes is determined by the energy spectrum $\varepsilon(k)$, where $\varepsilon(k)$ represents the energy in eddies with wave number k, and the total kinetic energy is given by

$$K = \int_0^\infty \varepsilon(k)dk. \tag{8}$$

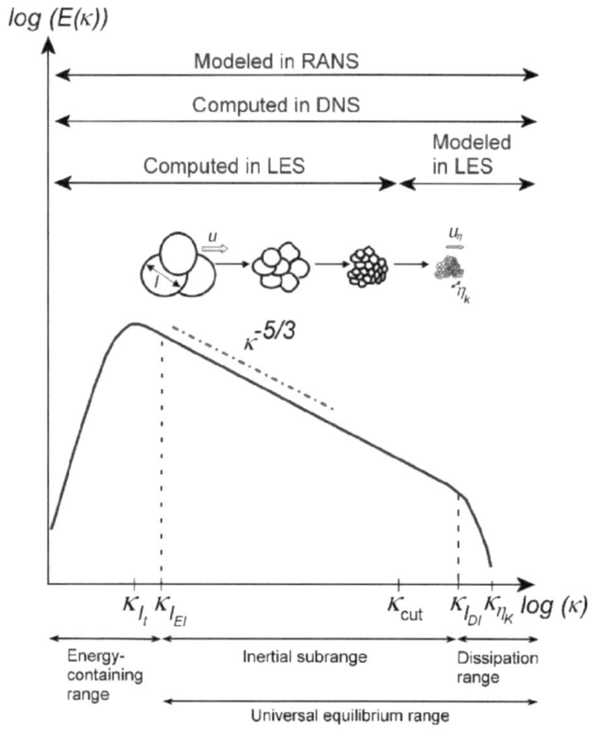

Figure 4.
Schematic representation of turbulence energy spectrum [18].

In the inertial region, where the energy is in equilibrium, it is expressed as

$$\varepsilon = \varepsilon(k, \varepsilon, \nu), \quad k \gg l^{-1}, \tag{9}$$

where l denotes the wavelength. Kolmogorov, using dimensional analysis $[K] = m^2/s^2$, $[\varepsilon] = m^2/s^3$, $[k] = 1/m$, $[\varepsilon(k)] = [K]/[k] = m^3/s^2$, and $\left[\varepsilon^{2/3}k^{-5/3}\right] = m^3/s^2$, derived the energy spectrum in the inertial region as

$$\varepsilon(k) = K_0\varepsilon^{2/3}k^{-5/3}, \quad \eta^{-1} \gg k \gg l^{-1}, \tag{10}$$

where K_0 is the Kolmogorov constant, taking values between $1.4 - 1.8$. Eq. (10) is known as 'Kolmogorov's $-5/3$ rule'. In the high-wavenumber region, dimensional analysis is not useful; thus, in the distribution region, it is generalized as

$$\varepsilon(k) = K_0\varepsilon^{2/3}k^{-5/3}f(\eta k), \quad k \gg l^{-1}. \tag{11}$$

The statistical approaches used in flow modeling are few and classical. In addition, new findings in the theory of stochastic processes and stochastic analysis are becoming important for modeling. Notably, stochastic Navier-Stokes equations with the addition of Brownian motion have been studied in recent years [19]. Earlier, Monin and Yaglom [20] were interested in the probability density function of particles moving according to the Navier-Stokes equation. Moreover, Pope [21] demonstrated the usefulness of the probability density function for the analysis of turbulent flows with complex responses. With his simplified modeling, the probability density function approach to turbulence has become central to Lagrange flow.

3. Numerical approaches to turbulence

The Navier-Stokes equations are widely used to analyze various phenomena, including air movements, ocean currents, water flow in pipes, and air flow around airplane wings. They have applications in power plants, vehicle design, and predicting air pollution, as well as modeling blood flow in veins. However, despite the many important applications, the lack of analytical solutions for the equations and the rapid development in computer technology have led to the use of numerical approaches for studying turbulence. Numerical simulation of turbulent flows has emerged as an important tool for interpreting turbulence. The main numerical approaches are direct numerical simulation, Reynolds-averaged numerical simulation, and large eddy simulation.

3.1 Direct numerical simulation

Direct Numerical Simulation (DNS) is the most accurate numerical approach to solving the Navier-Stokes equations since it analyzes all scales in turbulent flows. The chosen numerical method plays a crucial role in ensuring the reliability of DNS. The numerical solution of the Navier-Stokes equations primarily employs the finite difference method, finite volume method, finite element method, and spectral method.

Orszag and Patterson [22] pioneered the application of the spectral method in Direct Numerical Simulation (DNS) discretization. This method relies on Fourier series expansion in periodic directions and Chebyshev polynomial expansion in the wall direction. While highly accurate, it faces challenges when applied to complex geometries.

In contrast, the finite difference method, though less accurate than the spectral method, is applicable to complex regions [23] and is based on Taylor series expansion. Its accuracy can be enhanced by reducing the time step and mesh structure and by using higher-order approximations, but this comes at the cost of increased computational time.

The finite volume method considers the integral form of the equations, making it highly sensitive to mass (or energy) conservation. It is adaptable to non-uniform regions, and in comparisons between finite difference and volume methods, it has been demonstrated that the finite volume method is more accurate. It has been suggested that a coarse mesh can be employed to save time in simulations [24, 25].

For complex geometries or situations with intense turbulence, the use of Direct Numerical Simulation (DNS) may not be suitable due to constraints such as computer capacity and time. In scenarios where it is challenging to describe motion at all scales of turbulent flow, a more practical approach is to derive equations that model the influence of small scales on large scales, governing only the large scales. In such cases, DNS becomes highly valuable for validating turbulence models. An initial approach based on the idea of modeling the effect of small scales on large ones is the Reynolds-centered Navier-Stokes equations.

3.2 Reynolds-averaged Navier-Stokes equations

Irregularity is one of the most important features of turbulent flows. While it may be challenging to describe turbulent flow in full detail as a function of time and space, expressing it through a new system of equations based on the average values of flow variables (velocity and pressure) is a viable approach. In 1895, Osborne Reynolds developed a statistical method and derived the renowned equations, named after him, to describe the dynamics of mean flow.

The Reynolds equations are obtained by averaging the velocity $\mathbf{u}$ and pressure p in the Navier-Stokes equations, as defined by Eqs. (1) and (2), and decomposing them into fluctuation components:

$$\mathbf{u} = \langle \mathbf{u} \rangle + \mathbf{u}' \; p = \langle p \rangle + p'. \tag{12}$$

The mean components $\langle \mathbf{u} \rangle$ and $\langle p \rangle$ represent the average of the pressure and velocity fields with respect to time and are expressed as:

$$\langle \mathbf{u} \rangle(\mathbf{x}, t) = \frac{1}{T}\int_t^{t+T} \mathbf{u}(\mathbf{x}, \lambda), d\lambda, \;\; \langle p \rangle(\mathbf{x}, t) = \frac{1}{T}\int_t^{t+T} p(\mathbf{x}, \lambda), d\lambda \tag{13}$$

For the mean and fluctuations of $\mathbf{u}$ and p, when $T \to \infty$, the following properties hold:

$$\langle\langle \mathbf{u} \rangle\rangle = \langle \mathbf{u} \rangle, \langle\langle p \rangle\rangle = \langle p \rangle, \langle \mathbf{u}' \rangle = 0, \langle p' \rangle = 0 \tag{14}$$

When the decomposition in Eq. (12) is applied to the Navier-Stokes equations, the Reynolds-Averaged Navier-Stokes (RANS) equations are obtained.

$$\frac{\partial \langle u_i \rangle}{\partial x_i} = 0 \tag{15}$$

$$\frac{\partial\langle u_i\rangle}{\partial t}+\frac{\partial\langle u_iu_j\rangle}{\partial x_j}=-\frac{1}{\rho}\frac{\partial\langle p\rangle}{\partial x_i}+\nu\frac{\partial^2\langle u_i\rangle}{\partial x_j^2}. \tag{16}$$

These new equations closely resemble the original equations but describe the behavior of the variables $(\langle\mathbf{u}\rangle, \langle p\rangle)$ instead of $(\mathbf{u}, p)$. Moreover, in addition to the mean viscosity stress involving the gradient of $\mathbf{u}$, there is a second stress, known as the Reynolds stress. This stress represents the effect of the fluctuating components on the mean flow components $(\langle\mathbf{u}\rangle, \langle p\rangle)$. The Reynolds stress is defined as follows:

$$\tau_{ij}=\langle u_iu_j\rangle-\langle u_i\rangle\langle u_j\rangle. \tag{17}$$

Applying the properties of temporal averaging to Eq. (17), the subgrid stress will only be equal to the Reynolds stress, as follows:

$$\tau_{ij}=R_{ij}=\left\langle u_i'u_j'\right\rangle \tag{18}$$

The RANS equations depend on the variable $\langle\mathbf{u}\rangle$. To solve the Reynolds-Averaged Navier-Stokes (RANS) equations, the Reynolds stress must be modeled using the variables $(\langle\mathbf{u}\rangle, \langle p\rangle)$. Various Reynolds stress models exist in the literature, including algebraic models, one-equation models, and two-equation models [26, 27].

3.3 Large Eddy simulation (LES)

It is based on applying a spatial filter instead of temporal or ensemble averaging to the Navier-Stokes equations. The fundamental idea is to derive a new system of equations based on averaged variables, similar to the RANS equations. Unlike the RANS, the computational grid in Large Eddy Simulation (LES) is fine enough to resolve some of the turbulent structures but not as fine as needed for energy-dissipative motions as in the DNS. In the LES, the interaction of motion and large-scale eddies is calculated directly, while the effect of small scales on large ones is modeled. In **Figure 5**, it can be seen that LES provides a closer approximation to DNS, capturing more detailed flow structures compared to RANS [28].

The LES approach is a widely used method for the numerical simulation of turbulent flows. The LES equations are obtained by applying a filter to the Navier-Stokes equations. A homogeneous filtering process decomposes the velocity $\mathbf{u}(\mathbf{x}, t)$ into the sum of the filtered (resolved) $\overline{\mathbf{u}}$ and the subgrid $\mathbf{u}'$ components. The filtered component represents the motion of large eddies and is defined as follows:

$$\overline{u}_i(x,t)=\int u_i(\xi,t)G(x,\xi)d\xi \tag{19}$$

where $G(x, \xi)$ is a weight function that depends on the filter kernel, filter type, and filter width Δ. Commonly used filter types include box, top line, Gaussian, and cutoff filters. The box or top-line filter is defined in physical space and is often used with finite difference or finite volume methods. The Gaussian filter is used in both physical and spectral spaces, while the cutoff filter is defined in spectral space. The box filter is defined as follows:

$$G(x-\xi)=\begin{cases}\dfrac{1}{\Delta}, & |x-\xi|\leq\dfrac{\Delta}{2}\\ 0, & \text{otherwise}\end{cases} \tag{20}$$

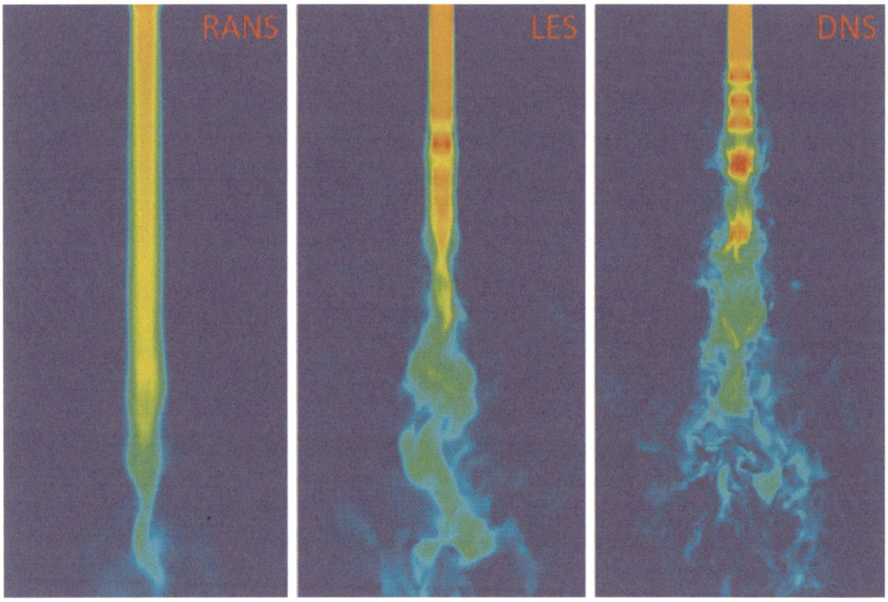

Figure 5.
CFD modeling of a turbulent jet using different approaches [28].

and corresponds to the local volume average in a three-dimensional region. The basic properties of the filtering operator are listed below:

- Conservation of Mass:

$$\overline{a} = a \leftrightarrow \int \int_{-\infty}^{\infty} G(\xi, t')d^3\xi = 1 \tag{21}$$

- Linearity:

$$\overline{v + \omega} = \overline{v} + \overline{\omega} \tag{22}$$

- Commutation with derivation:

$$\overline{\frac{\partial u}{\partial t}} = \int \frac{\partial u}{\partial t} G(x, \xi) d\xi \tag{23}$$

$$= \frac{\partial}{\partial t} \int uG(x, \xi) d\xi \tag{24}$$

$$= \frac{\partial \overline{u}}{\partial t} \tag{25}$$

Since the kernel function does not depend on the time variable t, there is a commutative property in the derivative with respect to time. Also, since the kernel space variable in the box filter does not depend on x, this property will also hold in the derivative with respect to x.

The filtered variables, which represent the largest scales in turbulence, can be decomposed as follows:

$$\mathbf{u} = \overline{\mathbf{u}} + \mathbf{u}', \quad p = \overline{p} + p' \tag{26}$$

where $\overline{\mathbf{u}}$ and $\overline{p}$ are termed resolved (filtered or large-scale) components, while $\mathbf{u}'$ and p' are unresolved (subgrid or small-scale) components, as described by Leonard [29]. In general, even if the flow is statistically stationary, the following properties are satisfied:

$$\overline{\overline{\mathbf{u}}} \neq \overline{\mathbf{u}}, \quad \overline{\overline{p}} \neq \overline{p}, \quad \overline{\mathbf{u}'} \neq 0, \quad \overline{p'} \neq 0 \tag{27}$$

By applying the filtering process to the incompressible Navier-Stokes equations, the general form of the LES equations is obtained:

$$\frac{\partial \overline{u}_i}{\partial x_i} = 0 \tag{28}$$

$$\frac{\partial \overline{u}_i}{\partial t} + \frac{\partial \overline{u_i u_j}}{\partial x_j} = -\frac{\partial \overline{p}}{\partial x_i} + \nu \frac{\partial}{\partial x_j}\left(\frac{\partial \overline{u}_i}{\partial x_j} + \frac{\partial \overline{u}_j}{\partial x_i}\right) \tag{29}$$

Here $\overline{p}$ is the filtered pressure. The filtered momentum equation introduces the nonlinear term $\overline{u_i u_j}$. In order to solve these equations numerically, we need to express the nonlinear term as a function of the variable $\overline{\mathbf{u}}$. If we rearrange Eq. (29) by adding and subtracting the term $\overline{u}_i \overline{u}_j$ to the nonlinear term and rewrite the equation, we get

$$\frac{\partial \overline{u_i}}{\partial x_i} = 0 \tag{30}$$

$$\frac{\partial \overline{u}_i}{\partial t} + \frac{\partial \overline{u}_i \overline{u}_j}{\partial x_j} = -\frac{1}{\rho}\frac{\partial \overline{p}}{\partial x_i} + \nu \frac{\partial^2 \overline{u}_i}{\partial x_j^2} - \frac{\partial \tau_{ij}}{\partial x_j}. \tag{31}$$

The filtration process results in a new system of equations similar to the original Navier-Stokes equations, governed by the variables $(\overline{u}, \overline{p})$, and introduces a new stress term called the subgrid stress:

$$\tau_{ij} = \overline{u_i u_j} - \overline{u}_i \overline{u}_j. \tag{32}$$

As seen from Eq. (32), this new stress requires knowledge of the $\mathbf{u}$ velocity field due to the $\overline{u_i u_j}$ term. However, the filtered Navier-Stokes equations contain filtered variables $(\overline{\mathbf{u}}, \overline{p})$. Therefore, the nonlinear term τ_{ij} must be modeled using these variables. The numerical solution of the filtered Navier-Stokes equations using a subgrid-scale model is called Large Eddy Simulation (LES).

The effect of the subgrid fluctuations is present in the subgrid-scale stress tensor $\tau_{ij} = \overline{u_i u_j} - \overline{u}_i \overline{u}_j$. If this stress tensor is rewritten using the decomposition $\mathbf{u}' = u - \overline{u}$:

$$\overline{u_i u_j} = \overline{\left(\overline{u}_i + u_i'\right)\left(\overline{u}_j + u_j'\right)} \tag{33}$$

$$= \overline{\overline{u}_i\overline{u}_j} + \overline{\overline{u}_i u_{j'}} + \overline{u_{i'}\overline{u}_j} + \overline{u_i' u_j'}. \quad (34)$$

The decomposition, known as the Leonard decomposition [29], is obtained, and the nonlinear stress tensor is expressed as a function of the filtered quantity $\overline{\mathbf{u}}$ and the fluctuation part $\mathbf{u}'$. The subgrid-scale stress tensor τ is decomposed as follows:

$$\tau_{ij} = L_{ij} + C_{ij} + R_{ij} = \overline{u_i u_j} - \overline{\overline{u}_i \overline{u}_j} \quad (35)$$

where

$$L_{ij} = \overline{\overline{u}_i\overline{u}_j} - \overline{u}i\overline{u}j, \quad (36)$$

$$Cij = \overline{\overline{u}_i u_j'} + \overline{\overline{u}_j u_i'}, \quad (37)$$

$$Rij = \overline{u_{i'} u_{j'}}. \quad (38)$$

The L term is called the Leonard tensor, representing interactions between large scales. C is the cross-stress tensor, capturing interactions between large and small scales. R is the Reynolds stress tensor, reflecting interactions between unresolved scales.

If we use this decomposition and rewrite the filtered momentum equation, we obtain:

$$\frac{\partial \overline{u}_i}{\partial t} + \frac{\partial \overline{u}_i \overline{u}_j}{\partial x_j} = -\frac{\partial \overline{p}}{\partial x_i} + \nu \frac{\partial}{\partial x_j}\left(\frac{\partial \overline{u}_i}{\partial x_j} + \frac{\partial \overline{u}_j}{\partial x_i}\right) - \frac{\partial \tau_{ij}}{\partial x_j}. \quad (39)$$

There are various subgrid-scale stress tensor models in the literature, with the main ones described below.

3.3.1 Eddy viscosity models

In large eddy simulation, τ_{ij} defined by Eq. (35) is called the subgrid-scale stress. Stress is the resistance force per unit area of the body against deformation. Let F be the external force, and A be the cross-sectional area; it is mathematically defined as follows:

$$\sigma = \frac{F}{A} \quad (40)$$

Applying a filter to the Navier-Stokes equations means that small-scale structures are removed from the motion. According to Kolmogorov's first hypothesis, the motion of small scales (or Kolmogorov scale) is universal, and therefore, it is possible to develop a model applicable to any flow. There are many alternatives for modeling the subgrid stress, with the most commonly used method based on the following assumption:

$$\tau_{ij} - \frac{\tau_{kk}}{3}\delta_{ij} = -2\nu_t \overline{S}_{ij} \quad (41)$$

Here, ν_t is the subgrid-scale eddy viscosity and $\overline{S}_{ij} = \frac{1}{2}\left(\frac{\partial \overline{u}_i}{\partial x_j} + \frac{\partial \overline{u}_j}{\partial x_i}\right)$ is the filtered strain rate tensor. These models are based on the modeling of eddy viscosity. In the literature, almost all eddy viscosity models are constructed to accurately analyze the kinetic energy distribution of filtered scales. Kinetic energy distribution for the filtered Navier-Stokes equations is given by:

$$\varepsilon_{sgs} = -\tau_{ij}\frac{\partial \overline{u}_i}{\partial x_j} = -\tau_{ij}\overline{S}_{ij} \tag{42}$$

where ε_{sgs} is the kinetic energy distribution, and τ is the subgrid-scale tensor. Some basic subgrid-scale stress models are listed below.

3.3.1.1 Smagorinsky model

A widely used subgrid-scale stress model is the Smagorinsky model [30], which is based on the concept of eddy viscosity. Smagorinsky modeled eddy viscosity as follows:

$$\nu_t = \left(C_S\overline{\Delta}\right)^2|S|, \quad |S| = 2\left(\overline{S}ij\overline{S}ij\right)^{1/2} \tag{43}$$

Here, C_S is the Smagorinsky coefficient, and $\overline{\Delta}$ is the filter width. The coefficient C_S typically varies between 0.17 and 0.21. The eddy viscosity ν_t is analogous to the kinematic viscosity ν and has units of L^2/T, where L represents length and T represents time.

According to Piomelli et al. [31], the Smagorinsky model is inadequate in representing energy transfer from small to large scales. To address this limitation, several models derived from the Smagorinsky model have been proposed. Comprehensive details on these models can be found in [32].

3.3.1.2 Dynamic model

One challenge of the Smagorinsky model lies in the variation of the C_S coefficient across different flow types. For instance, Smagorinsky coefficients range from 0.17 to 0.24 for high Reynolds number flows, while in laminar flow they are set to 0. The C_S coefficient is directly obtained from the resolved velocity field using the dynamic technique, first introduced by Germano [33]. The Navier-Stokes equation is filtered with a width $\overline{\Delta}$; consequently, the resolved velocity field is necessary to determine the new equation $\overline{u}_i$ and the subgrid tensor τ_{ij}.

Similarly, applying a second filter with a width of $\hat{\overline{\Delta}}$ results in a new equation that depends on the variable $\hat{\overline{u}}_i$. The new subgrid-scale tensor is given by:

$$T_{ij} = \widehat{\overline{u_i u_j}} - \hat{\overline{u}}_i\hat{\overline{u}}_j. \tag{44}$$

The Leonard term is obtained using the Germano identity [33]:

$$L_{ij} = \widehat{\overline{u}_i\overline{u}_j} - \hat{\overline{u}}i\hat{\overline{u}}j = Tij - \hat{\tau}ij. \tag{45}$$

The Leonard term L_{ij} can be expressed in terms of the resolved velocity field in large eddy simulation. With the same model coefficient, the stresses T_{ij} and τ_{ij} can be represented similarly:

$$\tau_{ij}^{d} \equiv \tau_{ij} - \frac{\tau_{kk}}{3}\delta_{ij} = -2\left(C_S\overline{\Delta}\right)^2 |\overline{S}|\overline{S}ij, \tag{46}$$

$$Tij^{d} \equiv T_{ij} - \frac{T_{kk}}{3}\delta_{ij} = -2\left(C_S\overline{\Delta}\right)^2 |\overline{S}|\overline{S}_{ij}. \tag{47}$$

Hence, the Leonard stress L_{ij} can be written as:

$$L_{ij} - \frac{L_{kk}}{3}\delta_{ij} = 2\left(C_S\overline{\Delta}\right)^2 |\overline{S}|\hat{\overline{S}}ij - 2\left(C_S\hat{\overline{\Delta}}\right)^2 |\hat{\overline{S}}|\hat{\overline{S}}ij. \tag{48}$$

C_S is determined by the least squares method:

$$C_S^2 = \frac{L_{ij}^{d}M_{ij}}{M_{ij}M_{ij}}, \quad M_{ij} = 2\overline{\Delta}^2|\overline{S}|\hat{\overline{S}}ij - 2\overline{\Delta}^2|\overline{S}|\overline{S}ij. \tag{49}$$

However, it was found that this procedure was numerically unstable, leading to extremely large fluctuations when the Smagorinsky coefficient was negative. Therefore, an additional average is taken to minimize the error:

$$C_S^2 = \frac{\left\langle L_{ij}^{d}M_{ij}\right\rangle}{\left\langle M_{ij}M_{ij}\right\rangle}. \tag{50}$$

Here, $\langle . \rangle$ denotes the average in homogeneous directions. Many studies assume that the test filter width is equal to twice the filter width, that is, $\hat{\overline{\Delta}} = 2\overline{\Delta}$. Backscattering—that is, negative C_s from small scales to the resolved flow—is possible in this model.

3.3.2 Kinetic energy subgrid-scale model

This model is based on the kinetic energy of subgrid scales:

$$k^{sgs} = \frac{1}{2}\left(\overline{u_i u_i} - \overline{u}_i\overline{u}_i\right). \tag{51}$$

The eddy viscosity is defined as follows:

$$\nu_t = C_\nu \overline{\Delta}\left(k^{sgs}\right)^{1/2} \tag{52}$$

where C_ν is the model coefficient and $\overline{\Delta}$ is the filter width. Subgrid-scale kinetic energy is obtained from the transport equation which is given by the following,

$$\frac{\partial k^{sgs}}{\partial t} + \frac{\partial}{\partial x_i}\left(\overline{u}_i k^{sgs}\right) = P^{sgs} - D^{sgs} + \frac{\partial}{\partial x_i}\left(\nu_t \frac{\partial k^{sgs}}{\partial x_i}\right). \tag{53}$$

The three terms on the right-hand side represent energy production, energy dissipation, and diffusion of k^{sgs}, respectively. The energy production term is defined as:

$$P^{sgs} = \tau_{ij}\left(\frac{\partial \overline{u}_i}{\partial x_j}\right). \tag{54}$$

The model for the subgrid stress tensor τ_{ij} is as follows [27, 34]:

$$\tau_{ij} = -2\nu_t \overline{S}ij + \frac{2}{3}k^{sgs}\delta ij \tag{55}$$

The dissipation term is also modeled as,

$$D^{sgs} = C_e \frac{\left(k^{sgs}\right)^{3/2}}{\overline{\Delta}} \tag{56}$$

where C_e is the model coefficient [34].

3.3.3 Scale similarity and mixed models

Bardina [35] proposed the scale similarity hypothesis, which states that the statistical structure of tensors on subgrid scales is similar to that of the smallest resolved scales. Filtering the subgrid velocity field, denoted as $u_i' = u_i - \overline{u}_i$, yields the largest subgrid scales,

$$\overline{u'}_i = \overline{u}_i - \overline{\overline{u}}_i. \tag{57}$$

When the subgrid-scale stress is decomposed as including the terms $\overline{u}_i$ and u_i', and Eq. (57) is applied, the stress is represented as

$$\tau_{ij} - \frac{\tau_{kk}}{3}\delta_{ij} = C_B\left[\overline{\overline{u}_i\overline{u}_j} - \overline{\overline{u}}_i\overline{\overline{u}}_j - \frac{\delta_{ij}}{3}\left(\overline{\overline{u}_i\overline{u}_j} - \overline{\overline{u}}_i\overline{\overline{u}}_j\right)\right]. \tag{58}$$

The subgrid-scale stress is stated as follows: where L_{ij} is the Leonard tensor in the dynamical model:

$$\tau_{ij} = C_B L_{ij}. \tag{59}$$

The mixed model is used when the scale similarity model is combined with the Smagorinsky model to illustrate the dissipative effect of small scales:

$$\tau_{ij} = C_B\left[\overline{\overline{u}_i\overline{u}_j} - \overline{\overline{u}}_i\overline{\overline{u}}_j\right] - 2\nu_t\overline{S}_{ij}. \tag{60}$$

The scale similarity model does not account for the dispersion attribute of subgrid scales. Therefore, the Smagorinsky model enables the mixed model to exhibit the dispersion property as well.

3.3.4 The wall-adapting local Eddy viscosity (WALE) model

The traceless symmetric part of the square of the velocity gradient tensor is used by the WALE model, and it is defined as:

$$S_{ij}^d = \frac{1}{2}\left(\overline{g}_{ij}^2 + \overline{g}_{ji}^2\right) - \frac{1}{3}\delta_{ij}\overline{g}_{kk}^2 \tag{61}$$

where $g_{ij} = \frac{\partial u_i}{\partial x_j}$. The eddy viscosity ν_t is given by

$$\nu_t = (C_W \Delta)^2 \frac{\left(S_{ij}^d S_{ij}^d\right)^{3/2}}{\left(S_{ij}^d S_{ij}^d\right)^{5/2} + \left(S_{ij}^d S_{ij}^d\right)^{5/4})}, \tag{62}$$

where $C_W = 0.33$ [36].

3.3.5 σ-model

The σ-model, a subgrid-scale model introduced for enclosed fluids [37], is based on the singular values of the resolved velocity gradient tensor. With $\sigma_1(\mathbf{x}, t) \geq \sigma_2(\mathbf{x}, t) \geq \sigma_3(\mathbf{x}, t) > 0$ the singular values of $\nabla \overline{u}$, sigma model is given by the following eddy viscosity:

$$\nu_t = C_\sigma \overline{\Delta}^2 \left(\frac{\sigma_3(\sigma_1 - \sigma_2)(\sigma_2 - \sigma_3)}{\sigma_1^2} \right), \tag{63}$$

where C_σ is a suitable constant and $\overline{\Delta}$ the filter width.

3.3.6 Vortex-based subgrid-scale model

Misra and Pullin [38] proposed a vortex-based model for the subgrid stress tensor τ_{ij}. This model originates from the work of Lundgren [39], who demonstrated that the helical eddy's energy expression has an exponential form [39]. Later on, these eddies came to be known as Lundgren eddies. An extended two-dimensional eddy is used as the model for the subgrid-scale motion. This eddy rotates in stochastically distinct ways around the primary axis. Using this information, the following model is proposed:

$$\tau_{ij} = \left(\delta_{ij} - e_i^\vartheta e_j^\vartheta\right) K \equiv \left(\delta_{ij} - e_i^\vartheta e_j^\vartheta\right) \int_{k_c}^{\infty} \varepsilon(k) dk. \tag{64}$$

In this expression, k is the wave number, $\varepsilon(k)$ is the energy spectrum, k_c is the cutoff point of the wave numbers, and $\left(\delta_{ij} - e_i^\vartheta e_j^\vartheta\right)$ is the orientation tensor. The subgrid-scale stress, which only includes the Reynolds stress and ignores the other two terms Leonard and cross stress, is calculated as follows: $\tau_{ij} = \overline{u_i' u_j'}$. In addition, the cross stress was also modeled in the following ways by Chung and Pullin [40]:

$$\overline{u_i' \overline{u}_j} = -K_s \left[e_j^\vartheta e_k^\vartheta \frac{\partial \overline{u}_k}{\partial x_l} \left(\delta_{li} - e_l^\vartheta e_i^\vartheta\right) \right] \tag{65}$$

where $K_s = \gamma \Delta K^{1/2}/2$ and e^ϑ represents the axis of the subgrid vortex.

3.3.7 Çinlar subgrid-scale model

The Çinlar SGS model is based on the Çinlar velocity field for the subgrid-scale and makes use of the viscous effects with the resolved strain tensor. The generalized form of the Çinlar velocity field is constructed as a random sum of vortices given by

$$u'(x,t) = \int_{-\infty}^{t} \int_{Q} N(ds, dz, da, db) e^{-c_q(t-s)} a\, v\left(\frac{x-z}{b}\right) \tag{66}$$

where $x \in \mathbb{R}^2$, $t \in \mathbb{R}$, $q = (z, a, b)$ represents the type of an eddy and includes its center z in space, its amplitude a as well as its radius b and N is Poisson random measure on the Borel sets of $\mathbb{R} \times Q$ with mean measure $\mu(dt, dq) \equiv \mu(dt, dz, da, db) = \lambda dt\, dz \alpha(da) \beta(db)$ where λ is the arrival rate per unit time-unit space, and α and β are probability distributions for the amplitudes and radii of eddies, respectively. The notation u' is used to indicate that we aim to model the subgrid scales with Eq. (66). The decay parameter is explicitly given by $c_q(x) = c\left|\frac{x-z}{b}\right|^{2\gamma}$ where $c > 0$ and $\gamma > 0$ [41]. $v = (v_1, v_2)$ is the deterministic velocity field, called the basic eddy, on $\mathbb{R}^2$ and $Q = \mathbb{R}^2 \times \mathbb{R} \times (0, \infty)$ be the set of types of eddies. The basic vortex v is taken as a rotation on $\mathbb{R}^2$, with center at 0.

For modeling Reynolds stress, which represents the interaction of the small scales, the covariance tensor of the velocity field is computed as

$$R^{ij}(x,t) = \frac{\lambda}{c} \int_{\mathbb{R}^2} dz \int_{\mathbb{R}} \alpha(da) a^2 \int_{\mathbb{R}+} \beta(db) \frac{b^2 \exp\left(-c|z|^{2\gamma}|t|\right)}{|z|^{2\gamma} + \left|z + \frac{x}{b}\right|^{2\gamma}} v_i(z) v_j\left(z + \frac{x}{b}\right). \tag{67}$$

The covariance in space and time (0,0) is used as an approximation for the Reynolds stress R_{ij}. The vortex structures are assumed to have random orientations in three dimensions, which is the same viewpoint as Misra and Pullin [38]. For orienting Çinlar vortices in a three-dimensional flow, we can approximate $R_{ij} \approx (\delta_{ij} - e_i e_j) R^{11}(0, 0)$. Assuming that the vortices are aligned with the resolved vorticity vector $\overline{\omega} = \nabla \times \overline{u}$ we model $e_i \approx \overline{\omega}_i / |\overline{\omega}|$ as in ([38], p. 2445). As a result, we set

$$R_{ij} \approx \left(\delta_{ij} - \frac{\overline{\omega}_i \overline{\omega}_j}{|\overline{\omega}|^2}\right) \frac{\lambda}{c} \mathbb{E}(a^2) \mathbb{E}(b^2) \frac{3\pi}{8} \tag{68}$$

as the Reynolds stress, and model its parameters with the resolved velocity field as given in [42]. As a result, Reynolds stress is modeled as

$$R_{ij} \equiv \frac{3\pi}{32} \frac{C_3^2 C_4}{C_1} \overline{u}^2 \overline{\Delta}^2 e^{-2C_1|\overline{S}|} \cdot \left(\delta_{ij} - \frac{\overline{\omega}_i \overline{\omega}_j}{|\overline{\omega}|^2}\right) \left[\theta(\theta + 1) - \frac{C_3^{-\theta}}{e^{1/C_3} \Gamma_{1/C_3}(\theta)} \left(\frac{1}{C_3} + \theta + 1\right)\right] \tag{69}$$

where the radius of the largest eddy in dissipation range B is taken to be equal to half of the grid size as $\overline{\Delta}/2$, and $\theta := C_2 \overline{\Delta} / (|\overline{S}| \overline{u})$ will be used. For the cross stress C_{ij}, we use Clark's cross stress model [43], and Leonard stress L_{ij} does not need to be modeled as it depends only on the resolved velocity field. These altogether form the subgrid-scale tensor τ_{ij} [42].

The hierarchical representation of various turbulence models is shown in **Figure 6** [44]. In terms of physical complexity, DNS is at the top of the hierarchy, as seen in the graphic, followed by LES and RANS approaches. Consequently, the choice of

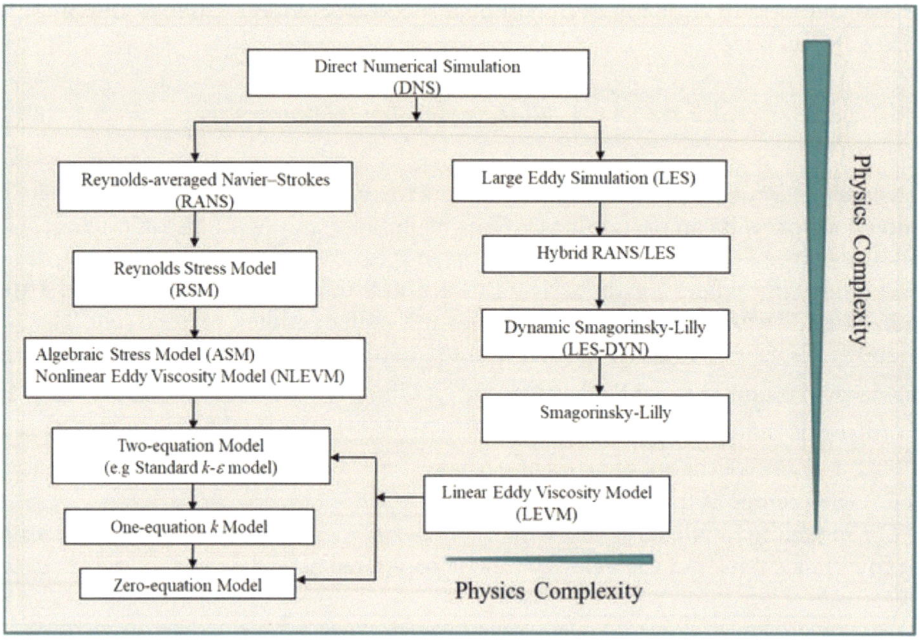

Figure 6.
Illustrative hierarchy of simulation methods and turbulence models [44].

turbulence model depends on the specific requirements of the simulation, including the desired level of detail and computational resources. LES models provide a sophisticated tool for studying complex turbulent flows, offering a balance between resolution and efficiency.

4. Conclusions

Turbulence is a well-known phenomenon that occurs frequently in nature and is essential to many applications. LES is useful as a turbulent system simulation technique in a wide range of applications, including aerodynamics, geophysics, astronomy, oceanography, combustion, and the medical sciences. Their profound understanding of complex fluid dynamics is invaluable in the optimization of designs and processes. In order to forecast an aircraft's aerodynamic performance, complex flow phenomena such as turbulent waves and vortical structures are studied using LES simulations in aerospace engineering. Chalot et al. employed LES in the design of business jets and military aircraft [45].

Another important type of turbulent flow is turbulent combustion. There are huge challenges in this area because turbulent combustion is very complex. For example, liquid fuel injection in aircraft engines involves liquid fuel atomization, droplet fragmentation, and evaporation, chemical reactions, and turbulence interactions. The review article by Fureby [46] provides much more discussion on the LES of turbulent combustion.

LES simulation is frequently used in automotive engineering to assess drag coefficients and improve vehicle aerodynamics. Three hybrid RANS-LES models were subjected to testing on DrivAer notchback and fastback automobile bodies in the

Ekman et al. study, and the outcomes were compared with wind tunnel measurements [47]. Pollutant dispersion in a three-dimensional roadway canyon was examined using LES simulations by Tominaga and Stathopoulos [48], who compared the results with experimental observations.

Cardiac valves are also a crucial topic in LES application areas. Aortic valves often degenerate and lead to regurgitation or stenosis, which, if left untreated, can lead to patient death. In cases where the aortic valve needs to be replaced with a prosthesis, one of the main engineering concerns for the design of an artificial aortic valve is hemodynamics. A more realistic model based on the Fluid-Structure Interaction (FSI) formulation can be used to validate an artificial aortic valve. In addition, the LES can be used to monitor the progression of the disease and to assess the extent of damage to the aortic wall caused by the turbulent flow of blood. The LES is therefore of crucial importance for the modeling of human cardiovascular diseases and the application of potential solutions [49].

Author details

Rukiye Kara
Department of Mathematics, Mimar Sinan Fine Arts University, Istanbul, Turkey

*Address all correspondence to: rukiye.kara@msgsu.edu.tr

References

[1] Ghose P, Sahoo TK, Sahu AK. Pulverized coal combustion computational modeling approach: A review. Proceedings of the Institution of Mechanical Engineers, Part A: Journal of Power and Energy. 2023; **237**(4):797-818

[2] Cuoci A. Modellazione numerica dei flussi turbolenti, Corso di Fluidodinamica degli Incendi. Italia: Politecnico di Milano, Dipartimento di Chimica, Materiali, Ingegneria Chimica “Giulio Natta”; 2015

[3] Goc KA, Lehmkuhl O, Park GI, Bose ST, Moin P. Large eddy simulation of aircraft at affordable cost: A milestone in computational fluid dynamics. Flow. 2021;**1**:E14

[4] Xiao H, Jenny P. A consistent dual-mesh framework for hybrid LES/RANS modeling. Journal of Computational Physics. 2012;**231**(4):1848-1865

[5] Bose ST, Park GI. Wall-modeled large-eddy simulation for complex turbulent flows. Annual Review of Fluid Mechanics. 2018;**50**:535-561

[6] Friess C, Manceau R, Gatski TB. Toward an equivalence criterion for hybrid RANS/LES methods. Computers and Fluids. 2015;**122**:233-246

[7] Li J, Zhong C, Pan D, Zhuo C. A gas-kinetic scheme coupled with SST model for turbulent flows. Computers and Mathematics with Applications. 2019; **78**(4):1227-1242

[8] Duraisamy K, Zhang ZJ, Singh AP. New approaches in turbulence and transition modeling using data-driven techniques. In: 53rd AIAA Aerospace Sciences Meeting. Kissimmee, Florida: AIAA SciTech; Jan 2015

[9] Gamahara M, Hattori Y. Searching for turbulence models by artificial neural network. Physical Review Fluids. 2017; **2**(5):054604

[10] Ma M, Lu J, Tryggvason G. Using statistical learning to close two-fluid multiphase flow equations for bubbly flows in vertical channels. International Journal of Multiphase Flow. 2016;**85**:336

[11] Batchelor GK. An Introduction to Fluid Dynamics. San Diego: Cambridge University Press; 1967

[12] Landau LD, Lifschitz EM. Fluid Mechanics. 2nd ed. Oxford, England: Pergamon Press; 1982

[13] Leray J. Sur le mouvement d’un liquide visqueux emplissant l’espace. Acta Mathematica. 1934;**63**(1):193-248

[14] Darrigol O. Between hydrodynamics and elasticity theory: The first five births of the Navier-Stokes equation. Archive for History of Exact Sciences. 2002;**56**: 95-150

[15] Navier CLMH. Mémoire sur les lois du mouvement des fluides. Mémoires de l’Académie Royale des Sciences. 1823;**6**: 389-416

[16] Stokes GG. On the effect of the internal friction of fluids on the motion of pendulums. Transactions of the Cambridge Philosophical Society. 1845;**8**: 287-305

[17] Versteeg HK, Malalasekera W. An Introduction to Computational Fluid Dynamics—The Finite Volume Method. London, England: Pearson Education Limited; 2007

[18] Tay-Wo-Chong L. Numerical simulation of the dynamics of turbulent

swirling flames [Ph.D. thesis]. München, Garching, Germany: Technische Universit.; 2012

[19] Flandoli F. Remarks on 3D stochastic Navier-Stokes equations. Seminar on Stochastic Analysis, Random Fields and Application V, Progress in Probability. 2008;**59**(1):123-134

[20] Monin AS, Yaglom AM. On the laws of small-scale turbulent flow of liquids and gases. Uspekhi Matematicheskikh Nauk. 1963;**18**(5(113)):93-114

[21] Pope SB. Computations of Turbulent Combustion: Progress and Challenges, in 23rd International Symposium on Combustion. Pittsburgh: The Combustion Institute; 1990

[22] Orszag SA, Patterson GS Jr. Numerical simulation of three-dimensional homogeneous isotropic turbulence. Physical Review Letters. 1972;**28**(2):76

[23] Rai MM, Moin P. Direct simulations of turbulent flow using finite-difference schemes. Journal of Computational Physics. 1991;**96**(1):15-53

[24] Leonard BP. Comparison of truncation error of finite difference and finite volume formulation of convection terms. Applied Mathematics Modelling. 1994;**18**:46-50

[25] Rouboa A, Monteriro E. Heat transfer in multiblock grid during solidification-performance of finite differences and finite volume methods. Journal of Materials Processing Technology. 2008;**204**: 451-458

[26] Pope SB. Turbulent Flows. New York: Cambridge University Press, Cornell University; 2000

[27] Sagaut P. Large Eddy Simulation for Incompressible Flows: An Introduction. Berlin, Heidelberg: Springer; 2005

[28] Rodriguez S. LES and DNS turbulence modeling. In: Applied Computational Fluid Dynamics and Turbulence Modeling. Cham: Springer; 2019

[29] Leonard A. Energy cascade in large-eddy simulations of turbulent fluid flows. Advances in Geophysics A. 1974; **18**:237-248

[30] Smagorinsky J. General circulation experiments with the primitive equations. Monthly Weather Review. 1963;**91**(3):99-164

[31] Piomelli U, Zang TA, Speziale CG, Hussaini MY. On the large-eddy simulation of transitional wall-bounded flows. Physics of Fluids A: Fluid Dynamics. 1990;**2**(2):257-265

[32] Rogallo RS, Moin P. Numerical simulation of turbulent flows. Annual Review of Fluid Mechanics. 1984;**16**: 99-137

[33] Germano M, Piomelli U, Moin P, Cabot W. A dynamic subgrid-scale eddy viscosity model. Physics of Fluids A. 1991;**3**(7):1760-1765

[34] Schumann U. Subgrid scale model for finite difference simulations of turbulent flows in plane channels and annuli. Journal of Computational Physics. 1975;**18**:376-404

[35] Bardina J, Ferziger JH, Reynolds WC. Improved subgrid scale models for large eddy simulation. AIAA Paper. 1980;**80**:1357

[36] Ducros F, Nicoud F, Poinsot T. Wall-adapting local eddy-viscosity models for simulations in complex geometries.

Numerical Methods for Fluid Dynamics VI. 1998;**6**:293-299

[37] Nicoud F, Toda HB, Cabrit O, Bose S, Lee J. Using singular values to build a subgrid-scale model for large Eddy simulations. Physics of Fluids. 2011;**23**:085106

[38] Misra A, Pullin DI. A vortex-based subgrid stress model for large-eddy simulation. Physics of Fluids. 1997;**9**: 2443-2454

[39] Lundgren TS. Strained spiral vortex model for turbulent fine structure. Physics of Fluids. 1982;**25-12**:2195-2203

[40] Chung D, Pullin DI. Large-eddy simulation and wall modelling of turbulent channel flow. Journal of Fluid Mechanics. 2009;**631**:281-309

[41] Çağlar M. Velocity fields with power-law spectra for modeling turbulent flows. Applied Mathematical Modelling. 2007;**31**:1934-1946

[42] Kara R, Çağlar M. Çinlar subgrid scale model for large eddy simulation. Applied Mathematics and Computation. 2018;**322**:89-99

[43] Clark TL. A small-scale dynamic model using a terrain following coordinate transformation. Journal of Computational Physics. 1977;**24**:186-215

[44] Lü X et al. Improving the energy efficiency of buildings based on fluid dynamics models: A critical review. Energies. 2021;**14-17**:5384

[45] Chalot F, Levasseur V, Mallet M, Petit G, Reau N. LES and DES simulations for aircraft design. In: Proceeding in ARC Aerospace Research Central, 45th AIAA Aerospace Sciences Meeting and Exhibit. Reno, Nevada; Jan 2007. p. 723

[46] Fureby C. Towards the use of large eddy simulation in engineering. Progress in Aerospace Sciences. 2008;**44**(6): 381-396

[47] Ekman P, Wieser D, Virdung T, Karlsson M. Assessment of hybrid RANS-LES methods for accurate automotive aerodynamic simulations. Journal of Wind Engineering and Industrial Aerodynamics. 2020;**206**: 104301

[48] Tominaga Y, Stathopoulos T. CFD modeling of pollution dispersion in a street canyon: Comparison between LES and RANS. Journal of Wind Engineering and Industrial Aerodynamics. 2011; **99**(4):340-348

[49] Kara R, Vergara C. Assessing turbulent effects in ascending aorta in presence of bicuspid aortic valve. Computer Methods in Biomechanics and Biomedical Engineering. 2023. Published online November 10, 2023. DOI: 10.1080/10255842.2023.2279938

Chapter 3

A Numerical Study of the Minimum-Dissipation Model for Large-Eddy Simulation in OpenFOAM

Jing Sun and Roel Verstappen

Abstract

The minimum-dissipation model (QR) has been utilized in studying turbulent channel flows at Reynolds numbers (Re_τ) of up to 2000, as well as in investigating the flow past a circular cylinder at a Reynolds number (Re) of 3900, and flow over periodic hills at Re = 10,595. In our investigations, we have employed both symmetry-preserving discretizations and standard second-order accurate discretization methods within the OpenFOAM framework. The outcomes are compared with Direct Numerical Simulation (DNS) and experiment, indicating a favorable alignment between the QR results and the reference data. The findings suggest that the static QR model achieves comparable performance to dynamic models while cutting computational costs by a factor of three. The model coefficient C = 0.024 produces the most precise predictions, and as the mesh resolution increases, the influence of the subgrid model decreases, dropping to less than 20% of the molecular viscosity at the finest mesh. Furthermore, the QR model can predict the mean and root-mean-square velocity accurately up to $Re_\tau = 2000$ without a wall damping function. The characteristics of turbulence strongly rely on spatial discretization methods. Various comparisons demonstrate the QR model conjugated with symmetry-preserving discretization performs better than the standard OpenFOAM discretization. Within the realm of OpenFOAM discretization, central difference schemes outperform other approaches.

Keywords: minimum-dissipation model, turbulence, large-eddy simulation, symmetry-preserving discretization, OpenFOAM

1. Introduction

Turbulent flows are prevalent in numerous engineering applications, yet simulating them using Direct Numerical Simulation (DNS) can be prohibitively expensive, especially for high-Reynolds-number flows. The Unsteady Reynolds-Averaged Navier-Stokes (RANS) method, while reducing computational costs, introduces significant simplifications, leading to decreased accuracy due to its treatment of large-scale unsteadiness. To tackle these challenges, large-eddy simulation (LES) has been developed. LES directly resolves the larger-scale, unsteady turbulent motions while modeling the effects of smaller-scale motions. This is achieved by representing the unresolved scale of motion using a subgrid model.

Among the subgrid models available for LES, the Smagorinsky model stands out as the most widely used one [1]. While the Smagorinsky model demonstrates satisfactory performance in decaying homogeneous turbulence simulations [2], it tends to dissipate eddies inappropriately for laminar and transitional flows. A way to improve the Smagorinsky model's performance is by dynamically computing the model coefficient [3], although this approach comes with increased computational costs. Another alternative is the WALE model [3], which addresses behavior near walls by utilizing the square of the velocity gradient tensor. Additionally, there is the Vreman model, which remains insensitive to pure shear but may lead to eddy dissipation in cases of backscatter and solid body rotation [4].

Minimum-dissipation models offer a straightforward alternative to Smagorinsky-type approaches for parameterizing subfilter turbulent fluxes in large-eddy simulation (LES). The QR model, proposed by Verstappen [5, 6], is the first minimum-dissipation model introduced in this context. The QR model possesses several desirable characteristics: it is more computationally efficient than dynamic models, effectively deactivates in laminar and transitional flows, and maintains consistency with the exact subfilter stress tensor on isotropic grids. Subsequently, the anisotropic minimum-dissipation model (AMD) was developed by Rozema et al. specifically for flows on anisotropic grids [7]. Abkar and Moin utilized the AMD model to examine high-Reynolds-number rough-wall boundary layer flow [8]. Zahiri et al. further extended the application of the AMD model by implementing it into OpenFOAM and conducting tests on both single-phase and multi-phase flows [9].

However, there have been limited investigations into the QR model, particularly within open-source software. In this study, we introduce the QR model into OpenFOAM and conduct simulations in scenarios involving high Reynolds numbers and complex geometries, marking the first instance of such exploration.

The discretization error results from discretizing the partial differential equations into algebraic equations. Verstappen and Veldman [10] put forward a proposition to precisely maintain the symmetry properties of the underlying differential operators. This involves representing the convective operator with a skew-symmetric matrix and the diffusive operator with a symmetric, positive-definite matrix. Trias et al. [11] extended this concept to unstructured collocated grids and introduced an approach based on fully conservative regularization of the convective term to address the issue of checkerboard spurious modes. Expanding upon these concepts, Komen et al. [12] devised a symmetry-preserving second-order time-accurate PISO-based method for solving incompressible flow equations on collocated grids. Hopman et al. made the code available to the public [13].

This chapter starts by introducing the QR model, the symmetry-preserving discretization, and OpenFOAM standard discretization methods. Then, simulations are conducted for channel flow, flow over periodic hills, and flow over a cylinder. The model coefficient is optimized and an investigation of the numerical discretization schemes is carried out. The obtained results are compared with the references [14–21]. The chapter concludes with a summary.

2. Numerical schemes

2.1 Minimum-dissipation model

The dynamics of large eddies in incompressible fluid flow are governed by the following momentum and continuity equations

$$\partial_t v + (v \cdot \nabla)v + \nabla p - 2\nu\nabla \cdot S(v) = -\nabla \cdot \tau(v),\ \nabla \cdot v = 0, \tag{1}$$

In this context, v represents the spatially filtered velocity, ν denotes the kinematic molecular viscosity, and p represents the pressure. $S(v) = \left(\nabla v + \nabla v^T\right)/2$ refers to the symmetric portion of the velocity gradient. The subgrid tensor $\tau(v)$ can be expressed as:

$$\tau_{ij} = \frac{1}{3}\tau_{kk}\delta_{ij} + \left(\tau_{ij} - \frac{1}{3}\tau_{kk}\delta_{ij}\right) = \frac{2}{3}k_{sgs}\delta_{ij} + \left(\tau_{ij} - \frac{1}{3}\tau_{kk}\delta_{ij}\right).$$

where $k_{sgs} = \frac{1}{2}\tau_{kk}$ is the subgrid kinetic energy. The subgrid stress tensor τ_{ij} is split into an isotropic part $\frac{1}{3}\tau_{kk}\delta_{ij}$ and anisotropic part $\tau_{ij} - \frac{1}{3}\tau_{kk}\delta_{ij}$. The isotropic part can be included into the pressure. The eddy-viscosity model describes the anisotropic part of the subgrid tensor as

$$\tau_{ij} - \frac{1}{3}\tau_{kk}\delta_{ij} = -2\nu_e S(v) \tag{2}$$

Note that the trace of $S(v)$ is zero due to the divergence-free condition, $\nabla \cdot v = 0$. The coefficient ν_e is termed the eddy viscosity of the model. This subgrid model is time irreversible (for $\nu_e > 0$), meaning it contributes to dissipation forward in time. In the classical Smagorinsky eddy-viscosity model [1], the eddy viscosity is set equal to:

$$\nu_e = C_S^2\Delta^2\sqrt{4q}. \tag{3}$$

Here $q(v) = \frac{1}{2}tr\left(S^2(v)\right)$ represents the second invariant of the strain-rate tensor $S(v)$. It's worth noting that $|S(v)| = \sqrt{2tr\left(S(v)^2\right)} = \sqrt{4q}$. Various Smagorinsky coefficient values C_s have been suggested for different scenarios, spanning from $C_s =$ 0.1-0.17 for channel flow and temporal mixing layers, to $C_s = 0.20 - 0.22$ for decaying homogeneous isotropic turbulence [2].

The first minimum-dissipation model, introduced by Verstappen [5], is based on the invariants of the strain-rate tensor, and is designed to deactivate in laminar flow and flows exhibiting negative eddy dissipation. This model operates under the assumption that the eddy-viscosity model should prevent the residual field, denoted as $v' = v - \overline{v}$ where the bar represents the LES filter, from being dynamically significant. This criterion is formalized by constraining the subgrid kinetic energy using Poincaé's inequality. According to Poincaré's inequality, there exists a constant C_Δ, contingent solely on Ω_Δ, such that for every function v in the Sobolev space $W^{1,2}(\Omega_\Delta)$

$$\int_{\Omega_\Delta} \|v - \overline{v}\|^2 dx \leq C_\Delta \int_{\Omega_\Delta} \|\nabla v\|^2 dx \tag{4}$$

where the residual field $v' = v - \bar{v}$ contains the eddies of size smaller than the length of the filter Δ, and $\| \cdot \| = \sqrt{\langle , \rangle}$ is the standard norm of the inner product $\langle , \rangle$ on the space of real-valued $L^2(\Omega_\Delta)$ functions, where Ω_Δ is a domain with diameter Δ_j; here it will be taken equal to a grid cell. The Poincaré constant C_Δ, which remains unaffected by v, is equal to the inverse of the smallest non-zero eigenvalue of the dissipative operator $-\Delta = -\nabla \cdot \nabla = \nabla^T \nabla$ on a grid cell Ω_Δ [22]. Note that $-\nabla = \nabla^T$ is utilized for the $L^2(\Omega_\Delta)$ inner product and periodic domain Ω_Δ. For convex domains, the Poincaré constant is denoted as $C_\Delta = (\Delta/\pi)^2$ [23]. Poincaré's inequality dictates that the kinetic energy of the residual field v' is bounded by a constant multiplied by the velocity gradient energy

$$\int_{\Omega_\Delta} \frac{1}{2} \|v'\|^2 dx \leq C_\Delta \int_{\Omega_\Delta} \frac{1}{2} \|\nabla v\|^2 dx \tag{5}$$

The equation governing the evolution of velocity gradient energy can be obtained by computing the inner product of L^2 with $\nabla^2 v$. Integration by part yields

$$\frac{d}{dt} \int_{\Omega_\Delta} \frac{1}{2} \|\nabla v\|^2 dx = -\nu \int_{\Omega_\Delta} \|\nabla^2 v\|^2 dx + \int_{\Omega_\Delta} (v \cdot \nabla) v \cdot \Delta v dx - \nu_e \int_{\Omega_\Delta} \|\nabla^2 v\|^2 dx. \tag{6}$$

The boundary terms arising from integration by parts vanish since Ω_Δ is a periodic box. The second term on the right-hand side of Eq. (6) signifies the generation of velocity gradient energy by the convective term. It can be represented as $r(v) = -tr(S^3(v))/3 = -det(S(v))$, which corresponds to the third invariant of the strain-rate tensor $S(v)$ (for further details, see Ref. [5–7]). We make the assumption that the molecular and eddy viscosity remains constant over a grid cell. The third term in Eq. (6) accounts for the dissipation caused by eddy viscosity and can be expressed as $q(\omega) = tr(S^2(\omega))/2$, where ω denotes the vorticity, $\omega = \nabla \times v$.

Incorporating $r(v)$ and $q(\omega)$ into the evolution equation of velocity gradient energy (Eq. (6)) leads to

$$\frac{d}{dt} \int_{\Omega_\Delta} \frac{1}{2} \|\nabla v\|^2 dx = -\nu \int_{\Omega_\Delta} \|\nabla^2 v\|^2 dx + 4 \int_{\Omega_\Delta} r(v) dx - 4\nu_e \int_{\Omega_\Delta} q(\omega) dx. \tag{7}$$

Let us consider that the eddy viscosity is chosen in such a way that the last two terms on the right-hand side of Eq. (7) counterbalance each other

$$\int_{\Omega_\Delta} r(v) dx = \nu_e \int_{\Omega_\Delta} q(\omega) dx \tag{8}$$

Then, we obtain

$$\frac{d}{dt} \int_{\Omega_\Delta} \frac{1}{2} \|\nabla v\|^2 dx = -\nu \int_{\Omega_\Delta} \|\nabla^2 v\|^2 dx. \tag{9}$$

Utilizing Poincaré's inequality and Gronwall's lemma on the right-hand side of the aforementioned equation yields:

$$\int_{\Omega_\Delta} \|v'\|^2 (x,t) dx \leq C_\Delta \int_{\Omega_\Delta} \|\nabla v\|^2 (x,t) dx \leq C_\Delta e^{\frac{-2\nu t}{C_\Delta}} \int_{\Omega_\Delta} \|\nabla v'\|^2 (x,0) dx.$$

The energy of the subgrid-scale decays at least as fast as $C_\Delta e^{\frac{-2\nu t}{C_\Delta}}$, for any filter length Δ. Therefore, we are able to keep the subfilter component v' under control through the utilization of Eq. (8). The minimum amount of eddy dissipation must adhere to the dissipation condition (9).

The expression for $q(\omega)$ in Eq. (8) can be equivalently written in terms of $q(v)$:

$$\int_{\Omega_\Delta} q(\omega)dx = \frac{1}{4}\int_{\Omega_\Delta} |\nabla\omega|^2 dx = \frac{1}{4}\int_{\Omega_\Delta} \omega\cdot(-\Delta)\omega dx = \frac{\int_{\Omega_\Delta} \omega\cdot(-\Delta)\omega dx}{\int_{\Omega_\Delta} \omega\cdot\omega dx}\cdot\int_{\Omega_\Delta} q(v)dx. \tag{10}$$

Thus, Eq. (8) becomes

$$\nu_e = \frac{\int_{\Omega_\Delta} \omega\cdot\omega dx}{\int_{\Omega_\Delta} \omega\cdot -\Delta\omega dx}\cdot\frac{\int_{\Omega_\Delta} r(v)dx}{\int_{\Omega_\Delta} q(v)dx} \tag{11}$$

The first fraction on the right-hand side above is at most C_Δ, which corresponds to the reciprocal of the smallest eigenvalue of $-\Delta$ on Ω_Δ.

Therefore, we establish $\nu_e\int_{\Omega_\Delta} q(v)dx = C_\Delta\int_{\Omega_\Delta} r(v)dx$. This equation guarantees that the subgrid scales remain dynamically insignificant, implying that their energy is bounded by Eq. (9), where the subgrid scales energy $\int_{\Omega_\Delta}\|v'\|^2(x,t)dx$ decays at least as rapidly as $C_\Delta e^{\frac{-2\nu t}{C_\Delta}}$, for any filter length Δ. Consequently, the minimum eddy viscosity required to counteract the nonlinear production is determined by

$$\nu_e = C_\Delta\frac{\overline{|r(v)|}}{\overline{q(v)}}. \tag{12}$$

To ensure non-negativity of the eddy viscosity, we employ the absolute value of $r(v)$. The quantities $\overline{q(v)}$ and $\overline{r(v)}$ represent the grid cell averages of the second and third invariant of the strain-rate tensor, respectively. In practical applications, grid cell averaging is typically approximated using midpoint integration. This characterizes the QR model

$$\tau - \frac{1}{3}tr(\tau)I = -2\nu_e S(v) = -2C_\Delta\frac{|r(v)|}{q(v)}S(v) \tag{13}$$

2.2 Numerical schemes in OpenFOAM

2.2.1 Spatial and temporal discretization

The discretization schemes typically maintain a second-order accuracy level. OpenFOAM employs a collocated arrangement where all variable values are computed and stored at the center x_P of the control volume V_P. These values are depicted by a piecewise constant profile, commonly known as the mean value

$$\phi(x_P) = \phi_P \approx \frac{1}{V_P}\int_{V_P} \phi(x)dV, \tag{14}$$

where ϕ_P denotes a discretized quantity. By employing the Divergence theorem (or Gauss theorem), the volume integrals present in the governing equations are transformed into surface integrals. Consequently, the problem shifts toward interpolating cell-centered values to face-centered values. The face values involved in the convective and diffusive fluxes must be determined through some form of interpolation from the centroid of the control volumes to their respective faces. The Gauss linear interpolation scheme in OpenFOAM is given by $\phi_f = \alpha\phi_P + (1-\alpha)\phi_N$, where $\alpha = \frac{f-N}{P-N}$, which yields a central difference scheme on a uniform mesh. The solver we use is pimpleFOAM if no other specification is provided.

If no specific instruction is provided, the continuous temporal derivative $\partial/\partial t$ is discretized using an implicit backward scheme, as indicated by Eq. (15)

$$\frac{3\phi^{n+1} - 4\phi^n + \phi^{n-1}}{2\Delta t} \approx \frac{\partial}{\partial t}(\phi), \tag{15}$$

where the $n + 1$ represents the value at the subsequent time level $t + \Delta t$, n denotes the value at the current time level, and $n - 1$ signifies the value at the preceding time level $t - \Delta t$.

2.3 Symmetry-preserving discretization scheme

2.3.1 Navier-stokes equation

The governing equation for incompressible Navier-Stokes (NS) is expressed as

$$\partial_t \mathbf{u} + (\mathbf{u}\cdot\nabla)\mathbf{u} - \frac{1}{Re}\nabla\cdot\nabla\mathbf{u} + \nabla p = 0, \quad \nabla\cdot\mathbf{u} = 0. \tag{16}$$

Here, u stands for the velocity (no filter applied), p is the pressure, and Re represents the Reynolds number.

When there are no external sources, for instance, body or boundary forces, the change rate of the total energy remains unaffected by both pressure disparities and convective transport. Instead, it is solely dictated by dissipation. This fundamental physical characteristic can be easily inferred from the symmetric properties of the differential operators in the NS Eq. (16).

The total energy of the flow (u, u) is defined using the conventional scalar product. Its temporal evolution can be derived by differentiating (u, u) with respect to time and then reformulating $\partial_t \mathbf{u}$ with the assistance of Eq. (16). This process yields:

$$\frac{d}{dt}(\mathbf{u},\mathbf{u}) = -((\mathbf{u}\cdot\nabla)\mathbf{u},\mathbf{u}) - (\mathbf{u},(\mathbf{u}\cdot\nabla)\mathbf{u}) + \frac{1}{Re}((\nabla\cdot\nabla\mathbf{u},\mathbf{u}) + (\mathbf{u},\nabla\cdot\nabla\mathbf{u})) - (\nabla p,\mathbf{u}) - (\mathbf{u},\nabla p).$$

By performing integration by parts on the linear and trilinear forms on the right-hand side, disregarding any boundary contributions, we get

$$(\mathbf{u},\nabla)^* = -(\mathbf{u},\nabla) \quad \text{and} \quad \nabla^* = -\nabla \tag{17}$$

As a result of these (skew-) symmetries, the convective- and pressure-dependent terms cancel, leading to a simplification in the change rate of the total energy to

$$\frac{d}{dt}(\mathbf{u},\mathbf{u}) = -\frac{2}{Re}(\nabla\mathbf{u},\nabla\mathbf{u}) \leq 0 \tag{18}$$

In the discrete context, the evolution of energy also adheres to Eq. (18), where u is substituted with the discrete velocity, and ∇ is replaced by its discrete counterpart. This holds true provided that the discretization of the differential operator upholds the (skew-) symmetries outlined in Eq. (17). Under this condition, the energy of any discrete solution remains conserved in the absence of viscosity, and diminishes over time in the presence of dissipation. Essentially, a symmetry-preserving spatial discretization of the NS equation exhibits unconditional stability and conservativeness.

2.3.2 Symmetry-preserving discretization

Consider the discretization of a first-order derivative in one spatial dimension. The Lagrangian interpolation (minimizing local truncation error) violates the skew-symmetry of the convective operator on nonuniform grids. Consequently, quantities conserved in the continuous formulation, such as kinetic energy, are not preserved at the discrete level. To stabilize the discretization, the energy undergoes systematic damping; for instance, in upwind methods, where the convective term introduces artificial dissipation that dampens the kinetic energy. Alternatively, explicit damping may be required to maintain stability. However, artificial dissipation disrupts the delicate balance between convective transport and physical dissipation, particularly at the smallest scales of motion, thereby distorting the essence of turbulence. Hence, symmetry-preserving discretization is employed.

Consider the continuing equation and linear momentum in one spatial dimension

$$\partial_t u + \overline{u}\partial_x u - \partial_{xx} u/\,Re + \partial_x p = 0, \qquad \partial_x u = 0. \tag{19}$$

For simplicity, the convective transport velocity $\overline{u}$ is assumed to be constant. In matrix-vector notation, the spatial discretization of Eq. (19) can be expressed as:

$$\Omega_0\frac{d\mathbf{u}_h}{dt} + \mathbf{C_0}(\overline{u})\mathbf{u}_h + \mathbf{D}_0\mathbf{u}_h + \Omega_0\mathbf{G}_0\mathbf{p}_h = 0, \qquad \mathbf{M}_0\mathbf{u}_h = 0 \tag{20}$$

where the diagonal matrix Ω_0 is built of the spacing of mesh: $(\Omega_0)_{i,i} = \frac{1}{2}(x_{i+1} - x_{i-1})$, the discrete velocities $u_i(t) \approx u(x_i, t)$ constitute the vector $\mathbf{u_h}$; the tridiagonal matrices $\mathbf{C_0}(\overline{u})$ and $\mathbf{D_0}$ represent the convective and diffusive operate, respectively.

The continuity equation $\partial_x u = 0$ is discretized as $\frac{u_{i+1}-u_{i-1}}{x_{i+1}-x_{i-1}} = 0$. This yields $\mathbf{M}_0\mathbf{u}_h = 0$.

The discretization of the pressure gradient is facilitated by utilizing the symmetry relation outlined in Eq. (17). According to Eq. (17), the continuous gradient operator equals the negative transpose of the divergence, implying that any velocity and pressure fields adhere to the relation $(\nabla p, \mathbf{u}) = -(p, \nabla\cdot\mathbf{u})$. This relation also applies to the discrete pressure $\mathbf{p}_h$ and the discrete pressure gradient $\mathbf{G}_0\mathbf{p}_h$, expressed as

$$\left(\mathbf{G}_0\mathbf{p}_h\right)^*\Omega_0\mathbf{u}_h = \mathbf{p}_h^*\mathbf{G}_0^*\Omega_0\mathbf{u}_h = -\mathbf{p}_h^*\mathbf{M}_0\mathbf{u}_h, \quad (21)$$

if the gradient operator is approximated by

$$\mathbf{G}_0 = -\Omega_0^{-1}\mathbf{M}_0^*. \quad (22)$$

Note that the gradient matrix, describing the integration of the pressure over the control volumes, is represented by $-\mathbf{M}_0^*$. Since the discrete pressure gradient also inherits boundary conditions from the discrete divergence, there is no need to explicitly specify boundary conditions for the pressure.

In the absence of diffusion, that is for $\mathbf{D_0} = 0$, the energy $\|\mathbf{u_h}\|^2 = \mathbf{u_h^*}\Omega_0\mathbf{u_h}$ of any solution $\mathbf{u_h}$ of the dynamic system of Eq. (20) is conserved if and only if the right-hand side of

$$\frac{d}{dt}\|\mathbf{u_h}\|^2 = -\mathbf{u_h^*}\left(\mathbf{C_0}(\overline{u}) + \mathbf{C_0^*}(\overline{u})\right)\mathbf{u_h} + \mathbf{u}_h^*\left(\mathbf{M}_0^*\mathbf{p}_h\right) + \left(\mathbf{M}_0^*\mathbf{p}_h\right)^*\mathbf{u}_h$$

is zero. This property remains valid (for any $\mathbf{u}_h$) if and only if the coefficient matrix $\mathbf{C}_0(\overline{u})$ exhibits skew-symmetry,

$$\mathbf{C}_0(\overline{u}) + \mathbf{C}_0^*(\overline{u}) = 0, \quad (23)$$

In other words, the discrete operator must retain the skew-symmetry observed in the continuous convective derivative $(\mathbf{u}\cdot\nabla)$.

To meet the skew-symmetry condition stated in Eq. (23), the interpolation weights of the neighboring discrete variables should be set to 1/2. Consequently, symmetry-preserving discretization yields

$$\overline{u}\partial_x u(x_i) \approx \overline{u}\frac{u_{i+1} - u_{i-1}}{x_{i+1} - x_{i-1}} = \left(\Omega_0^{-1}\mathbf{C_0}(\overline{u})\mathbf{u_h}\right)_i \quad (24)$$

The elements of the tridiagonal matrix $\mathbf{C_0}(\overline{u})$ are defined as follows: $\mathbf{C_0}(\overline{u})_{i,i-1} = -\frac{1}{2}\overline{u}$, $\mathbf{C_0}(\overline{u})_{i,i} = 0$, and $\mathbf{C_0}(\overline{u})_{i,i+1} = \frac{1}{2}\overline{u}$. It has been rigorously proven that Eq. (24) yields second-order accurate solutions on both uniform and nonuniform grids [24].

Diffusion is discretized in the same vein. The resulting coefficient matrix $\mathbf{D}_0$ is positive-definite, like the underlying differential operator $-\partial_{xx}$

$$\mathbf{D}_0 = \frac{1}{Re}\Delta_0^*\Lambda_0^{-1}\Delta_0,$$

where the difference matrix Δ_0 is defined as $(\Delta_0\mathbf{u}_h)_i = u_i - u_{i-1}$, and the non-zero entries of the diagonal matrix Λ_0 are given by $(\Lambda_0)_{i,i} = \delta x_i$. Consequently, the symmetric component of $\mathbf{C_0}(\overline{u}) + \mathbf{D_0}$ is solely influenced by diffusion, and hence is positive-definite. The energy of any solution $\overline{\mathbf{u}}_h$ of the semi-discrete system Eq. (20) evolves in a manner analogous to the continuous case; compare Eq. (18) to

$$\frac{d}{dt}\left(\mathbf{u}_h^*\Omega_0\mathbf{u}_h\right) \overset{(20)+(23)}{=} -\mathbf{u}_h^*\left(\mathbf{D}_0 + \mathbf{D}_0^*\right)\mathbf{u}_h \leq 0.$$

The right-hand side equals zero if and only if $\mathbf{u}_h$ lies within the null space of $\mathbf{D}_0 + \mathbf{D}_0^*$. Thus, in summary, as the energy $\|\mathbf{u}_h\|^2 = \left(\mathbf{u}_h^*\Omega_0\mathbf{u}_h\right)$ remains constant over time, a stable solution can be achieved on any grid. Similarly, higher-order symmetry-preserving discretization can be derived in a comparable manner (refer to [10] for details). Taking all ingredients together yields the symmetry-preserving discretization of the NS equation.

2.3.3 Symmetry-preserving discretization of Navier-Stokes equation

In three spatial dimensions, the incompressible Navier-Stokes equations in their semi-discrete form are expressed as

$$\Omega \frac{d\mathbf{u}_h}{dt} + \mathbf{C}(\mathbf{u}_h)\mathbf{u}_h + \mathbf{D}\mathbf{u}_h - \mathbf{M}^*\mathbf{p}_h = 0, \qquad \mathbf{M}\mathbf{u}_h = 0 \tag{25}$$

Global conservation laws involve integrals across the flow domain, which are transformed into scalar products upon discretization of the flow. For example, the change of the total mass of the flow is discretized as the scalar product of the constant vector 1 (with dimensions equal to the number of grid cells) and the discrete mass flux $\mathbf{M}\mathbf{u}_h$. Due to this scalar product equals zero ($\mathbf{M}\mathbf{u}_h = 0$) the total mass is conserved.

To determine the total momentum, the scalar product of velocity vector $\mathbf{u_h}$ with the vector $\Omega\mathbf{1}$ is taken, where the constant vector now encompasses as many entries as there are control volumes for the discrete velocity components $u_{i,j}$ and $v_{i,j}$). The evolution of total momentum straightforwardly follows from Eq. (25)

$$\frac{d}{dt}(\mathbf{1}^*\Omega\mathbf{u}_h) = -\mathbf{1}^*(\mathbf{C}(\mathbf{u}_h) + \mathbf{D})\mathbf{u}_h + \mathbf{1}^*\mathbf{M}^*\mathbf{p}_h = 0. \tag{26}$$

Thus, momentum conservation is ensured if $(\mathbf{C}(\mathbf{u}_h) + \mathbf{D})^*1 = 0$, and the law of mass conservation is consistently discretized as $\mathbf{M1} = \mathbf{0}$. The former condition can be divided into two conditions: one for convective discretization, $\mathbf{C}^*(\mathbf{u}_h)\mathbf{1} = 0$, and one for diffusive discretization, $\mathbf{D}^*\mathbf{1} = \mathbf{0}$. Moreover, we can omit the *'s, yielding $\mathbf{C}(\mathbf{u}_h)\mathbf{1} = 0$ and $\mathbf{D1} = \mathbf{0}$, since the convective matrix $\mathbf{C}(\mathbf{u}_h)$ is skew-symmetry and the diffusive matrix D is symmetric. Thus, it suffices to ensure that the constant vector lies in the null space of the approximate, convective, and diffusive operators.

The discretization is configured to ensure that the evolution of the (kinetic) energy $\mathbf{u}_h^*\Omega\mathbf{u}_h$ of any solution to Eq. (25) adheres to

$$\frac{d}{dt}(\mathbf{u}_h^*\Omega\mathbf{u}_h) = -\mathbf{u}_h^*(\mathbf{C}(\mathbf{u}_h)) + \mathbf{C}^*(\mathbf{u}_h))\mathbf{u}_h - \mathbf{u}_h^*(\mathbf{D} + \mathbf{D}^*)\mathbf{u}_h + \mathbf{u}_h^*(\mathbf{M}^*\mathbf{p})_h + \left(\mathbf{M}^*\mathbf{p}_h\right)^*\mathbf{u}_h,$$

where the right-hand side is negative for all $\mathbf{u}_h$'s, except those within the null space of $\mathbf{D} + \mathbf{D}^*$. The convective component cancels out due to $\mathbf{C}(\mathbf{u})_h$'s skew-symmetry, while the pressure terms nullify on the staggered grids (thus, cannot destabilize the spatial discretization) because the discrete pressure gradient correlates with the transpose of M, as seen in Eq. 22.

In summary, energy is conserved for inviscid flow, while for viscous flow, the energy $\|\mathbf{u_h}\|^2 = \mathbf{u_h^*}\Omega\mathbf{u_h}$ does not escalate in time. This indicates that the symmetry-preserving discretization Eq. 25 is stable and upholds conservation laws for mass, momentum, and energy.

2.3.4 Addressing the pressure-velocity coupling on the collocated grid

In a collocated grid setup, however, the actual velocity $\mathbf{u_c}$ is stored in the cell center. The coupling between velocity and pressure introduces an additional error term proportional to the third-order derivative of pressure $\tilde{\mathbf{p}}_c'$ into the momentum equation. This issue is commonly referred to as the checkerboard problem [25]. Trias et al. [11] proposed a method to eliminate the checkerboard spurious mode without introducing any nonphysical dissipation. The concept behind this approach involves utilizing a linear shift operator to transform a cell-centered velocity $\mathbf{u_c}$ into a staggered one $\mathbf{u_s}$ and employing a fully conservative regularization of the convective term to restrain the generation of small motions.

The linear shift operator serves to establish a relationship between the staggered velocity field and the cell-centered counterparts and vice versa. In this context, the subscript c denotes variables centered on the collated mesh, while s indicates variables staggered on the faces. The cell-to-face linear shift operator, denoted as $\Gamma_{c\rightarrow s} \in \mathbb{R}^{m\times 3n}$, converts a cell-centered velocity field into a staggered one

$$\mathbf{u}_s = \Gamma_{c\rightarrow s}\mathbf{u}_c. \tag{27}$$

Conversely, the cell-centered fields are linked to the staggered ones through the linear shift transformation $\Gamma_{s\rightarrow c} \in \mathbb{R}^{3n\times m}$,

$$\mathbf{u}_c = \Gamma_{s\rightarrow c}\mathbf{u}_s. \tag{28}$$

Note the general relation $\Gamma_{s\rightarrow c}\Gamma_{c\rightarrow s} = I$ only holds approximately, meaning $\mathbf{u}_c \approx \Gamma_{s\rightarrow c}\Gamma_{c\rightarrow s}\mathbf{u}_c$.

The face-to-cell shift operator $\Gamma_{s\rightarrow c}$ is constrained by Eq. (22) to eliminate the contribution of the pressure gradient term to the global kinetic energy. The expression for this is as follows

$$\Gamma_{s\rightarrow c} = (I_3 \otimes \Omega_c)^{-1}\Gamma^{*}_{c\rightarrow s}\Omega_s, \tag{29}$$

where $I_3 \in \mathbb{R}^{3\times 3}$ is the identity matrix.

The linear shift operator is given by

$$\Gamma_{c\rightarrow s} = N_s\Pi, \tag{30}$$

where matrices $\Pi \in \mathbb{R}^{3m\times 3n}$ and $N_s \in \mathbb{R}^{m\times 3m}$ are defined as $\Pi = I_3 \otimes \Pi_{c\rightarrow s}$ and $N_s = (N_{s,1}, N_{s,2}, N_{s,3})$, where $\Pi_{c\rightarrow s} \in \mathbb{R}^{m\times n}$ is the operator that interpolates a cell-centered scalar field to the faces, and $N_{s,i} \in \mathbb{R}^{m\times m}$ are diagonal matrices containing the x_i-spatial components of the face-normal vectors.

A classical fractional step projection method [26–28] is employed to address the velocity-pressure coupling. The staggered velocity field, denoted as $\mathbf{u}_s$, a velocity $\mathbf{u}_s^p$ can be uniquely decomposed into a curl-free vector represented as the gradient of a scalar field, $\mathbf{G}\tilde{\mathbf{p}}_c'$ and a solenoidal vector, $\mathbf{u}_s^{n+1}$. This decomposition is formulated as

$$\mathbf{u}_s^p = \mathbf{u}_s^{n+1} + \mathbf{G}\tilde{\mathbf{p}}_c', \tag{31}$$

taking the divergence of Eq. (31) yields a discrete Poisson equation for $\tilde{\mathbf{p}}_c'$,

$$\mathbf{M}\mathbf{u}_s^p = \mathbf{M}\mathbf{u}_s^{n+1} + \mathbf{M}\mathbf{G}\tilde{\mathbf{p}}_c' \Rightarrow \mathbf{M}\mathbf{u}_s^p = \mathbf{M}\mathbf{G}\tilde{\mathbf{p}}_c'. \tag{32}$$

The cell-center predicted velocity field, $\mathbf{u}_c^p$, is computed with the projection method and then corrected to obtain the velocity at the next time step, $\mathbf{u}_\mathbf{c}^{\mathbf{n+1}}$. Assuming $\mathbf{u}_c^p \approx \Gamma_{s\to c}\Gamma_{c\to s}\mathbf{u}_c^p$ is satisfied, the overall correction algorithm can be explicitly written by combining the expression of Eq. (31) and the linear shift operator, i.e.,

$$\mathbf{u}_c^{n+1} = \mathbf{u}_c^p + \Gamma_{s\to c}\Omega_s^{-1}\mathbf{M}^*\mathbf{L}^{-1}\mathbf{M}\Gamma_{c\to s}\mathbf{u}_c^p \tag{33}$$

$$= \mathbf{u}_c^p + (I_3 \otimes \Omega_c)^{-1}\Gamma_{c\to s}\mathbf{M}^*\mathbf{L}^{-1}\mathbf{M}\Gamma_{c\to s}\mathbf{u}_c^p, \tag{34}$$

where the $\Omega_s \in \mathbb{R}^{m\times m}$ and $\Omega_c \in \mathbb{R}^{n\times n}$ are diagonal matrices with staggered and cell-centered control volumes, respectively. The discrete Laplacian operator $\mathbf{L} \in \mathbb{R}^{n\times n}$, defined as$L \equiv -\mathbf{M}\Omega^{-1}\mathbf{M}^*$, is symmetric and negative-definite. The checkerboard problem stems from the unrealistic component of the cell-centered velocity field that cannot be eliminated by the pseudo-projection matrix,

$$\mathbf{u}_c^{\ominus} = \mathbf{u}_c^{\ominus} + \left((I_3 \otimes \Omega_c)^{-1}\Gamma_{c\to s}\mathbf{M}^* L^{-1}\mathbf{M}\Gamma_{c\to s}\right)\mathbf{u}_c^{\ominus}, \tag{35}$$

where $\mathbf{u}_c^{\ominus}$ denotes the spurious modes. These'unrealistic' velocity modes persist unless explicitly removed. Trias et al. [27] suggested regularizing (smoothing approximation) the convective term to address the origin of these 'unrealistic' velocity modes while maintaining the numerical solution devoid of nonphysical oscillations.

2.3.5 Constructing the discrete operators on unstructured collocated grids

A skew-symmetric convective operator $\mathbf{C}(\mathbf{u_s})$ is obtained in two stages [10]. Firstly, attention is given to the off-diagonal elements. The matrix $\mathbf{C}(\mathbf{u_s})$-diag$(\mathbf{C}(\mathbf{u_s}))$ is skew-symmetry when the interpolation weights of the adjacent discrete variables are set equal to 1/2; hence the discrete normal velocity $[\mathbf{u}_s]_f \approx \mathbf{u}_f \cdot \mathbf{n}_f$, situated at the centroid of the cell faces f, is expressed as

$$[\mathbf{u}_s]_f = [\mathbf{\Gamma}_{c\to s}\mathbf{u}_c]_f = \frac{1}{2}\left([\mathbf{u}_c]_{c1} + [\mathbf{u}_c]_{c2}\right) \cdot \mathbf{n}_f, \tag{36}$$

where $c1$ and $c2$ represent the cells neighboring the face f. The entries of the matrix $[C(\mathbf{u}_s)]_k$ correspond to half of the flux through face f between the adjacent cells i and j, formulated as

$$[C(\mathbf{u}_s)]_k = \frac{1}{2}[\mathbf{u}_s]_f A_f, \tag{37}$$

where A_f denotes the face area of f. In addition, for skew-symmetry, the diagonal elements should be zero, that is

$$[C(\mathbf{u}_s)]_{k,k} = \frac{1}{2}\sum_{f\in F_f(i)} [\mathbf{u}_s]_f A_f = 0, \tag{38}$$

where the $F_f(i)$ represents the set of faces neighboring face f. This criterion is met due to the vanishing discrete divergence of $\mathbf{u}_s$. Thus, the convective operator on unstructured collocated grids becomes

$$[C(\mathbf{u}_s)\mathbf{u}_c]_k = \sum_{f \in F_f(k)} \frac{1}{2}\left([\mathbf{u}_c]_{c1} + [\mathbf{u}_c]_{c2}\right)[\mathbf{u}_s]_f A_f, \tag{39}$$

at cell i.

By integrating the continuity equation given in Eq. (16) over an arbitrary cell k with volume $(\Omega_c)_{kk}$, we obtain

$$\int_{(\Omega_c)_{k,k}} \nabla \cdot \mathbf{u} dV = \int_{\partial(\Omega_c)_{k,k}} \mathbf{u} \cdot \mathbf{n} dS = \sum_{f \in F_f(k)} \int_{S_f} \mathbf{u} \cdot \mathbf{n} dS. \tag{40}$$

A discretization of Eq. (40) that is second-order accurate is

$$\sum_{f \in F_f(k)} \int_{S_f} \mathbf{u} \cdot \mathbf{n} dS \approx \sum_{f \in F_f(k)} [\mathbf{u}_s]_f A_f. \tag{41}$$

Hence, the divergence operator is taken as

$$[\mathbf{M}\mathbf{u}_s]_k \equiv \sum_{f \in F_f(k)} [\mathbf{u}_s]_f A_f = 0. \tag{42}$$

According to Eq. (22) the integrated pressure gradient operator, $\Omega_s\mathbf{G}$, equals the negative of the transpose of the divergence operator $-\mathbf{M}^*$. Hence, the discretization of the pressure gradient at the face f follows from Eq. (42)

$$[\Omega \mathbf{G} \mathbf{p}_c]_f = (p_{c1} - p_{c2})A_f, \tag{43}$$

The discrete gradient inherits boundary conditions from the discrete divergence, thus negating the need for specifying boundary conditions for the pressure. Finally, we compute the pressure from a Poisson equation that results from the incompressibility constraint.

The Laplacian operator is represented by the matrix,

$$\mathbf{L} = -\mathbf{M}\Omega_s^{-1}\mathbf{M}^*. \tag{44}$$

This negative-definite and symmetric matrix resembles the continuous Laplacian operator $\Delta \equiv \nabla \cdot \nabla = -\nabla^T \cdot \nabla$.

The approach used to discretize the Laplacian operator in the pressure Poisson equation is also employed to discretize the diffusive term in NS equations. The diffusive operator is seen as the product of two first-order differential operators, a divergence and a gradient. The divergence operator is discretized and the gradient operator becomes the transpose of the discrete divergence

$$\mathbf{D}_c = -\frac{1}{Re}\mathbf{M}\mathbf{G} = \frac{1}{Re}\left(\mathbf{M}\Omega_s^{-1}\right)\mathbf{M}^*. \tag{45}$$

This procedure results in a symmetric, positive-definite, approximation of the diffusive operator $-\nabla \cdot \nabla$. The collocated diffusive operator for a cell-centered variable ϕ_c is expressed as,

$$[\mathbf{D}_c \phi_c] = \frac{1}{Re} \sum_{f \in F_f(k)} \frac{(\phi_{c1} - \phi_{c2})A_f}{\delta n_f}, \tag{46}$$

where the length δn_f approximates the distance between the centroid of the neighboring cells $c1$ and $c2$, calculated as $\delta n_f = |\mathbf{n}_f \cdot \overleftarrow{c1c2}|$. Afterwards, the volume of the cell associated with the face-normal velocity at face f is denoted as $(\Omega_s)_f = \delta n_f A_f$, where the $F_f(k)$ is the set of faces boarding the face f.

2.3.6 Solver

The RKSymFoam solver's primary algorithm comprises three hierarchical iterative levels in the case of implicit time discretization [12, 13]:

1. The outer loop iterates over each RungeKutta (RK) stage i, ranging from 1 to s (see Section 2.3.8 for details).

2. An intermediate loop manages to update the nonlinear convective term.

3. An inner PISO loop handles the pressure-velocity coupling.

For explicit temporal discretization, a single projection step suffices for the pressure-velocity coupling, and there's no need to update the convective term iteratively.

2.3.7 Temporal discretization

To maintain the favorable conservation and stability properties in discrete time, the time discretization in Eq. (25) should lead to a skew-symmetric operator, which can only be achieved through an implicit temporal integration. In high-Reynolds-number flow simulations, the implicit method may incur higher computational costs compared to the explicit counterpart. Yet in an explicit temporal discretization scheme for convection-diffusion equations, the time step Δt is typically constrained by convective stability conditions such as $U\Delta t < \Delta y$ (where U represents the maximum velocity magnitude and Δy denotes the spatial grid size), diffusive stability conditions akin to $2\Delta t < Re\,\Delta y^2$. The implicit method does not impose such stringent restrictions, allowing for a time step that can be much larger. To strike a balance between accuracy and stability, ensuring that the simulation remains globally second-order accurate, the Crank-Nicolson and diagonal-implicit Runge–Kutta methods are chosen.

2.3.8 Butcher tables specify Runge-Kutta methods

After applying the spatial discretization the momentum equations Eq. (16) become

$$\frac{d\mathbf{u}_c}{dt} = \mathbf{F}_c(t, \mathbf{u}_c) - (\nabla p)_c \quad \text{with} \quad \mathbf{u}_c(t^n) = \mathbf{u}_c^n. \tag{47}$$

Here, the subscript c denotes the cell-centered values, while the pressure gradient term remains to be discretized. Runge-Kutta (RK) schemes, specified by a Butcher table [29], are used to discretize Eq. (47) in time. Using a_{ij} to denote the stage weights for stage i, and c_i as the quadrature nodes for the schemes with $c_i = \sum_{j=1}^{s} a_{ij}$, where $i = 1, \ldots, s$, and $t^i = t^n + c_i \Delta t$, s denotes the number of stages, and b_j denotes the primary weights of the utilized RK scheme, satisfying $\sum_{j=1}^{s} b_j = 1$. Explicit schemes are characterized by having only non-zero entries in the lower triangle of the $a_{i,j}$ portion of the table. Entries on or above the diagonal lead to implicit methods as they involve $\mathbf{u}_c^{n+1}$ on the right-hand side of Eq. (47).

The intermediate solution $\tilde{\mathbf{u}}_c^i$ for stage i at time t^i is expressed as

$$\tilde{\mathbf{u}}_c^i = \mathbf{u}_c^n + \Delta t \left(\sum_{j=1}^{i} a_{ij} \mathbf{F}_c \left(t^j, \tilde{\mathbf{u}}_c^j \right) - c_i (\nabla \tilde{p})_c^i \right), \tag{48}$$

and the final solution $\mathbf{u}_c^{n+1}$ at time $t^{n+1} = t^n + \Delta t$ by

$$\mathbf{u}_c^{n+1} = \mathbf{u}_c^n + \Delta t \left(\sum_{j=1}^{i} b_j \mathbf{F}_c \left(t^j, \tilde{\mathbf{u}}_c^j \right) - (\nabla p)_c^{n+1} \right). \tag{49}$$

We first discuss the pressure-velocity coupling for an explicit time integration. Here we concentrate on explaining the forward-Euler method as an example, as higher-order explicit RK methods essentially involve a series of forward-Euler stages. We compute an intermediate velocity field $\mathbf{u}_c^p$ through the following predictor step

$$\frac{\mathbf{u}_c^p - \mathbf{u}_c^n}{\Delta t^n} = \mathbf{H}\left(\mathbf{u}_s^n\right)\mathbf{u}_c^n - \mathbf{G}_c \mathbf{p}_c^p. \tag{50}$$

Here $\mathbf{H}\left(\mathbf{u}_s^n\right)\mathbf{u}_c^n \equiv -\Omega_c^{-1}\left(\mathbf{C}\left(\mathbf{u}_s^n\right)\mathbf{u}_c^n + \mathbf{D}\mathbf{u}_c^n\right)$, where $\mathbf{H}\left(\mathbf{u}_s^n\right) \in \mathbb{R}^{3n\times 3n}$, and Δt^n denotes the time difference between time levels n and $n+1$. As the intermediate velocity is not divergence-free, we correct the final values for the time step $n+1$ by adding the following corrections to the intermediate values: $\mathbf{u}_c^{n+1} = \mathbf{u}_c^p + \mathbf{u}_c'$ and $\mathbf{p}_c^{n+1} = \mathbf{p}_c^p + \tilde{\mathbf{p}}_c'$, where the velocity and pressure correction are related by

$$\mathbf{u}_c' = -\Delta t \mathbf{G}_c \tilde{\mathbf{p}}_c'. \tag{51}$$

After obtaining the pressure correction $\tilde{\mathbf{p}}'$ from the Poisson equation Eq. (32), we can calculate the velocity correction $\mathbf{u}'$ using Eq. (51). Subsequently, the new velocity and pressure fields for the next time step can be computed using the velocity and pressure corrections $\mathbf{u}_c'$ and $\tilde{\mathbf{p}}_c'$.

We now consider the pressure-velocity coupling for implicit time integration. The RKSymFoam solver employs the PISO approach for implicit time integration. Let us illustrate this with the backward Euler method. The PISO algorithm comprises a predictor step followed by n_{corr} corrector steps (or inner iterations), as outlined below [30]:

1. Predictor step: The initial intermediate velocity $\mathbf{u}_c^1$ is calculated using the predictor equation

$$\frac{\mathbf{u}_c^1 - \mathbf{u}_c^n}{\Delta t^n} = \mathbf{H}\left(\mathbf{u}_s^p\right)\mathbf{u}_c^p - \mathbf{G}_c \mathbf{p}_c^p, \tag{52}$$

where $\mathbf{H}\left(\mathbf{u}_s^p\right)\mathbf{u}_c^p = \mathbf{H}\left(\mathbf{u}_s^n\right)\mathbf{u}_c^n$ is utilized to refrain from an implicit treatment of the convective and diffusive terms. The resulting initial intermediate velocity may not be divergence-free, necessitating subsequent corrector steps.

2. Corrector steps: In each correct step, a new pressure field $\mathbf{p}_c^k$ and a corresponding revised velocity $\mathbf{u}_c^{k+1}$ (which is divergence-free $\mathbf{M}\Gamma_{c\to s}\mathbf{u}_c^{k+1} = 0_{c,n}$) are computed. To enhance the stability of the time-discrete momentum equation during the corrector step, the operator is partitioned into a diagonal component $A^d(\mathbf{u}_s^n)$ acting on $\mathbf{u}_c^{k+1}$ and an off-diagonal component $\mathbf{H}^{od}(\mathbf{u}_s^n)$ acting on $\mathbf{u}_c^k$, resulting in

$$\mathbf{u}_c^{k+1} = \mathbf{B}^{-1}\left(\Delta t\mathbf{H}^{od}\left(\mathbf{u}_s^n\right)\mathbf{u}_c^k + \mathbf{u}_c^n\right) - \Delta t\mathbf{B}^{-1}\mathbf{G}_c\mathbf{p}_c^k, \quad \text{with} \quad k = 1, 2, \ldots, n_{corr}, \tag{53}$$

where $\mathbf{B} = \mathbf{I} - \Delta t^n A^d\left(\mathbf{u}_s^n\right) \in \mathbb{R}^{3n\times 3n}$. A Poisson equation for pressure $\mathbf{p}_c^k$ can be derived by taking the divergence of Eq. (53) and employing $\mathbf{M}\Gamma_{c\to s}\mathbf{u}_c^{k+1} = 0_{c,n}$. Thus we get

$$\mathbf{M}\mathbf{B}_S^{-1}\Omega_s^{-1}\mathbf{M}^*\mathbf{p}_c^k = \frac{1}{\Delta t}\mathbf{M}\Gamma_{c\to s}\mathbf{B}^{-1}\left(\Delta t\mathbf{H}^{od}\left(\mathbf{u}_s^n\right)\mathbf{u}_c^k + \mathbf{u}_c^n\right), \quad k = 1, 2, \ldots, n_{corr}, \tag{54}$$

where $\mathbf{B}_s^{-1} \approx \Gamma_{c\to s}\mathbf{B}^{-1}\Gamma_{s\to c}$, is a square diagonal matrix. The discrete Laplacian in the implicit scheme is $\mathbf{L} = -\mathbf{M}\mathbf{B}_s^{-1}\Omega_s^{-1}\mathbf{M}^*$. The pressure correction $\tilde{\mathbf{p}}_c^{\prime k}$ is obtained from the following Poisson equation

$$\mathbf{M}\mathbf{B}_S^{-1}\Omega_s^{-1}\mathbf{M}^*\mathbf{p}_c^{\prime k} = \frac{1}{\Delta t}\mathbf{M}\Gamma_{c\to s}\mathbf{B}^{-1}\left(\Delta t\mathbf{H}^{od}\left(\mathbf{u}_s^n\right)\mathbf{u}_c^k + \mathbf{u}_c^n - \mathbf{G}\mathbf{p}_c^n\right), \quad k = 1, .., n_{corr}, \tag{55}$$

After obtaining the pressure field $\mathbf{p}_c^k$ from the Poisson equation Eq. (54), the corresponding updated velocity can be computed using Eq. (53). This iterative process continues for n_{corr} iterations until an inner iteration criterion is met. The inner PISO iteration procedure concludes by updating the velocity, which is utilized to solve the non-linearity in the convective part of the operator H, with the new velocity $\mathbf{u}_c^{n_{corr}}$ for the subsequent outer iteration.

3. Verification and validation

The test cases encompass channel flow across a range of friction Reynolds numbers (Re_τ), including 180, 550, 1000, and 2000, along with flow over periodic hills and flow past a circular cylinder. Both static and dynamic minimum-dissipation models are subjected to numerical examinations in channel flow scenarios. The investigation delves into the sensitivity of the QR model concerning spatial discretization schemes, mesh resolution, and Reynolds number variations, comparing standard OpenFOAM discretization against symmetry-preserving discretization methods. This section focuses on the main findings; for comprehensive details, see [31].

3.1 Optimal QR model coefficient

The simulations were performed using the QR model with different model constants for a channel flow at $Re_\tau = 180$ and periodic hills, aiming to identify the optimal model constant. **Figure 1** depicts the normalized mean streamwise velocity against the distance to the wall. It demonstrates that the model coefficient C = 0.024 aligns precisely with DNS data. Additionally, utilizing five different error measurements (mean square, absolute, maximum absolute, and slope errors), the constant

C = 0.024 was identified as yielding the smallest error. This model coefficient yields accurate results under mesh convergence. Furthermore, the various comparisons carried out for flows over periodic hills demonstrate that the model coefficient of C = 0.024 is again the best one. In literature, the Smagorinsky model coefficient is typically considered optimal within the range of $C_s = 0.1$ to $C_s = 0.2$. Consequently, the ideal coefficient for the QR model falls within the range of C_s^2. Note that the model coefficient utilized in the QR model is not squared as in the Smagorinsky model.

3.2 Static QR model compared to dynamic models in channel flow at $\mathrm{Re}_\tau = 180$

The static QR model is compared with four dynamic models. The dynamic procedure is based on the Lilly-Germano identity [2]. All subgrid models depicted in **Figure 2** closely approximate the DNS results in the near-wall region, particularly the dynamic QR model, which agrees well with the DNS data. A detailed examination of the figure highlights that in the logarithmic wall range $y^+ > 30$, the static QR, dynamic minimum-dissipation (DMD), and dynamic QR (DQR) models closely match the DNS results. Conversely, the hybrid dynamic (HDM) and dynamic Smagorinsky (DSM) models nearly overlap and deviate from the DNS data in the logarithmic wall region.

In this case, both the static QR and DQR models exhibit similar performance, with negligible differences in errors. On the other hand, the DSM and HDM demonstrate comparatively lower accuracy, while the DMD model exhibits the largest error.

In summary, this study underscores the reliability of the static QR model. A suitably chosen model coefficient can yield results akin to a dynamic model while cutting computational costs by a factor of three.

3.3 Sensitivity to expansion ratio in channel flow at $\mathrm{Re}_\tau = 180$

This investigation focuses on the expansion ratio of the mesh in the wall-normal direction. The expansion ratio is defined as follows

$$\text{Expansion ratio} = \frac{\Delta y_c^+}{\Delta y_w^+}, \tag{56}$$

where Δy_c^+ represents the size of the cell at the channel center, while Δy_w^+ denotes the size of the first cell adjacent to the wall. **Figure 3** illustrates the normalized mean streamwise and rms velocity against wall distance in wall unit at $Re_\tau = 180$,

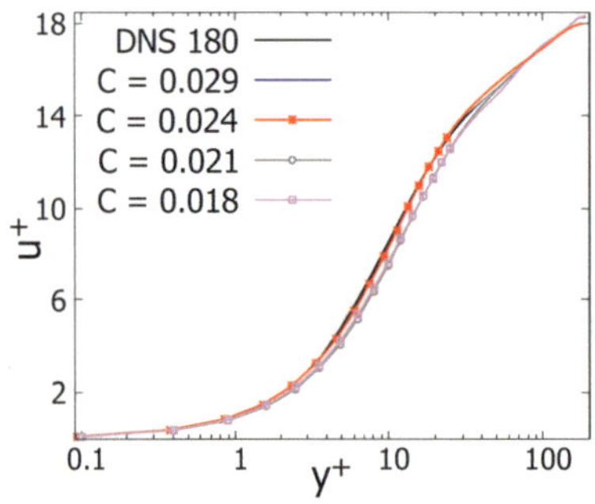

Figure 1.
Normalized mean streamwise velocity versus wall-normal distance in channel flow simulation at $Re_\tau = 180$, for different coefficient C in the QR model.

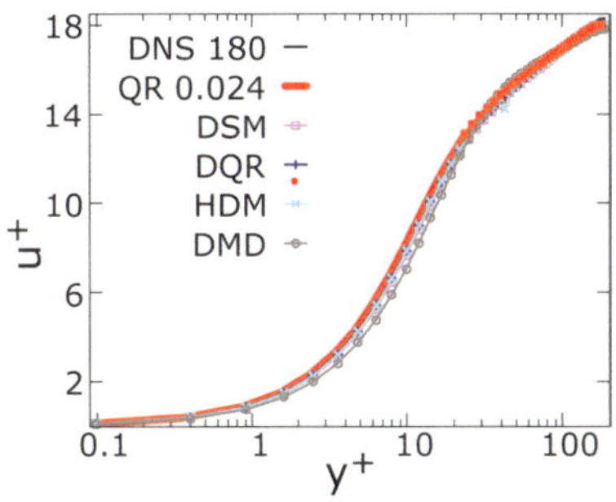

Figure 2.
Average streamwise velocity u^+ obtained from static QR and dynamic models compared to DNS data.

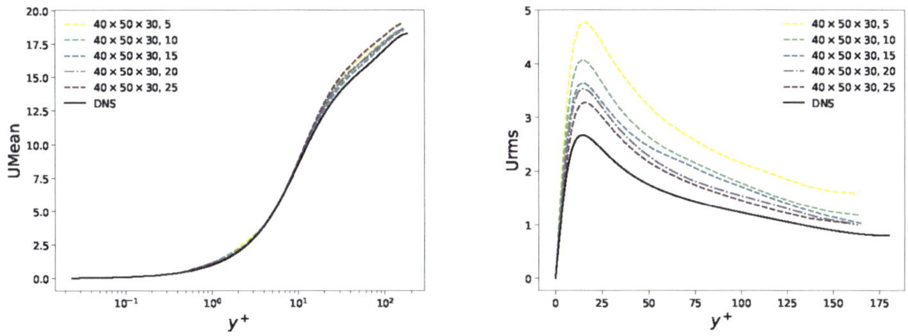

Figure 3.
Normalized mean streamwise and rms velocity using QR model with C = 0.023 at $Re_\tau = 180$ for five grid expansion ratios: 5,10,15, 20, and 25.

employing a grid resolution of $40 \times 50 \times 30$. The normalized uniform grid spacing in the streamwise direction is $\Delta x^+ = 18$ and in spanwise direction is $\Delta z^+ = 12$. The expansion ratios in the wall-normal direction are (approximately) 5, 10, 15, 20, and 25, respectively. The first normalized wall-normal grid height is $\Delta y_w^+ = 1.728$. The time step for simulation is $\Delta t^+ = 36$. As the expansion ratio increases, the near-wall mesh refines, while the mesh, in the channel center coarsens.

For the mean streamwise velocity u^+, reducing the expansion ratio from 15 to 5 leads to increasingly pronounced overestimation of the mean velocity beyond $y^+ > 10$. However, this overestimation exacerbates further with expansion ratios of 20 and 25.

For the velocity fluctuation in the streamwise direction $u'u'$, lowering the expansion ratio from 25 to 5 results in notably exaggerated peaks in the near-wall region and relatively diminished overestimation in the central channel. For the velocity fluctuation in the wall-normal $v'v'$ and spanwise $w'w'$ directions (not shown here), reducing the expansion ratio from 20 to 5 causes overestimation of peak values near the wall and comparatively smaller underestimations in the central channel. The highest expansion ratio, 25, underestimates the near-wall peak but provides more accurate values in the channel center.

3.4 Grid Independence study in channel flow at $\mathrm{Re}_\tau = 550$

This section of the survey investigates how different grid configurations affect the mean and root-mean-square velocities at a Reynolds number of $Re_\tau = 550$, utilizing a

QR model with C = 0.023. Four grid setups are considered: 128^3, 64^3, 32^3, and 16^3. **Figure 4** illustrates the normalized mean streamwise velocity as a function of wall distance in wall units, with the channel dimensions being $2\pi \times 2 \times \pi$. The expansion ratio in the wall-normal direction is approximately 10, and the simulation employs a time step of $\Delta t^+ = 55$.

Initially, the analysis focuses on the results of the simulation conducted using the 16^3 grid points (depicted by the blue line on the left side of **Figure 4**). The first mesh point is situated around $y^+ = 7$, indicating a significant distance from the wall. Despite the coarse mesh, the obtained results surprisingly exhibit reasonable agreement with the overall trend observed in the other simulations.

A noteworthy observation is the discernible trend that emerges with increasing mesh resolution: the mean velocity profile progressively aligns more closely with DNS data, particularly evident when utilizing 128^3 mesh points. As resolution increases, the proximity of the initial mesh points to the wall decreases, resulting in a greater concentration of mesh in the near-wall region. Consequently, the 128^3 case demonstrates complete agreement with DNS data in the near-wall region.

3.5 Symmetry-preserving compared to standard OpenFOAM discretization

In this section, we illustrate the effect of the numerical schemes by comparing the symmetry-preserving discretization with the Pimple algorithm conjugated with Gauss linear schemes in three benchmarks, that is, periodic channel flow, periodic hill flow, and flow over a cylinder.

The channel flow outcomes depicted in **Figure 5**, show that the results obtained with symmetry-preserving discretization closely resemble those of the DNS reference. Notably, in the region where $y^+ \geq 10$, the symmetry-preserving scheme with the QR model precisely aligns with the DNS results.

The symmetry-preserving method enhances velocity fluctuation predictions in the region where $y^+ \leq 220$. In other areas of the computational domain, both methods yield comparable results. For the Reynolds shear stress $u'v'$, the two methods give the same accurate prediction.

Additionally, it has been observed that the contribution of the subgrid-scale model to the diffusive flux in both methods is lower than 20% of the contribution of molecular viscosity [31].

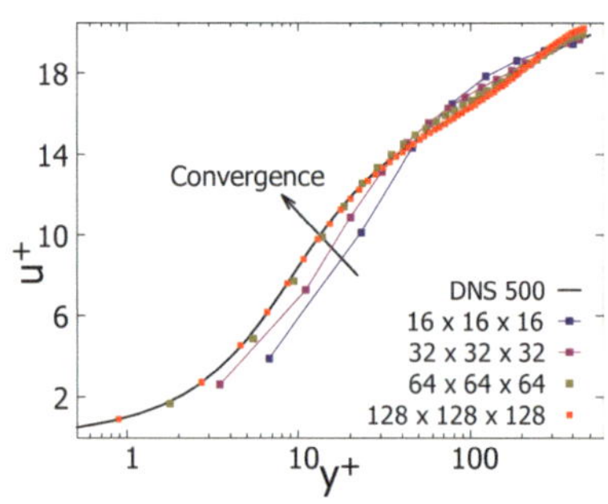

Figure 4.
Comparison of streamwise velocities using QR Model (C = 0.023) at $Re_\tau = 550$ on various mesh resolutions (128^3, 64^3, 32^3, 16^3) with DNS results at $Re_\tau = 500$.

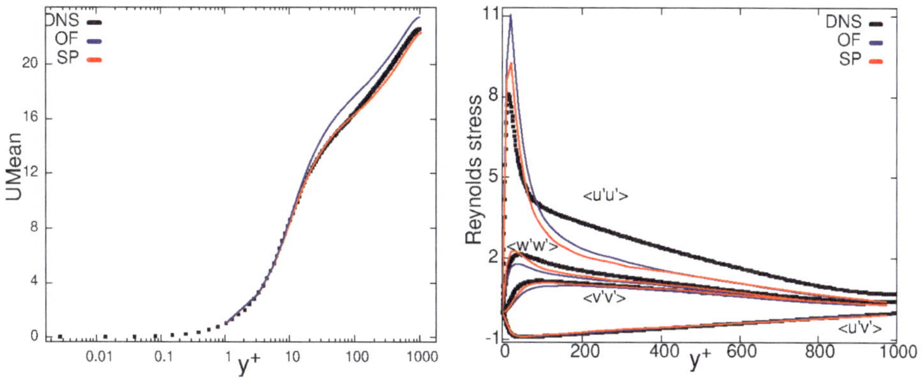

Figure 5.
The mean streamwise velocity obtained from the QR model with standard OpenFOAM and symmetry-preserving (SP) schemes for fully developed turbulent channel flow at $Re_\tau = 1000$.

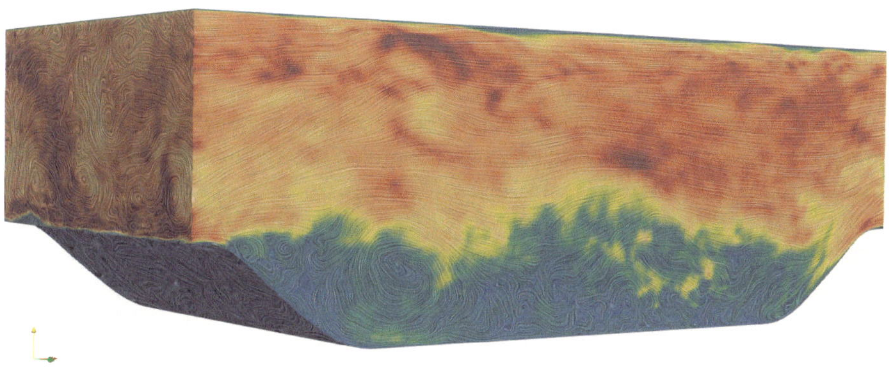

Figure 6.
The side view of the periodic hill flow.

3.6 Application to periodic hill flow

The phenomenon of flow separation from curved surfaces followed by reattachment is frequently encountered in engineering applications. To evaluate the effectiveness of the proposed minimum-dissipation model in OpenFOAM for simulating separated flows, three-dimensional flow simulations over periodic hills at $Re = 10{,}595$ as, illustrated in **Figure 6**, have been conducted.

The mean velocity and Reynolds stress calculated by the QR model, without incorporating a wall damping function, are compared with experimental data [32] and three numerical experiments [33]. It is important to note that all large-eddy simulations are based on second-order accurate numerical techniques both in space and time. Additionally, the QR model grid comprises six times fewer grid points compared to the grid used in the reference CFD data. The QR simulation mesh comprises 2.56 million grid points and necessitates 6 hours for simulation on two nodes, each equipped with 128 cores. This mesh size proves adequate for modeling this case, as further refinement does not yield improved results.

3.6.1 Separation and reattachment lengths

The separation and reattachment points are obtained at Re = 11,230, that is, the bulk velocity $u_b = 1.06$, using Gauss linear spatial discretization and backward temporal discretization. The separation point is accurately pinpointed by numerically solving boundary layer equations under pressure-adverse conditions. The QR model predicts the separation point, where the wall shear stress reaches zero, to be approximately $x/H \approx 0.175$, slightly smaller than the reference value of $x/H \approx 0.19$. This deviation is expected, as the separation point tends to shift upstream with increasing Reynolds numbers. The precise location of the separation point significantly influences the point of reattachment. Recirculation initiates around $x/H \approx 0.27$ and extends until approximately $x/H \approx 4.71$, with the main recirculation bubble spanning a length of roughly 4.48. The reattachment position, where the dividing streamline reconnects with the wall, is determined to be at $x/H = 4.71$, closely aligning with the reference value of 4.69 [32].

3.6.2 Sensitivity to the numerical schemes of standard OpenFOAM

A significant discrepancy has been identified in the prediction of the streamwise velocity component $\langle u \rangle / u_b$ within the range $1.5 < y/H < 2.8$, despite employing the same Reynolds number as the reference case (Re = 10, 595).

In an effort to understand the cause of this discrepancy, various interpolation numerical schemes for solving the divergence term and pressure gradient have been evaluated. These schemes include central difference, filtered and limited central difference, linear upwind blended with central difference, least squares, and detached eddy discretization schemes. However, the outcomes of these different numerical schemes fail to accurately predict the mean velocity, as illustrated in **Figure 7**. Consequently, it has been determined that increasing the Reynolds number to better align with the reference data is necessary. Furthermore, despite minor differences observed between the simulations, the central difference schemes are ultimately selected due to their satisfactory performance in previous validation cases. These discrepancies, while present, are deemed insignificant in distinguishing one technique from another.

3.6.3 Sensitivity to the Reynolds number

In order to address the underprediction of the streamwise velocity component $\langle u \rangle / u_b$ within the range of $1.5 < y/H < 2.8$ while maintaining the Reynolds number

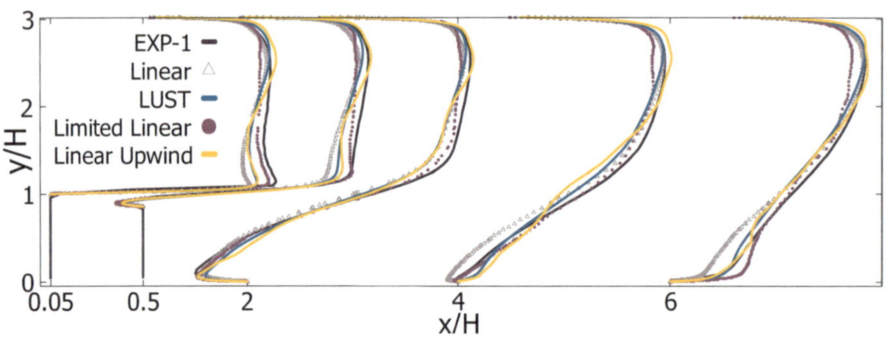

Figure 7.
Streamwise velocity at five different positions with varying finite volume discretization schemes.

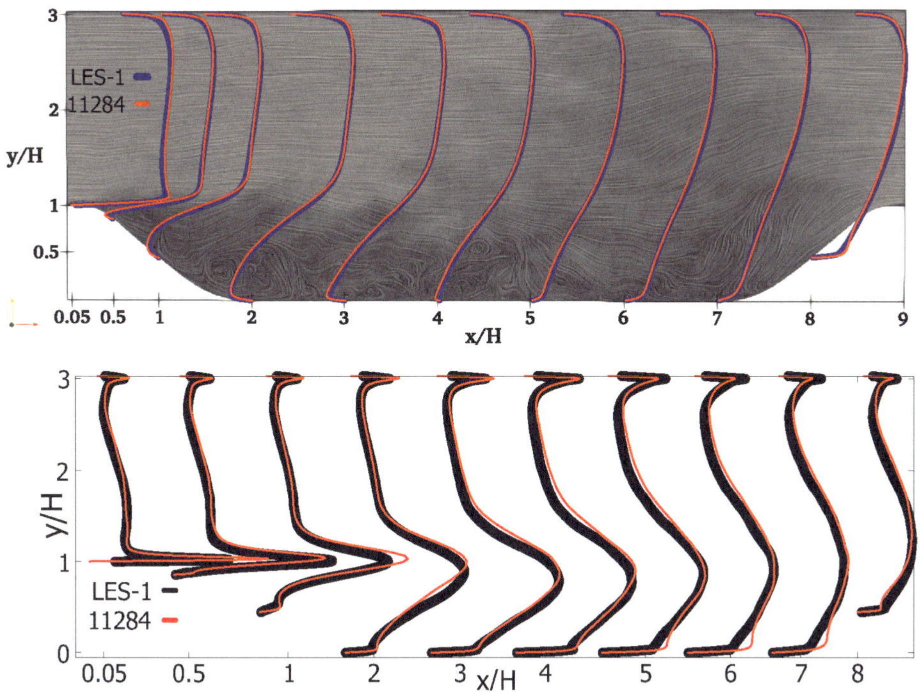

Figure 8.
Streamwise velocity (upper) and Reynolds stress (lower) at ten distinct locations, with a 6.5% increment in Reynolds number compared to the B-spline data. 11,284 refers to the QR results and LES-1 indicates the reference B-spline results.

equal to the reference value of *Re* = 10,595, simulations at higher Reynolds numbers were explored. The most notable revelation from the analysis, as depicted in **Figure 8**, is that the mean velocity and Reynolds stress obtained with a 6.5% higher Reynolds number align with the B-spline reference. Furthermore, a 7% increase in Reynolds number corresponds to agreement with experimental data (not illustrated here). These findings confirm that augmenting the Reynolds number by 3–7% successfully aligns the simulations with the reference data. Note that despite the experimental data and three LES simulations used for comparison being conducted at the same Reynolds number of *Re* = 10,595, they exhibit discrepancies among themselves.

3.7 Application to flow over a cylinder

3.7.1 Comparison between standard OpenFOAM spatial discretization

To examine the impact of different numerical schemes employed in standard OpenFOAM spatial discretization on turbulence, simulations are conducted for both periodic hills and flow over a cylinder.

Figure 9 displays the mean streamwise velocity u resulting from three distinct numerical discretizations applied to flow over a cylinder. Notably, discrepancies in mean streamwise velocity u are observed in the vicinity of the wake $(x/D < 2.02)$. Specifically, the upwind-blended scheme predicts a U-shaped mean velocity profile at $x/D = 1.06$, transitioning to a markedly lower V-shaped profile at $x/D = 1.54$ and 2.02, compared to the other two schemes.

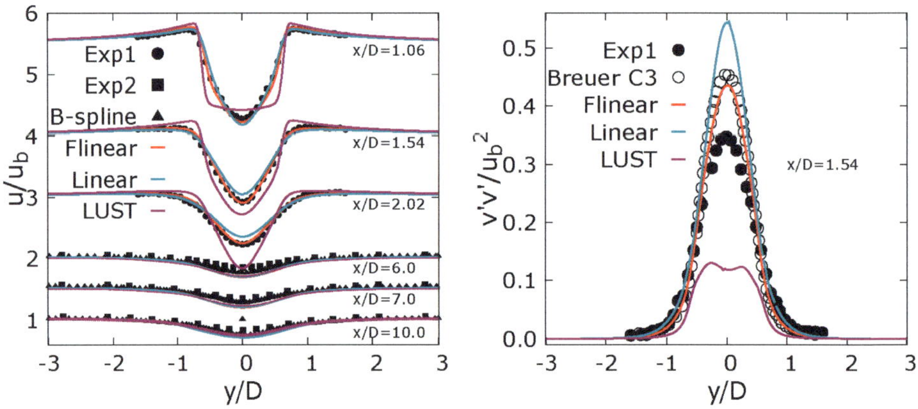

Figure 9.
The mean streamwise velocity at six locations (left) and the Reynolds stress in the cross-section at $x/D = 1.54$ *(right) are depicted for flow over a cylinder at* $Re_D = 3900$. *The QR model is utilized with filtered central difference (Flinear), pure central difference (Linear), and upwind-biased (LUST) schemes.*

Analysis from the central difference simulation reveals that the onset of turbulence in the separating shear layers occurs nearer to the cylinder, leading to the formation of a shorter vortex region and the development of a V-shaped velocity profile. Consequently, the shear layers exhibit reduced length, and the recirculation region diminishes in size [31]. In the downstream region, variations in numerical schemes appear to have minimal impact on mean velocity. However, simulations of periodic hills demonstrate that the selection of numerical schemes substantially influences both mean and root-mean-square variables.

4. Conclusions

A comprehensive comparison between the minimum-dissipation model (QR) and experimental/DNS/LES results across channel flow, flow past a circular cylinder, and flow over periodic hills reveals generally favorable agreement, particularly on coarse meshes.

The findings suggest that the static QR model is comparable to dynamic models while reducing computational costs. Notably, a model coefficient of C = 0.024 produces the most accurate predictions, with the contribution of the subgrid model diminishing to less than 20% of the molecular viscosity as mesh resolution increases to its finest level.

The characteristics of turbulence exhibit dependence on spatial discretization methods. Various comparisons highlight the superiority of the QR model when paired with symmetry-preserving discretization over standard OpenFOAM discretization techniques. Among standard OpenFOAM discretization methods, central difference schemes emerge as the most effective. Finally, for the periodic hill flow, augmenting the Reynolds number by 3–7% successfully aligns the simulations with the reference data.

Acknowledgements

This work was supported by the Chinese Scholarship Council and the University of Groningen under Grant (CSC No. 202006870021). The computing time was granted by the Dutch Research Council (NWO) under project 2022.009.

The first version was published as a preprint titled 'Minimum-dissipation model for large-eddy simulation in OpenFOAM – A study on channel flow, periodic hills, and flow over a cylinder' on arXiv.org in 2023.

Author details

Jing Sun*† and Roel Verstappen†
University of Groningen, Groningen, the Netherlands

*Address all correspondence to: j.sun@rug.nl

† These authors contributed equally.

References

[1] Smagorinsky J. General circulation experiments with the primitive equations: I. The basic experiment. Monthly Weather Review. 1963;**91**(3):99-164

[2] Lilly D, K. On the application of eddy viscosity concept in the inertial sub-range of turbulence. NCAR Manuscript. 1966;**123**:1-19

[3] Ducros Frédéric, Nicoud Franck, Poinsot Thierry. Wall-Adapting Local Eddy-Viscosity Models for Simulations in Complex Geometries. Numerical Methods for Fluid Dynamics VI. 1998;**6**:293-299

[4] Vreman AW. An eddy-viscosity subgrid-scale model for turbulent shear flow: Algebraic theory and applications. Physics of Fluids. 2004;**16**(10):3670-3681. DOI: 10.1063/1.1785131. ISSN 1070-6631

[5] Verstappen R. When does eddy viscosity damp subfilter scales sufficiently? Journal of Scientific Computing. 2011;**49**(1):94-110

[6] Verstappen R. How much eddy dissipation is needed to counterbalance the nonlinear production of small, unresolved scales in a large-eddy simulation of turbulence? Computers & Fluids. 2018;**176**:276-284

[7] Rozema W, Bae HJ, Moin P, Verstappen R. Minimum-dissipation models for large-eddy simulation. Physics of Fluids. 2015;**27**(8): 085107

[8] Abkar M, Bae HJ, Moin P. Minimum-dissipation scalar transport model for large-eddy simulation of turbulent flows. Physical Review Fluids. 2016;**1**(4):041701

[9] Zahiri A-P, Roohi E. Anisotropic minimum-dissipation (AMD) subgrid-scale model implemented in OpenFOAM: Verification and assessment in single-phase and multi-phase flows. Computers & Fluids. 2019; **180**:190-205

[10] Verstappen RWCP, Veldman AEP. Symmetry-preserving discretization of turbulent flow. Journal of Computational Physics. 2003;**187**(1): 343-368

[11] Trias FX, Lehmkuhl O, Oliva A, Pérez-Segarra CD, Verstappen RWCP. Symmetry-preserving discretization of Navier–stokes equations on collocated unstructured grids. Journal of Computational Physics. 2014;**258**:246-267. ISSN 0021-9991

[12] Komen EMJ, Hopman JA, Frederix EMA, Trias FX, Verstappen RWCP. A symmetry-preserving second-order time-accurate PISO-based method. Computers & Fluids. 2021;**225**:104979

[13] Hopman Jannes, Edo Frederix. A RKSymFoam Github page. Netherlands. 2023. Available from: https://github.com/janneshopman/RKSymFoam [Accessed: June 13, 2023]

[14] Hoyas S, Jiménez J. Scaling of the velocity fluctuations in turbulent channels up to $Re_\tau = 2003$. Physics of Fluids. 2006;**18**(1):011702

[15] Hoyas S, Jiménez J. Reynolds number effects on the Reynolds-stress budgets in turbulent channels. Physics of Fluids. 2008;**20**(10):101511

[16] Jimenez J, Hoyas S. Turbulent fluctuations above the buffer layer of wall-bounded flows. Journal of Fluid Mechanics. 2008;**611**:215-236

[17] Del Juan Carlos Álamo and Javier Jiménez. Direct numerical simulation of the very large anisotropic scales (vlas) in a turbulent channel. San Diego. In: APS Division of Fluid Dynamics Meeting Abstracts. 54, KF–003, 2001. 54, KF–003, 2001

[18] Del Alamo JC, Jiménez J. Spectra of the very large anisotropic scales in turbulent channels. Physics of Fluids. 2003;**15**(6):L41-L44

[19] Del Alamo JC, Jiménez J, Zandonade P, Moser RD. Scaling of the energy spectra of turbulent channels. Journal of Fluid Mechanics. 2004;**500**: 135-144

[20] Ong L, Wallace J. The velocity field of the turbulent very near wake of a circular cylinder. Experiments in Fluids. 1996;**20**(6):441-453

[21] Kravchenko AG, Moin P. Numerical studies of flow over a circular cylinder at $\mathrm{Re}_D = 3900$. New York: Physics of Fluids. 2000;**12**(2):403-417

[22] Courant R, Hilbert D. Methods of Mathematical Physics: Partial Differential Equations. New York: John Wiley & Sons; 2008

[23] Payne LE, Weinberger HF. An optimal poincaré inequality for convex domains. Archive for Rational Mechanics and Analysis. 1960;**5**(1):286-292

[24] Manteufel TA, White Jr AB. The numerical solution of second-order boundary value problems on nonuniform meshes. Mathematics of Computation. 1986;**47**:511

[25] Shashank JL, Iaccarino G. A co-located incompressible Navier-stokes solver with exact mass, momentum and kinetic energy conservation in the inviscid limit. Journal of Computational Physics. 2010;**229**(12):4425-4430

[26] Chorin AJ. Numerical solution of the Navier-stokes equations. Mathematics of Computation. 1968;**22**(104):745-762

[27] Yanenko NN. Economical implicit schemes (method of fractional steps). In: Doklady Akademii Nauk. Vol. 134. Moscow: Russian Academy of Sciences; 1960. pp. 1034-1036

[28] Blair J, Perot. An analysis of the fractional step method. Journal of Computational Physics. 1993;**108**(1): 51-58

[29] Butcher JC. A history of Runge-Kutta methods. Applied Numerical Mathematics. 1996;**20**(3):247-260

[30] Issa RI. Solution of the implicitly discretized fluid flow equations by operator-splitting. Journal of Computational Physics. 1986;**62**(1): 40-65

[31] Sun J, Verstappen R. Minimum-dissipation model for large-eddy simulation in OpenFOAM-A study on channel flow, periodic hills and flow over cylinder, Netherlands. arXiv. 2023, arXiv:2309.04415. DOI: 1048550/ arXiv.2309.04415

[32] Ch R, Pfleger F, Manhart M. New experimental results for a LES benchmark case. In: Direct and Large-Eddy Simulation VII: Proceedings of the Seventh International ERCOFTAC Workshop on Direct and Large-Eddy Simulation, Held at the University of Trieste. September 8–10, 2008. Netherlands: Springer; 2010. pp. 69-74

[33] Temmerman L, Leschziner MA. Large eddy simulation of separated flow in a streamwise periodic channel constriction. In: Second Symposium on Turbulence and Shear Flow Phenomena. London, UK: Begel House Inc.; 2001

Chapter 4

Recent Advancements in Large Eddy Simulations of Compressible Real Gas Flows

Neelakantan Padmanabhan

Abstract

This chapter explores some of the recent advancements in the field of computational fluid dynamics, specifically with respect to large eddy simulations. We start by introducing some fundamental concepts of scales of turbulent fluid flows. We then discuss the available simulation methods, major challenges of each method and the advantages of large eddy simulation (LES) over other methods. The focus of the chapter then shifts to real gas flows. Governing equations for LES of compressible flows and modifications for real gas flows are presented. Some of the filtering methods are discussed including physical space, Fourier space and proper orthogonal decomposition (POD) based filtering. The chapter then focuses on modeling of sub-filter / subgrid scale terms. A brief discussion of traditional modeling approaches is provided followed by a discussion of the current research and advancements. This discussion includes stochastic modeling using the filtered mass density function (FMDF) approach and machine learning based models for subgrid scales. For each method, a brief a background of the method, their applicability to different scenarios, their advantages, and disadvantages are presented.

Keywords: subgrid scale modeling, compressible flows, reacting flow, stochastic methods, machine learning

1. Introduction

Turbulence in fluid flows is a three dimensional phenomena characterized by chaos. However, over a century of research in the field has shown that patterns exist in such chaos. Turbulence occurs at large Reynolds numbers, where the inertial forces of the fluid are larger than the viscous forces, causing vortices (eddies) to be generated due to gradients in velocity. These vortices rapidly bend, stretch and collapse into smaller eddies. As a result there exist multiple length scales of turbulent flows with vortex sizes ranging from a few kilometers (in atmospheric turbulence) to a few hundred nanometers. Further, an energy cascade phenomenon is observed where the largest scales of the flow that contain the most energy, break down into smaller scales with lesser energy. This continues until the smallest realizable scale called the Kolmogorov scale [1], beyond which the energy dissipates as heat (**Figure 1**). The largest

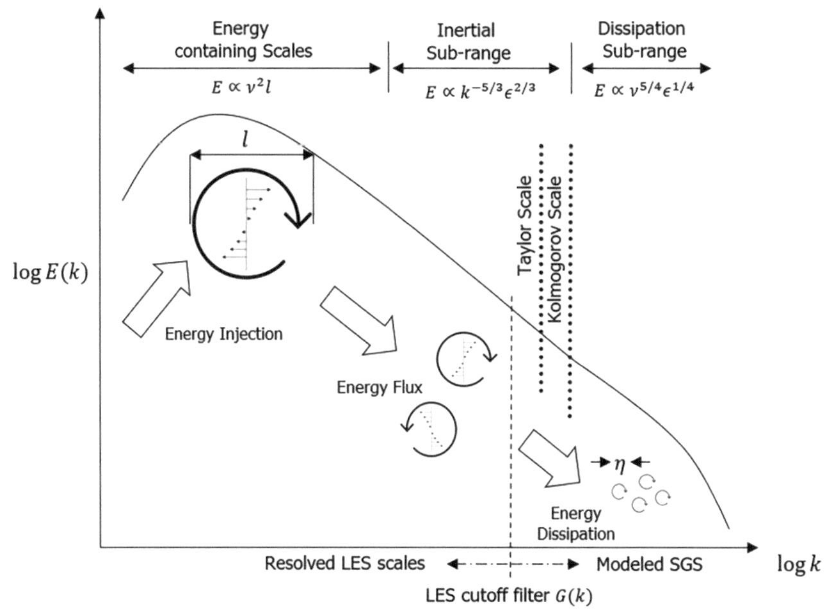

Figure 1.
Spectra of turbulent flow: This figure illustrates the length scales (from integral l to Kolmogorov η) of turbulent flows and the energy cascade phenomena. The turbulent kinetic energy (TKE) in inertial sub-range typically exhibits a universality $E(k) \sim k^{-5/3}$*. LES cutoff filter G(k) separates the resolved and modeled scales.*

flow scales are observed to be dependent on the physical constraints of the flow (initial and boundary conditions), while some degree of universality in structure and statistics is observed in the smaller scales. In computational fluid dynamics (CFD) three broad simulation techniques exist. The first is called Direct Numerical Simulation (DNS), where the computational domain is set such that the simulation captures all the length and time scales of the flow. This approach, solves the full form of Navier Stokes equations by setting the size of mesh, equivalent to the Kolmorogov scales and the number mesh points equivalent to the size of the largest eddy to capture all the length scales [2]. Needless to say, this approach is extremely computationally intensive and impractical for a majority of flows of interest. The second approach is called Reynolds Averaged Navier Stokes (RANS), where an averaging operation is applied to the flow variables in the governing equations. This results in generation of mean flow statistics where turbulence at all scales are modeled [3]. While computationally inexpensive, the effects of turbulence are oversimplified. The third approach is called Large Eddy Simulation (LES), where a filtering operation is applied to the flow variables in the governing equation, such that the largest scales of the flow are captured by the simulation, while the smaller (universal) scales are modeled. This approach results in a more accurate representation of turbulence, while being applicable to many complex flows of interest.

Research in LES focuses primarily on modeling the flow physics occurring at the length scales that are not explicitly captured by the simulation. These scales are called sub-filter or subgrid scale (SGS). Unlike an averaging operation that decomposes the statistics of the flow into mean and fluctuating parts, the flow characteristics of subgrid scales are dependent entirely on the size of the filter used in the simulation. In a sense, a trade-off between the accuracy and computational expense is set by the filter size. Over many decades, a rich collection of models have been developed for the SGS fluctuations occurring in various flow scenarios including incompressible flows, compressible flows, chemically reacting flows, and boundary layer flows [4, 5]. While modeling turbulence in incompressible flows is a challenge, the difficulty increases when compressibility is introduced. From a computation point of view, there are additional equations that need to be solved to account for density variations. The boundary conditions must be designed properly to allow proper dissipation of generated acoustic waves and noise without reflection. Further, the flow physics at the subgrid scales become more localized. Modeling difficulty increases in flows with chemical reactions since the flow physics vary over a wider range of length and time scales. For instance in combustion, the flame fronts are very thin compared to the rest of the flow scales yet strong interactions exist between these scales [6], and the rates of many chemical reactions of interest are in order of a few hundred nano-seconds. These challenges are further exacerbated when the ambient pressure is increased from $1atm$ as many modern day combustion devices such as rocket engines, diesel engines and gas turbines operate at elevated ambient pressures $(10atm < P < 100atm)$. At these pressures many of the simplifications such as the ideal gas equation of state, constant property models or simple forms of heat and mass flux models, are no longer applicable. Various regimes of flow fields are encountered such as large ranges of chemical time scales, transport rates, Reynolds and Mach numbers with weak and strong compressibility effects. Thus, the models developed must be consistent with the fundamental flow physics. This chapter provides a high level overview of a few modeling techniques in literature that are applicable to LES of compressible real gas flows. This chapter is not intended to be a detailed review of all the existing applicable methods.

2. Governing equations

In LES, filters of prescribed widths are applied to the primitive variables in the governing equations. A filtering operation of the form,

$$\overline{\Phi} = \int_{\Omega} \Phi(x', t) G(x', x) dx', \tag{1}$$

is applied to the variables over the domain, Ω. In the above equation $G(x)$ denotes the filter function. The term $\overline{\Phi}$ represents the filtered (resolved) value of variable and $\Phi' = \Phi - \overline{\Phi}$ the fluctuating subgrid component. For compressible flows, a density weighted Favre filtering operation is performed, $\tilde{\Phi} = \overline{\rho\Phi}/\overline{\rho}$, with $\Phi'' = \Phi - \tilde{\Phi}$. For reacting flows, the resulting governing equations for mass, momentum, energy, and scalar conservation equations, respectively can be expressed as,

$$\frac{\partial \overline{\rho}}{\partial t} + \frac{\partial \overline{\rho}\tilde{u}_i}{\partial x_i} = 0, \tag{2}$$

$$\frac{\partial \overline{\rho}\tilde{u}_i}{\partial t} + \frac{\partial \overline{\rho}\tilde{u}_i\tilde{u}_j}{\partial x_j} = -\frac{\partial P(\overline{\Psi})}{\partial x_i} - \frac{\partial}{\partial x_i}\left[\overline{P(\Psi)} - P(\overline{\Psi})\right] + \frac{\partial \tau_{ij}(\overline{\Psi})}{\partial x_j} - \frac{\partial}{\partial x_i}\left[\overline{\tau_{ij}(\Psi)} - \tau_{ij}(\overline{\Psi})\right] - \frac{\partial T_{ij}}{\partial x_j}, \tag{3}$$

$$
\begin{aligned}
&\frac{\partial \bar{\rho}\tilde{e}_t}{\partial t}+\frac{\partial \bar{\rho}\tilde{u}_j\tilde{e}_t}{\partial x_j}=-\frac{\partial P(\overline{\Psi})\tilde{u}_j}{\partial x_i}-\frac{\partial}{\partial x_i}\left[\overline{P(\Psi)u_j}-P(\overline{\Psi})\tilde{u}_j\right]+\\
&\frac{\partial \tilde{u}_i\tau_{ij}(\overline{\Psi})}{\partial x_j}-\frac{\partial}{\partial x_i}\left[\overline{u_i\tau_{ij}(\Psi)}-\tilde{u}_i\tau_{ij}(\overline{\Psi})\right]-\frac{\partial Q_j(\overline{\Psi})}{\partial x_j}-\frac{\partial}{\partial x_j}\left[\overline{Q_j(\Psi)}-Q_j(\overline{\Psi})\right]+\\
&\frac{\partial}{\partial x_j}\left[\sum_{\gamma=1}^{N_s}H_{,\gamma}(\overline{\Psi})J_{j,\gamma}(\overline{\Psi})\right]-\frac{\partial}{\partial x_j}\left[\sum_{\gamma=1}^{N_s}\overline{H_{,\gamma}(\Psi)J_{j,\gamma}(\Psi)}-\sum_{\gamma=1}^{N_s}H_{,\gamma}(\overline{\Psi})J_{j,\gamma}(\overline{\Psi})\right]-\\
&\frac{\partial}{\partial x_j}\left(\bar{\rho}\widetilde{u_je_t}-\bar{\rho}\tilde{u}_j\tilde{e}_t\right)+\overline{S_e},
\end{aligned}
\tag{4}
$$

$$
\begin{aligned}
&\frac{\partial \bar{\rho}\tilde{Y}_\gamma}{\partial t}+\frac{\partial \bar{\rho}\tilde{Y}_\gamma\tilde{u}_j}{\partial x_j}=-\frac{\partial J_{j,\gamma}(\overline{\Psi})}{\partial x_j}-\frac{\partial}{\partial x_j}\left[\overline{J_{j,\gamma}(\Psi)}-J_{j,\gamma}(\overline{\Psi})\right]-\\
&\frac{\partial}{\partial x_j}\left[\bar{\rho}\widetilde{Y_\gamma u_j}-\bar{\rho}\tilde{Y}_\gamma\tilde{u}_j\right]+\overline{S_{Y_\gamma}}.
\end{aligned}
\tag{5}
$$

In these equations, Ψ represents the primitive variables (density ρ, velocities u_i, energy e_i, and mass fraction of species Y_γ), $\overline{\Psi}$ represents the filtered values of the primitive variables and $\Phi(\overline{\Psi})$ represents the variables that are calculated from the filtered primitive variables. For instance, $\overline{P(\Psi)}$ represents the filtered pressure, whereas $P(\overline{\Psi})$ represents the pressure calculated from the filtered primitive variables available from LES and the equation of state. Additional closure terms are required for the source terms $\overline{S_e}, \overline{S_{Y_\gamma}}$, the viscous shear stress tensor, the heat and the mass flux vectors. The gravity term typically appearing in the Navier Stokes equations is neglected for high pressure flows.

2.1 Modifications for real gas flows

For real gas flows at elevated pressures, a few terms in the governing equations are modified. A brief summary of the modifications is provided in this section. Complete formulation of these equations are presented in the works of Palle [7], Foster and Miller [8].

2.1.1 Equation of state

While ideal gas equation of state is predominantly used for compressible flows at atmospheric pressures, a real gas equation is required for higher pressures to account for effects of compressibility, non-equilibrium thermodynamics, variations in material properties and effects of inter-molecular forces. A cubic form of Peng Robinson equation of state is typically used for closure of pressure term in real gases due to its simplicity.

$$
P(\overline{\Psi})=\frac{RT(\overline{\Psi})}{v(\overline{\Psi})-B_m(\overline{\Psi})}-\frac{A_m(\overline{\Psi})}{v(\overline{\Psi})^2+2v(\overline{\Psi})B_m(\overline{\Psi})-B_m(\overline{\Psi})^2},
\tag{6}
$$

where, $A_m(\overline{\Psi}), B_m(\overline{\Psi})$ represent resolved mixture parameters, R the molar gas constant and $v(\overline{\Psi}), T(\overline{\Psi})$ the molar volume and temperature calculated from the resolved primitive variables.

2.1.2 Property models

Flow properties at atmospheric pressures are typically assumed as constant or simple functions of temperature. However, for real gas flows with the aforementioned equation of state, realistic property models related to the equation of state through the Gibbs function must be derived. These property models include departure functions that represent the thermodynamic deviations of the properties from those at low pressures. For instance, modified the molar heat capacity term can be expressed as,

$$C_p(\overline{\Psi}) = C_p^0(\overline{\Psi}) + f(\Phi(\overline{\Psi})), \tag{7}$$

where $C_p^0(\overline{\Psi})$ represents the low pressure heat capacity and $f(\Phi(\overline{\Psi}))$ the deviation term modeled as a function of resolved variables. Similarly, property models for molar specific enthalpy $H(\overline{\Psi})$, partial molar volume $v(\overline{\Psi})$, partial molar enthalpy $H_{,\gamma}(\overline{\Psi})$, and the mixture molar internal energy $u(\overline{\Psi}) = H(\overline{\Psi}) - P(\overline{\Psi})v(\overline{\Psi})$ must be derived. The mass diffusion coefficients, mixture viscosity and thermal conductivity are typically obtained from look up tables.

2.1.3 Diffusion terms

In flows at atmospheric pressures, the Fourier form of heat diffusion $\left(Q_j = -\alpha \partial T/\partial x_j\right)$ and Fickian form of mass diffusion $\left(J_j = -D_m \partial Y_\gamma/\partial x_j\right)$ models are generally applicable. However, for flows at elevated pressures, generalized multi-component forms of the heat and mass flux vectors derived from non-equilibrium thermodynamics and fluctuation theory [9–11] are required. The modifications must include the effects of Soret, Dufour and cross diffusion for higher pressures. The general form for heat diffusion can be expressed as,

$$\begin{aligned} Q_j(\overline{\Psi}) &= -f(\alpha, D_m, \Phi(\overline{\Psi}))\frac{\partial T(\overline{\Psi})}{\partial x_j} \\ &-f(\alpha, D_m, \Phi(\overline{\Psi}))\frac{\partial P(\overline{\Psi})}{\partial x_j} \\ &-f(\alpha, D_m, \Phi(\overline{\Psi}))\frac{\partial X_\gamma(\overline{\Psi})}{\partial x_j}, \end{aligned} \tag{8}$$

where, α, D_m represent the thermal and mass diffusivities, $\Phi(\overline{\Psi})$ the filtered variables including molar fluxes (X_γ), molecular weights and mixture parameters. The cross diffusion effects are incorporated using non equal values of species mass diffusion coefficients. The Soret effect is associated with terms proportional to $\partial T(\overline{\Psi})/\partial x_j$ and $\partial X_\gamma(\overline{\Psi})/\partial x_j$, while the Dufour effect is associated with terms proportional to $\partial P(\overline{\Psi})/\partial x_j$ and $\partial X_\gamma(\overline{\Psi})/\partial x_j$. A similar form is used for the mass diffusion term. These terms produce similar effects as the Fourier and Fickian models at low pressures.

2.1.4 Reaction terms

Chemical kinetics of varying level of complexities can be included in simulation of reacting flows. However, to strike a balance between computational efficiency and accuracy, reduced form of pressure dependent kinetics must be used for reacting flows at elevated pressures. Pressure dependent mechanisms for high pressure reacting flows are proposed in the studies of Sohn et al. [12] and Chung et al. [13].

2.2 Unclosed terms

The governing equations (Eqs. (3)-(5)) contain many unclosed subgrid terms that require modeling.

- $T_{ij} = \overline{\rho}\widetilde{u_i u_j} - \overline{\rho}\tilde{u}_i\tilde{u}_j$—subgrid turbulent stress
- $\overline{S_e}, \overline{S_{Y_\gamma}}$—filtered reaction source
- $\left[\overline{\tau_{ij}(\Psi)} - \tau_{ij}\left(\overline{\Psi}\right)\right]$—subgrid viscous stress
- $\left[\overline{P(\Psi)} - P\left(\overline{\Psi}\right)\right]$—subgrid pressure
- $\left[\overline{Q_j(\Psi)} - Q_j\left(\overline{\Psi}\right)\right]$—subgrid heat flux vector
- $\left[\overline{P(\Psi)u_j} - P\left(\overline{\Psi}\right)\tilde{u}_j\right]$—subgrid pressure flux
- $\left[\overline{u_i\tau_{ij}(\Psi)} - \tilde{u}_i\overline{\tau_{ij}}\left(\overline{\Psi}\right)\right]$—subgrid viscous diffusion
- $\left[\overline{\rho}\widetilde{u_j e_t} - \overline{\rho}\tilde{u}_j\tilde{e}_t\right]$—subgrid convective energy
- $\left[\overline{\sum_\gamma H_{,\gamma}(\Psi)J_{j,\gamma}(\Psi)} - \sum_\gamma H_{,\gamma}\left(\overline{\Psi}\right)J_{j,\gamma}\left(\overline{\Psi}\right)\right]$—subgrid molecular enthalpy flux
- $\left[\overline{\rho}\widetilde{Y_\gamma u_j} - \overline{\rho}\tilde{Y}_\gamma\tilde{u}_j\right]$—subgrid scalar flux
- $\left[\overline{J_{j,\gamma}(\Psi)} - J_{j,\gamma}\left(\overline{\Psi}\right)\right]$—subgrid mass flux vector

Of these, the subgrid turbulent stress, and filtered reaction terms are usually given the most importance in LES modeling. The rest of the chapter discusses some of the modeling techniques in literature that are used to close these terms. The effects of subgrid pressure, subgrid molecular enthalpy flux, the subgrid shear-stress tensor, the subgrid heat and mass flux vectors are typically negligible for flows at atmospheric pressures. However, for reacting flows at elevated pressures the subgrid pressure, heat and mass flux vectors are not negligible as shown in the studies of Ma et al. [14], Foster and Miller [15], due to the presence of large scalar gradients. Closure for these terms are not presented here, but some of the methods described in this chapter may be used to model these terms as well.

2.3 Numerical approach

LES of compressible flows result in generation of acoustic waves and large instabilities in both low and high Mach number regimes. Choice of appropriate numerical technique, low diffusivity schemes for discretization of the equations, appropriate filters, boundary conditions for proper inflow and outflow of acoustic waves, entropy waves, and numerical instabilities with minimal reflection for high quality simulations are critical. Several methods to address the numerical implementations for compressible LES are reported in the studies of Jiang and Lai [16].

3. Filtering techniques

Filtering in LES may be performed either in a physical space or in a spectral space as illustrated in **Figure 1**. The filter functions are generally required to satisfy a few mathematical properties including normalization, linearity and commutativity [17].

3.1 Physical space filters

The simplest, physical space filter is a grid based filter called top hat / box filter that can be represented as, $G(x) = \frac{1}{\Delta}H\left(\frac{1}{2}\Delta - |x|\right)$, where H is the Heavyside function, and G a filter kernel that is spatially and temporally invariant. This is an implicit filter, where the grid / mesh size is used as the control. The flow characteristics larger than the grid are resolved, while those smaller than the grid require modeling.

3.2 Spectral space/Fourier mode based filters

In spectral space, a cut off wave number (or frequency) can be chosen and a filter kernel at that wave number can applied. Applying this low pass filter to the variable removes the small scales associated with high frequencies. To perform the filtering operation, Fourier transforms of the variable ($\hat{\Phi}$) and the filter kernel ($\hat{G}$) are first obtained. The filtering is then performed as a convolution operation, $\overline{\hat{\Phi}}(k, \omega) = \hat{G}(k, \omega) * \hat{\Phi}(k, \omega)$.

3.3 Proper orthogonal decomposition (POD) assisted filters

POD is a method primarily applied to development of reduced order models in CFD [18, 19]. For a given data, POD extracts the modes that contain the most dominant characteristics of the data. POD of a turbulent flow results in scale separation. The POD modes represent all the characteristic scales of motion. In this method, a singular value decomposition (SVD) of the turbulent flow field data $\Phi(x, t)$ is performed. This results in a decomposition of the form, $SVD(\Phi) = \Theta\Sigma\Xi^*$, where Θ represents the spatial modes (orthonormal basis functions) of the flow field data, Ξ the corresponding temporal coefficients and Σ the magnitude of significance of the modes. The instantaneous value of the field variable can be expressed as,

$$\Phi(x, t) = \sum_{n=1}^{\infty} \theta_n(x)\xi_n(t) \tag{9}$$

It is noted that not all ∞ modes are actually required to reconstruct the data. A finite number of modes ($n < N_c$) typically contain upto 80% of the energy of the flow field data. This is analogous to other LES filtering methods, where a set filter size captures the most energetic scales of flow. Here, the filter criteria is the number of modes chosen to reconstruct the data. The filtering operation may be defined as follows,

$$\Phi(x,t) = \overline{\Phi}(x,t) + \Phi'(x,t) = \sum_{n=1}^{N_c} \theta_n(x)\xi_n(t) + \sum_{n=N_c+1}^{\infty} \theta_n(x)\xi_n(t). \tag{10}$$

In this equation, N_c represents the number of modes used to represent the resolved (most energetic) scales of the data. The sub-filter scale modeling for POD assisted filters is discussed in Section 4.1.2.

4. Subgrid scale modeling

The unresolved terms identified in Section 2.2 must be closed using modeling techniques such that forward scatter (energy transfer from large scales to small scales) and possible back scatter (energy transfer from small scales to large scales) are captured. The general approach to modeling is to calculate the subgrid terms as a function of the appropriate resolved variables. This section presents some of the most predominantly used modeling approaches to close the subgrid turbulent stress and the reaction source terms. The general principles outlined here can further be used to model various other unclosed terms discussed in the preceding section.

4.1 Closure for turbulent stress

4.1.1 Eddy viscosity based models

Turbulent stress appears as a result of filtering applied to the momentum equation. While the large scale turbulent features cause mixing and transport the energy, the smaller scales are generally associated with diffusion. Thus, the subgrid turbulent stress is generally modeled similar to viscous diffusion with a special diffusivity factor. This form of model is called an eddy viscosity model or Smagorinsky model. The unclosed turbulent stress is expressed as a function of the resolved strain energy and a modeled diffusivity factor.

$$T_{ij} = -2\nu_{SGS}\tilde{S_{ij}} = C_s\Delta^2|\tilde{S}|\tilde{S_{ij}}, \tag{11}$$

where, ν_{SGS} represents the subgrid scale diffusivity, Δ the size of filter, $\tilde{S_{ij}} = \frac{1}{2}\left(\frac{\partial \tilde{u_i}}{\partial x_j} + \frac{\partial \tilde{u_j}}{\partial x_i}\right)$ the resolved strain rate tensor and C_s a model constant typically derived via dimensional analysis. While the form of the model is simple and consistent with the idea of energy cascade, a constant value of C_s makes it too restrictive. Further this model does not predict backscatter of energy. A number of improvements have been proposed for this model over the years. The most popular method is the usage of a dynamic model constant $C_D(x,t)$ that varies in space and time. The idea here is to perform LES at two different grid (filter) sizes (original grid size Δ_0, and test grid size $\Delta_T \geq 2\Delta_O$). The subgrid scales of the test simulation contain many of the resolved scales of the original simulation. Here it is assumed that the interaction between the turbulent eddies at the length scales between Δ_T and Δ_O must be similar to the interaction between the eddies at Δ_O and its subgrid scales. It is assumed that the same eddy viscosity type model is applicable in both levels. While the subgrid turbulent stress terms $\left(T_{ij_O}, T_{ij_T}\right)$ at each of these filter levels cannot be directly computed, a relationship exists between the two. This is expressed by the Germano identity [20] as, $L_{ij} = T_{ij_T} - \tilde{T}_{ij_0} = \widetilde{\tilde{u}_i\tilde{u}_j} - \tilde{\tilde{u}}_i\tilde{\tilde{u}}_j$. The term L_{ij} is called Leonard's stress tensor that represents the subgrid turbulent stress at length scales between test and the original grid filters. This can be explicitly calculated from the resolved velocities as a function of space-time and can be modeled similar to the Smagorinsky model as,

$$L_{ij} = C_D\left(\Delta_T^2 \tilde{\tilde{S}}, \tilde{\tilde{S}}_{ij} - \Delta_O^2 \widetilde{|\tilde{S}|\tilde{S}_{ij}}\right). \tag{12}$$

The dynamic model constant C_D is then obtained via least square minimization. This model is self contained and requires no prior knowledge of the model constant. This form of model is still predominantly used in LES of compressible real gas flows. However, the variable values of eddy viscosity may cause numerical instabilities and care must be taken when applying this model to highly localized flows.

4.1.2 Scale similarity based models

Scale similarity model proposed by Bardina et al. [21] assume similarities in characteristics between the eddies that are slightly larger than the grid size (smallest resolved scale) and the eddies are slightly smaller than the grid size (largest unresolved scale). With this hypothesis, the subgrid turbulent stress is expressed in terms of the smallest resolved velocities. This approach does not use an eddy viscosity assumption. This approach is also used to model the sub-fliter terms generated by POD based filtering presented in Section 3.3. Applying the scale similarity assumption between the smallest resolved mode ($n = N_c$) and the largest subfilter POD modes ($n > N_c + 1$), the drain of modal energy at the cut-off mode can be expressed in terms of a ratio of the singular values σ (magnitude of significance of the modes), $\beta_i = \sigma_{N_C}/\sigma_{N_C-1}$, which is assumed to be constant for all the sub-filter scales. The energy distribution of higher energy modes can be approximated in terms of coefficients $\alpha_i = \left(1 + \sum_{n=N_c+1}^{\infty} \beta_i^{n-N_c}\right)^{1/2}$. The subgrid turbulent stress is then computed as a function of the energy coefficient α_i and the resolved modes $\theta(x), \xi(t)$. A complete derivation of the turbulent stress for POD assisted filter is presented in the works of Shinde [22].

4.2 Closure for chemical source terms

Reaction source terms are highly non-linear and computationally stiff. Some of the popular models for closure of the reaction source terms in LES include flamelet libraries and linear eddy models. Each has its own advantages and limitations. However, both are approximations. Major improvements in closure of this term were made after introduction of the Probability Density Function (PDF) method by Pope [23] and Filtered Mass Density Function (FMDF) method by Jaberi et al. [24]. In these methods, the complete statistical information of the flow is provided by a filtered joint PDF for velocity and scalars of the flow. The moments of statistics of the flow are obtained from the solutions of a transport equation derived for the PDF. The primary advantage is that the chemical source term is closed and does not require any modeling. Thus, for combustion at high pressures this approach may generate more accurate results. For a scalar $\phi(x,t)$, the filtered mass density function F_L can be defined as,

$$F_L(\psi,x,t) = \int_{-\infty}^{\infty} \rho(x',t)\zeta[\psi,\phi(x',t)]G(x')dx', \tag{13}$$

where, $\zeta[\psi,\phi(x,t)] = \delta[\psi - \phi(x,t)] = \Pi_{\alpha=1}^{\sigma}\delta[\psi - \phi(x,t)]$ is the fine-grained density with δ the delta function and ψ the composition domain. The integral property of FMDF results in filtered density, $\int_{-\infty}^{\infty} F_L(\psi,x,t)d\psi = \int_{-\infty}^{\infty}\rho(x',t)G(x')dx' = \overline{\rho}(x,t)$. From the properties of FMDF, the filtered value of a scalar is obtained by integration over the composition space, as $Q(\tilde{x_i},t) = \int_{-\infty}^{\infty}\rho(x_{i'},t)Q(x_{i'},t)F_L(\psi,x_i,t)d\psi$. Any term that is a function of the variables within FMDF can be directly calculated. The transport equation for FMDF can be expressed as,

$$\begin{gathered}\frac{\partial F_L(\psi,x,t)}{\partial t} + \frac{\partial}{\partial x_i}\left[(u_i\tilde{,}\psi)\right]F_L(\psi,x,t)] = \\ \frac{\partial}{\partial \psi}\left[\left\langle\frac{1}{\rho}\frac{\partial J_{j,\alpha}}{\partial x_j},\psi\right\rangle F_L(\psi,x,t)\right] - \frac{\partial}{\partial \psi}[S_\alpha(\psi)F_L(\psi,x,t)/\overline{\rho}].\end{gathered} \tag{14}$$

The chemical reaction source term in the above equation is closed but two new terms appear that remain unclosed. The first unclosed term $u_i\widetilde{|\psi}$, is called SGS scalar convection and represents the motion of scalars at the subgrid scale due to the flow velocity. This term may be decomposed as, $u_i\widetilde{|\psi}F_L = \tilde{u_i}F_L + \left[u_i\widetilde{|\psi} - \tilde{u_i}\right]F_L$, where the second term is the effect of SGS convective flux. This is typically modeled using an eddy viscosity type model. The second unclosed term $\left\langle\frac{1}{\rho}\frac{\partial J_{j,\alpha}}{\partial x_j},\psi\right\rangle$, is the conditional scalar diffusion that represents the effects of molecular transport (diffusion) in physical space and molecular mixing in composition space. The resulting transport equation is then written as,

$$\frac{\partial F_L}{\partial t} + \frac{\partial \tilde{u_i}F_L}{\partial x_i} = \frac{\partial}{\partial x_i}\left[(\mu+\mu_t)\frac{\partial F_L/\overline{\rho}}{\partial x_i}\right] + \frac{\partial}{\partial \psi}\left[\left\langle\frac{1}{\rho}\frac{\partial J_{j,\alpha}}{\partial x_j},\psi\right\rangle - S_\alpha/\overline{\rho}\right]F_L. \tag{15}$$

This equation can be further integrated to obtain the transport equations for the SGS moments (mean, variance). The spatial transport of the FMDF is represented by a diffusion process governed by a stochastic differential equation

$$dX_i(t) = \tilde{u_i} + (1/\overline{\rho})\partial(\mu+\mu_t)/\partial x_i + \sqrt{2(\mu+\mu_t)/\overline{\rho}}, \tag{16}$$

and the composition evolution is governed by an equation,

$$\frac{d\phi_\alpha}{dt} = \Theta_\alpha - S_\alpha = \left\langle\frac{1}{\rho}\frac{\partial J_{j,\alpha}}{\partial x_j},\psi\right\rangle - S_\alpha. \tag{17}$$

This equation is typically solved via Lagrangian Monte Carlo approach. Here, the term Θ_α is the conditional scalar diffusion (CSD) which remains as the last unknown term in this transport eq. A few mixing models that stem from different physical concepts have been proposed for modeling the CSD. The most popular choice is the Interaction by exchange of mean (IEM) method that is linear and causes the composition of a Lagrangian particle within a filter volume to relax to a mean value. However, it is local only in the physical space and not in the composition space. Improvements to this model include interaction by exchange of conditional mean (IECM) and IEM plus mean drift but they also lack the locality in composition space property. Additional models include Mapping Closure [25] that maps the composition space of a single scalar to a Gaussian reference field and then solves for it evolution. This preserves the properties of mean, variance decay and is local in composition space but is applicable only to a single scalar. An improvement to this model is the Euclidean Minimum Spanning Tree (EMST) [26] that forms a minimum spanning tree based on composition of particles in an ensemble. The lengths of the vertices of the tree are dependent on the weights, and age properties of particles and enables treatment of multiple scalars. The mixing is performed for particles lying in the minimum spanning tree. This preserves the locality property but does not satisfy the linearity and independence properties. Evaluation of various mixing models are presented in the studies of Shetty et al. [27], Chen et al. [28] and Padmanabhan [29]. Most molecular mixing models however, are developed for pure mixing flows. Their application to reacting flows is based on an assumption that the reaction and diffusion are indirectly coupled as the molecular diffusion smooths toward the scalar means in state space. This assumption is not true for reacting flows and the models that worked for pure mixing flows may not work for reacting flows. Further, a common factor between all the models is the mixing frequency $\Omega_m = f(\overline{\rho}, \Delta_G, \tilde{u}_i, \mu, D_m)$ that is determined from the simulation parameters that represents the rate of change of composition of particles within a filter volume. The standard form of this term is only applicable to flows with constant diffusivity at atmospheric pressure. For flows at elevated pressures with generalized diffusion models, pressure dependent diffusivity corrections must be included in the mixing frequency term as shown in the study by Padmanabhan and Miller [30].

5. Current research in SGS modeling—Machine learning based models

Application of machine learning (ML) to turbulence modeling and control has been a topic of research since early 2000s [31, 32], but over the last several years major strides have been made in terms of development of new ML models, available libraries and collaborative open source codes that enable development of better turbulence closure models for LES. This section briefly covers a few popular supervised and semi-supervised techniques that have been successfully applied to modeling of SGS terms in LES. The SGS closures currently in literature typically pertain to turbulent stress and chemical source terms, but the general approach can be applied to model any of the unclosed terms identified in Section 2.2. Details of implementations of algorithms are not covered in this chapter, as it is beyond the scope of this work.

5.1 Artificial neural network (ANN)/deep learning (DL) based models

An ANN is a network of interconnected nodes in a layered structure (typically, an input, an output and one or more hidden layers). The nodes in the input layer hold numerical values (activations) that represent the features generated from a training dataset and the nodes in the output layer hold values that represent the generated prediction. The nodes in the hidden layers contain activations that map the input to the output. The core of ANN lies in automatic adjustments of these activations in a way that the machine is able to identify a correct mapping between a given input and output, not only for the set of data on which it is trained but also for a new set of data it has never encountered before. The structure of ANN is based on the Perceptron [33], where the activation of each node in a particular layer (except input layer) is a weighted sum of the activations of the nodes in the previous layer. This is expressed as,

$$a_j^L = \sigma\left(\sum_{k=1}^{N} w_{jk}^L a_k^{L-1} + b_k^L\right), \tag{18}$$

where N, a_k^L represent the number of nodes and activation of node k in layer L, w_{jk}^L the weight connecting node k in layer L and node j in layer $L-1$, b_k^L a bias term typically added to rescale this weighted sum to a positive value and σ a non-linear activation function (ReLU or Sigmoid function) that is used to force the output between 0 and 1. For CFD simulations, data or statistics computed at various mesh locations within the domain may be used as the input features. The activations of the input nodes are propagated forward using Eq. 18 until the last layer is reached. This is called forward propagation. A loss function is then defined to evaluate the error between the prediction in the output layer and the actual output y determined from a training dataset,

$$\mathcal{L} = \frac{1}{N}\sum_{k}^{N}\left(a_k^L - y_k\right)^2 \tag{19}$$

The weights and biases of the nodes in all the layers are then adjusted (from the last layer to first layer) using an optimization technique like gradient descent such that the loss function is minimized. This process is called back propagation. The updates to the weights and biases are applied as, $V = V - \alpha_R \partial\mathcal{L}/\partial V$, where $V = [w, b]$ denotes the weight, bias terms, α_R denotes the rate of descent and the derivatives $\partial\mathcal{L}/\partial V$ are determined using the functional dependencies between each layer and chain rule. This form of network can be trained to identify mapping between any input / output pairs. Identifying and generating the right set of inputs / outputs to train a model for optimal performance is the biggest challenge. Modeling approach for SGS terms using ANN can be classified into two categories:

1. Direct modeling of SGS terms using the resolved variables from LES as inputs. A DNS dataset can be explicitly filtered to compute the resolved and SGS terms. The SGS terms from this filtered DNS (FDNS) form the ground truth. A training dataset can then be created using the resolved FDNS variables, filter parameters and measured SGS terms. The resulting network can create a mapping between the resolved variables and the SGS terms. Implementations similar to this approach are presented in the studies of Gamhara and Hattori [34], Wang et al. [35], Park and Choi [36]. A similar approach for closure of reaction source term was proposed by Ranjan et al. [37] where DL was used to compute reaction rates from the filtered species mass fractions, temperatures and pressures. The training data were generated using a variety of methods including 1D linear eddy mixing (LEM), flamelet models and appropriately sliced DNS datasets.

2. Tuning of the model parameters of existing SGS models. Xie et al. [38], Meng et al. [39] generated coefficients used in dynamic SGS modeling, with the filtered flow variables and their gradients computed from LES as inputs, and the coefficients determined from filtered DNS as the ground truth. An illustration of this approach is presented in **Figure 2**.

These two approaches are often very cumbersome requiring large training datasets with equivalent DNS and LES states. Equivalency in filter kernel used in LES and explicit filter performed on DNS is not always guaranteed [40]. Time dependence is not typically captured in the network and training is typically performed with a snapshot data. This is acceptable with statistically stationary dataset but is error prone for instantaneously evolving flows. This approach has not shown to outperform the traditional models in terms of computational costs [41]. Further, the predictions by the models are not universal, but specific to the flow conditions used in the training dataset.

In addition to traditional form of NN, Physics Informed Neural Networks (PINN) can be applied to SGS modeling. In this approach, in addition to the traditional loss function, physics constrained loss functions can be imposed on the initial, boundary conditions and the governing equation. As a result the updated network parameters are constrained to satisfy the underlying governing equations. This enables the NN to generate a robust model with less data, and may enable faster computation for a well defined problem. An implementation of PINN for SGS modeling in lagrangian LES is presented in the study by Tian et al. [42].

Another powerful addition to NN, typically applied in image processing, is the idea of Convolutional Neural Network (CNN). In this approach, a convolutional filter is applied to the input, such that a feature map (activations) can be created. From these activations, certain known features of the data can be extracted. In image processing, this translates to extraction of known patterns within the image. These features are then pooled together and then passed to a fully connected neural network (FCNN), which then categorizes the input. In LES, CNN can be used to extract known flow patterns, SGS statistics or other quantities of interest. An implementation of this approach is presented in the study of Xing and Lapeyre [43] who applied CNN to

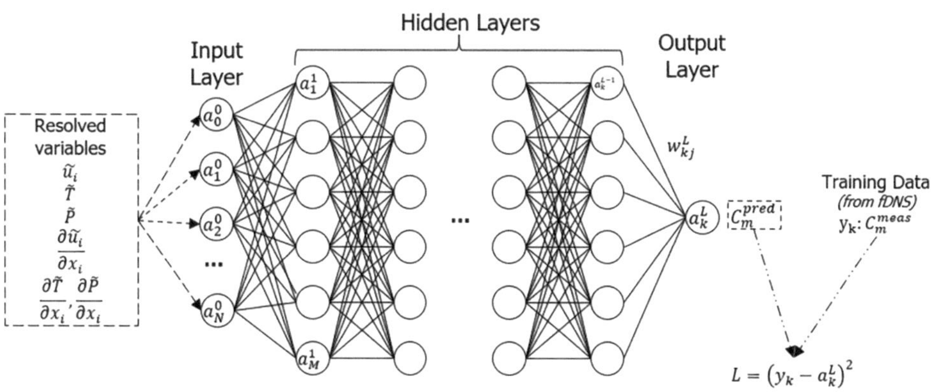

Figure 2.
An illustration of deep learning based model via tuning of existing SGS model is presented. The network inputs are the resolved variables measured from LES and the output is a model constant for an applicable eddy viscosity model.

predict the total flame surface density (unresolved due to thin flamefront) from the resolved progress variable c. This was then used to predict SGS flame wrinkling.

For high pressure real gas flows, there is a lack of modeling effort and studies evaluating the applicability of existing SGS models for the unclosed terms. Hence, the approach of direct modeling of SGS terms from resolved LES variables is more applicable to closure of the SGS pressure, molecular enthalpy, heat and mass flux vectors terms. While the concerns of equivalency in filter kernels and non-universality of the developed models exist, this approach does not disregard the effects of these terms. This would however require development of a DNS database for high pressure combustion [7, 44].

5.2 Reinforcement learning (RL)/deep reinforcement learning

Reinforcement learning is a ML technique where a model is not explicitly trained by a training set, instead it learns to improve itself by interacting with a dynamic environment. A brief overview of the method and a few core terminologies are provided in this section. An RL framework consists of an agent within an environment at a state s. The agent can take an action $a(s)$ to receive a reward $R(s)$. The rewards are cumulative in nature. The environment is typically stated as a Markov Decision Process (MDP). For deterministic systems, a transition model $T(s', s, a)$ can be developed that describes the new state s' the agent will be in, given it is at a particular state and takes an action. A policy for the agent can then be defined as a probability $\pi_\theta(s) = P(a, s)$, which determines the actions it must take, given it is in a particular state. Further, a value function can be defined for the agent which quantifies the expected rewards it can receive, for a given policy $V(s) = \mathbb{E}\left(\sum_i R(s_i, s, \pi_\theta)\right)$. The value function therefore is dependent on the path the agent takes. When defining a value function a discount factor γ is typically added to the rewards to prioritize the immediate rewards over the time delayed rewards. Policy and value functions can have any functional form including an ANN (in deep reinforcement learning).

The goal of an agent in RL, is ultimately to determine the best policy that maximizes the value function. When a transition function for a system is available, the policy design for an agent is straightforward and can be achieved via value iteration (updating the value function iteratively for every possible sequence to determine the best value) or via policy iteration (for each policy, determine the best value and choose the best policy that maximizes the value). This is called model based learning. For non-deterministic systems, it is not practical to derive a transition function because of the large action space. Instead, an optimal policy can be directly derived from the agent's interaction with its environment. This approach is called model free learning. One such method is Q learning where a function $Q(s, a)$ can be learnt for a state-action pair. This function represents the value (quality) of an action at a state. Once a Q function is learnt, the optimal value function and policy can be derived as the ones that return the highest Q value.

However, for continuous action spaces like a CFD simulation, Q learning is not always feasible because a Q value is required for every action and the method becomes unstable. In this case, policy gradient method is more suitable. The premise of policy gradient method is to directly optimize the policy instead of learning a Q function first. One of the most efficient policy gradient methods is proximal policy optimization (PPO). In PPO, the policy and value functions are each represented as an ANN. For each time iteration, the policy network takes states s as input and generates the probabilities $P(a, s)$. This information is stored as a tuple (s, a, R, P) in a dataset. The value function network then takes (s, a) pairs as input and generates $Q(s, a)$ that quantify the goodness of the decision. Future expected rewards for the decisions at each time step are then computed. The difference between the computed Q values and the total future rewards (for a certain set of actions), generates an advantage function $\hat{A}(s, a) = Q(s, a) - V_\pi(s)$. This advantage function is used to compute a loss, which is then back-propagated through the value function to improve it. A similar loss function is computed using the advantage function $\hat{A}$ and the stored probabilities $P(a, s)$, and back-propagated to improve the policy network. A clipped loss function used for the policy network increases the stability of training by constraining the maximum change of policy at each time step. Effectively the policy and value functions are trained and improved together.

An implementation of deep reinforcement learning to model SGS terms is presented in a study by Kurz et al. [45]. In this study, an agent interacts with an environment of LES of homogeneous isotropic turbulence (HIT), and adapts the Smagorinsky model coefficients $C_s(x_i, t)$ for each element at each time, in the domain. The framework for the model is stated as follows: the agent's states s are defined by the current flow state of the LES, the agent's actions space are defined as the element wise Smagorinsky model constants C_s, the policy $\pi_\theta(a, s_t)$ is defined by a CNN that takes resolved flow variables (momentum field, invariants of velocity) from LES as inputs and predicts applicable C_s, and the rewards $R(s)$ are defined by closeness of the turbulent energy spectra computed for the LES using the predicted C_s to the spectra measured from a filtered DNS solution, $R(s) = f(E_{FDNS}(k) - E_{LES}(k))$. The form of reward used in this case can also be extended to other flow configurations where the energy spectrum follows $K^{-5/3}$ trend. An illustration of this approach is presented in **Figure 3**. This approach is more efficient than a supervised learning model because the rewards are independent of the filters used. In another study by Novati et al. [46], a multi-agent reinforcement learning (MARL) approach is adopted for SGS turbulent stress tensor modeling. In this work a PPO method is adopted for policy improvement. N_{agents} are distributed across the computational domain each of which collect the state, action, reward and statistics information. These agents then send the information to a

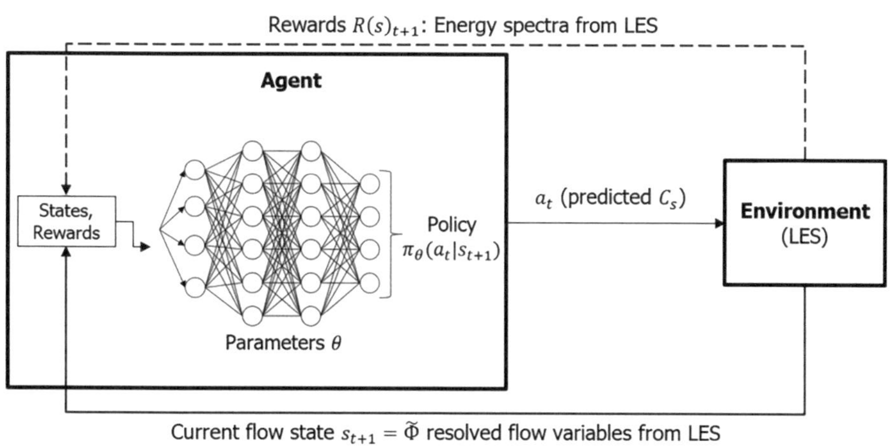

Figure 3.
At each time step, the agent computes C_s and passes to the environment. The environment then computes the SGS terms, the energy spectra and the corresponding reward. This information is then passed to the agent, which then adjusts its actions (C_s) based on the rewards it received.

master at the end of each simulation and receive updated policy parameters. The policy network takes invariants of gradients and Hessians of velocity and generates a Smagorinsky constant as output. Two types of rewards are used to control the model. One reward function ensures the sum of resolved and modeled contribution to the SGS stress is independent of the LES resolution (filter size), while the other reward function ensures the statistics between LES and DNS data are comparable. The MARL approach exploits the potential of cooperating agents resulting in an improved computation efficiency.

5.3 Super resolution

With the continuous development of newer ML methods and libraries, creative ML based modeling techniques are being applied in the field of turbulence modeling. One such approach is super resolution (SR) that aims at scaling up the data from a low resolution simulation to produce high resolution data. A trivial method to accomplish this would be to perform interpolation of data from the neighbors of a point. However, interpolation typically produces an average of the data which is not indicative of high resolution data. Instead, an ANN can be trained to learn this scaling based on prior information it collected from large sets of data. The ANN performs upscaling in a manner that fills the missing information. This approach does not violate the data processing inequality theory because the missing information is inherently present in the large set of input data. One approach to perform a direct mapping between low resolution LES and high resolution DNS data would be to use a super resolution convolution neural network (SRCNN), which uses a CNN for pattern recognition (flow structure / statistics in LES) and a SR model to fill the information gap. Some implementations of SR for modeling in LES include a Generative Adversarial Netowrk (GAN) based model by Nista et al. [47] that has a generator network to upscale the LES information and a discriminator to evaluate the quality of scaling, a physics guided super resolution network (PGSRN) by Chen et al. [48] that combines the idea of PINN and SR, and a meta learning deep CNN by Zhao et al. [49] that generates high resolution data without manually inputting filtered DNS information.

6. Future research challenges and concluding remarks

Research and progress in ML is seeing an unparalleled growth because of its applications in every field of science, technology, and business. While ML has tremendous potential, research focus must also be directed toward incorporation of fundamental physics and mathematical attributes of turbulence in the models. This is especially true for problems like combustion because of the complex flow physics and two-way coupling between turbulence and flames. Further, the vast majority of modeling efforts in literature only cover the SGS turbulent stress and chemical reaction terms. However, for flows at high pressures, additional unclosed SGS terms are observed to be significant and future modeling efforts in LES of real gas flows must take these terms into account. Many of the ML based models presented in this chapter use simple flow configurations and are intended to be a proof of concept. Application of these models to more complex flow scenarios must be evaluated. The ultimate goal of LES modeling is to develop accurate models with high computational efficiency. Thus, along with exploration of new innovative methods, improving the computational efficiency of the current methods must also be explored.

Author details

Neelakantan Padmanabhan
ZF Group, Livonia, USA

*Address all correspondence to: neel.p.3187@gmail.com

References

[1] Kolmogorov AH. The local structure of turbulence in incompressible viscous fluid for very large Reynolds' numbers. Proceedings of the USSR Academy of Sciences. 1941;**30**:301-305

[2] Moin P, Mahesh K. Direct numerical simulation: A tool in turbulence research. Annual Review of Fluid Mechanics. 1998;**30**(1):539-578

[3] Tennekes H, Lumley JL. A First Course in Turbulence. Cambridge, Massachusetts and London, England: MIT Press; 1972

[4] Georgiadis NJ, Rizzetta DP, Fureby C. Large-eddy simulation: Current capabilities, recommended practices, and future research. AIAA Journal. 2010; **48**(8):1772-1784

[5] Zhiyin Y. Large-eddy simulation: Past, present and the future. Chinese Journal of Aeronautics. 2015;**28**(1):11-24. DOI: 10.1016/j.cja.2014.12.007

[6] National Research Council. Research Directions in Computational Mechanics. Washington, DC: The National Academies Press; 1991. DOI: 10.17226/1909

[7] Palle S. On real gas and molecular transport effects in high pressure mixing and combustion. ProQuest. 2006

[8] Foster J, Miller RS. Fundamentals of High-Pressure Combustion. In: Lackner M, editer. Chapter in high-pressure processes in Chemical Engineering. United Kingdom: ProcessEng Engineering GmbH; 2010. pp. 53-75

[9] De Groot SR, Mazur P. Non-equilibrium Thermodynamics. New York: Dover Publications; 1984

[10] Keizer J. Statistical Thermodynamics of Nonequilibrium Processes. New York: Springer Verlag; 1987

[11] Okong'o N, Bellan J. Real-gas effects on mean flow and temporal stability of binary-species mixing layers. AIAA Journal. 2003;**41**(12):2429-2443

[12] Sohn CH, Chung SH. Effect of pressure on the extinction, acoustic pressure response, and NO formation in diluted hydrogen–air diffusion flames. Combustion and Flame. 2000;**121**(1–2): 288-300

[13] Chung SH, Lee SR, Mauss F, Peters N. Reduced kinetic mechanisms and NOx formation in diffusion flames of CH4/ C2H6 mixtures. In: Peters N, Rogg B, editors. Reduced Kinetic Mechanisms for Applications in Combustion Systems. Lecture Notes in Physics Monographs. Vol. 15. Berlin, Heidelberg: Springer; 1993. DOI: 10.1007/978-3-540-47543-9_17

[14] Ma Z, Korucu A, Miller RS. A priori analysis of subgrid scale pressure and heat flux in high pressure mixing and reacting shear layers. Combustion Theory and Modelling. 2015;**19**(6): 807-832

[15] Foster J, Miller RS. A priori analysis of subgrid mass diffusion vectors in high pressure turbulent hydrogen/oxygen reacting shear layer flames. Physics of Fluids. 2012;**24**(7):075114. DOI: 10.1063/ 1.4739065

[16] Jiang X, Lai C-H. Numerical Techniques for Direct and Large-Eddy Simulations. Boca Raton, FL, USA: CRC Press; 2016. DOI: 10.1201/9781420075793

[17] Lund TS. The use of explicit filters in large eddy simulation. Computers and Mathematics with Applications. 2003; **46**(4):603-616

[18] Berkooz G, Holmes P, Lumley JL. The proper orthogonal decomposition in the analysis of turbulent flows. Annual Review of Fluid Mechanics. 1993;**25**(1): 539-575

[19] Padmanabhan N. Reduced order model of a convection-diffusion equation using proper orthogonal decomposition. 2023. arXiv preprint arXiv:2303.07176. DOI: 10.48550/arXiv.2303.07176

[20] Germano M, Piomelli U, Moin P, Cabot WH. A dynamic subgrid-scale eddy viscosity model. Physics of Fluids A: Fluid Dynamics. 1991;**3**(7):1760-1765

[21] Bardina J, Ferziger J, Reynolds WC. Improved subgrid-scale models for large-eddy simulation. In: 13th Fluid and PlasmaDynamics Conference. Snowmass, CO, USA. 1980. p. 1357. DOI: 10.2514/6.1980-1357

[22] Shinde V. Proper orthogonal decomposition assisted subfilter-scale model of turbulence for large eddy simulation. Physical Review Fluids. 2020;**5**(1):014605

[23] Pope SB. PDF methods for turbulent reactive flows. Progress in Energy and Combustion Science. 1985;**11**(2):119-192

[24] Jaberi FA, Colucci PJ, James S, Givi P, Pope SB. Filtered mass density function for large-eddy simulation of turbulent reacting flows. Journal of Fluid Mechanics. 1999;**401**:85-121

[25] Pope SB. Mapping closures for turbulent mixing and reaction. Theoretical and Computational Fluid Dynamics. 1991;**2**(5–6):255-270

[26] Subramaniam S, Pope SB. A mixing model for turbulent reactive flows based on Euclidean minimum spanning trees. Combustion and Flame. 1998;**115**(4): 487-514

[27] Shetty D, Chandy A, Frankel S, Dinger M, Cai J, Tong C. Assessment of Mixing Models for LES/Filtered Mass Density Function in a Three-Stream Jet. In APS Division of Fluid Dynamics Meeting Abstracts. 2008;**61**:GR-003

[28] Chen W, Liang J, Zhang L, Guan Q. A numerical investigation of mixing models in LES-FMDF for compressible reactive flows. Energies. 2021;**14**(16):5180

[29] Padmanabhan N. On High Pressure Real Gas Turbulent Mixing Jets, 2017. All Dissertations. 1981. Available from: https://tigerprints.clemson.edu/all_dissertations/1981

[30] Padmanabhan N, Miller RS. Assessment of subgrid scale mixing models used in LES at high pressures. Journal of Turbulence. 2018;**19**(8): 683-715

[31] Brunton SL, Noack BR, Koumoutsakos P. Machine learning for fluid mechanics. Annual Review of Fluid Mechanics. 2020;**52**:477-508

[32] Duraisamy K, Iaccarino G, Xiao H. Turbulence modeling in the age of data. Annual Review of Fluid Mechanics. 2019;**51**:357-377

[33] Rosenblatt F. The perceptron: A probabilistic model for information storage and organization in the brain. Psychological Review. 1958;**65**(6):386

[34] Gamahara M, Hattori Y. Searching for turbulence models by artificial neural network. Physical Review Fluids. 2017; **2**(5):054604

[35] Wang Z, Luo K, Li D, Tan J, Fan J. Investigations of data-driven closure for subgrid-scale stress in large-eddy simulation. Physics of Fluids. 2018;**30**(12):125101. DOI: 10.1063/1.5054835

[36] Park J, Choi H. Toward neural-network-based large eddy simulation: Application to turbulent channel flow. Journal of Fluid Mechanics. 2021;**914**:A16

[37] Ranjan R, Panchal A, Karpe S, Menon S. Machine learning strategy for subgrid modeling of turbulent combustion using linear Eddy mixing based tabulation. In: Machine Learning and Its Application to Reacting Flows: ML and Combustion. Cham: Springer International Publishing; 2023. pp. 175-208

[38] Xie C, Wang J, Li H, Wan M, Chen S. Artificial neural network mixed model for large eddy simulation of compressible isotropic turbulence. Physics of Fluids. 1 August 2019;**31**(8): 085112. DOI: 10.1063/1.5110788

[39] Meng Q, Jiang Z, Wang J. Artificial neural network-based subgrid-scale models for LES of compressible turbulent channel flow. Theoretical and Applied Mechanics Letters. 2023;**13**(1): 100399

[40] Benjamin M, Domino SP, Iaccarino G. Neural networks for large eddy simulations of wall-bounded turbulence: Numerical experiments and challenges. The European Physical Journal E. 2023;**46**(7):55

[41] Nikolaou ZM, Minamoto Y, Chrysostomou C, Vervisch L. Machine-learning for stress tensor modelling in large Eddy simulation. In: Machine Learning and its Application to Reacting Flows: ML and Combustion. Cham: Springer International Publishing; 2023. pp. 89-116

[42] Tian Y, Woodward M, Stepanov M, Fryer C, Hyett C, Livescu D, et al. Lagrangian large eddy simulations via physics-informed machine learning. Proceedings of the National Academy of Sciences. 2023;**120**(34):e2213638120

[43] Xing V, Lapeyre CJ. Deep convolutional neural networks for subgrid-scale flame wrinkling modeling. In: Machine Learning and its Application to Reacting Flows: ML and Combustion. Cham: Springer International Publishing; 2023. pp. 149-174

[44] Foster J. On simulation and modeling of turbulent non-premixed reacting shear flames at low and high pressure. All Dissertations. 2012:1011

[45] Kurz M, Offenhäuser P, Beck A. Deep reinforcement learning for turbulence modeling in large eddy simulations. International Journal of Heat and Fluid Flow. 2023;**99**:109094

[46] Novati G, de Laroussilhe HL, Koumoutsakos P. Automating turbulence modelling by multi-agent reinforcement learning. Nature Machine Intelligence. 2021;**3**(1):87-96

[47] Nista L, Schumann CDK, Grenga T, Attili A, Pitsch H. Investigation of the generalization capability of a generative adversarial network for large eddy simulation of turbulent premixed reacting flows. Proceedings of the Combustion Institute. 2022;**39**(4):5279-5288. DOI: 10.1016/j.proci.2022.07.244

[48] Chen S, Sammak S, Givi P, Yurko JP, Jia X. Reconstructing high-resolution turbulent flows using physics-guided neural networks. In: 2021 IEEE International Conference on Big Data (Big Data). Orlando, FL, USA: IEEE; 2021. pp. 1369-1379. DOI: 10.1109/BigData52589.2021.9671524

[49] Zhao Q, Jin G, Zhou Z. Deep learning method for the super-resolution reconstruction of small-scale motions in large-eddy simulation. AIP Advances. 2022;**12**(12):125304. DOI: 10.1063/5.0127808

Chapter 5

Heat Transfer and Thermal Analysis with Computational Fluid Dynamics

Yingjie Tang

Abstract

This chapter provides general descriptions of heat transfer and thermal analysis within mechanical designs using computational fluid dynamics (CFD) tools. CFD modeling is capable of evaluating all heat transfer mechanisms: conduction, convection, and radiation, with predictions on temperature distributions in solid embodiment or fluids. This modeling process can be extremely useful in mechanical design to help further understanding of thermal management and avoid hotspots with extreme temperatures. For the prototype design of any mechanical machine, CFD-based heat transfer and thermal analysis will efficiently discover the thermal signatures and save costs from prototype manufacture and field tests. This book chapter will review the numerical modeling methodologies in CFD, the discretization approach, and later, with an example analysis of real physical applications, including heat exchangers, flow distributing systems, power generation/conversion systems, etc. The goal of this chapter is to provide a general research map and help readers use CFD tools in future mechanical designs.

Keywords: thermal analysis, conduction, convection, radiation, temperature distribution, optimized design

1. Introduction

Nowadays a few scientific methodologies are commonly used to solve flow-related problems, which include theoretical-based (analytical) fluid dynamics, experimental-based fluid dynamics, and numerical solutions using computational fluid dynamics (CFD). With the development of its governing equations, e.g., the Navier-Stokes equations, the theoretical-based analysis has been established for nearly 200 years since the initiation of the Newton fluid dynamics [1]. The methodology of theoretical-based analysis has been proven to be active in modern physics study, and is widely applied in many studies of scientific and engineering interests, such as modeling of atmosphere weather, ocean currents, internal or external flows, and greatly supporting mechanical designs in a number of areas [2, 3].

Experimental-based analysis of fluid dynamics is another useful tool for solving flow-related problems. Currently, it has been used in multiple industrial or academic research areas which are critical to validate the flow solutions and explain the deviations between engineering approximations and the theoretical-based results. Nowadays, experimental fluid dynamics is still the most trustable scientific approach for fluid analysis. However, experimental-based analysis normally is found with disadvantages due to its nature of expensive costs, especially for a full-scale measurement. During engineering practice, the experimental analysis is most likely limited by the size of the model, and if extreme conditions are difficult to achieve in specific cases (e.g., pressure, velocity, Reynolds number, temperature, and types of atmospheres that can be simulated, etc.). For some design instances with high-level requirements on their accuracy and reliabilities, such as an aircraft design in wind tunnel test, the full-scale measurement in the experiment may cause a dilemma in the industry because of its high cost in reproductions for design alternatives, both economically and temporally.

The method of computational fluid dynamics (CFD) was first developed to solve 2-D linear potential equations nearly 100 years ago [2]. CFD has experienced advanced progress in the last 70 years, and has become an unneglectable and cost-effective alternative approach for fluid analysis, as a supplement to theoretical and experimental fluid dynamics. With extreme blowouts of the computational power capability and significant developments in computational techniques in the last couple of decades, CFD has become one of the most preferable scientific design methods used in multiple engineering branches, because of its advantages in the following features [2, 4]: (a) Reduced analysis time in the mechanical design process, especially for design alternatives or iterations; (b) Ability to simulate or reproduce extreme flow conditions, which may be non-applicable in full-scale experiment measurements; (c) Ability to provide more detailed information on fluid analysis solutions, compared to experimental approaches which have sensory limitations; and (d) Economically cost-effective, less energy consumption, environmentally friendly with less carbon footprint.

Today, CFD analysis is widely used in many engineering branches to test alternative designs, such as aerodynamic, flow machinery, energy, automobile industries, and chemical process industries [2, 4, 5]. The outcome of such analysis includes flow field analysis, pressure-based flow restriction analysis, acoustic analysis, and last but not least, thermal analysis. Thermal analysis is a scientific and engineering analytical technique, targeting the profiles of time and temperature when physical changes occur upon a mass substance, for example, being heated or cooled. It is the consequence of a heat transfer process, i.e., the flow of 'heat', or thermal/internal energy, due to differences in temperature profiles within or between material substances. This transfer process of the heat energy would also lead to changes in the distributions of the subsequent temperature [6]. The main mechanisms of heat transfer include conduction, convection, and radiation, which in practice could be strongly related to the process of mass transfer as well [6]. In such case, the thermal analysis certainly bonded to the flow simulation most likely and becomes an important problem that can be resolved by CFD analysis.

Most thermal analysis applications in engineering design require a prediction of the changes in the substance temperature, or temperature distributions. For example, in a heat exchanger, there are two working fluids with temperatures either increasing or decreasing, along the time when they are passing through exchanging heat energy; in another example, an electric motor has generated heat from its source, e.g., electric windings, and the temperatures through its body are spatially different. In mechanical engineering designs, it may be a critical issue to have an accurate and reliable prediction on the temperature distributions: in the first example, the temperature gradients for both working fluids determine the performance and thermal efficiency of this heat exchanger; and in the motor application, the existence of the hotspots (locations with extremely high-temperature) could cause failures on the designed motor during its operations. Certain theoretical equations (i.e., transfer functions) or empirical equations based on a thermal network analysis could be useful to handle the thermal analysis and find out the temperature distribution for cases with simple or simplified geometries. However, for a complex mechanical design with alternative prototypes, it is possible to have highly complicated geometries which would make using such analytical or empirical equations difficult to get a practical solution without compromising its accuracy. Therefore, the utilization of numerical solution through discretization in CFD modeling becomes an advantageous complement for thermal analysis, if the industry is trying to avoid the high cost brought by expensive experimental studies.

This chapter will start with a general introduction to the heat transfer process, the general descriptions of the CFD-profiled major heat transfer mechanisms (conduction, convection, and radiation) will be discussed, followed by a review of the numerical modeling methodologies in CFD applied in thermal analysis, the discretization approach, and later with a study example on thermal analysis of industrial applications for alternative thermal designs. The goal of this chapter is to provide an instructional research map and help readers further understand the advantages as well as its limitations in the process of using CFD in thermal analysis.

2. Heat transfer mechanisms in thermal analysis

In physics, heat transfer is a form of energy transfer process where the heat energy, or thermodynamic form of work, gets exchanged within or between material substances or physical systems [7, 8]. The heat transfer generally requires a temperature difference in materials (solid/liquid/gas), and the spontaneous heat transfer process, as stated in the second law of thermodynamics, would cause the thermal form of energy to move from the higher temperature bodies (or parts of the body) to the lower temperature ones. In physical systems, the thermal energy is normally transferred from a 'heat source', generally at higher temperatures and probably with heat generation or energy conversion, to a 'heat sink' which is normally at lower temperatures. As mentioned before, there are three main mechanisms for the process of heat transfer, including thermal conduction, thermal convection, and thermal radiation. Some references also consider the energy transfer process by a material phase change or chemical reaction as other heat transfer mechanisms, but it should be noted that the heat energy transfer by a phase change or chemical reaction seldom acts alone and most likely performs the energy exchange together with these three main mechanisms [7]. All these mechanisms could be active individually with distinct features and characteristics, but on a common basis in the natural environment, most likely would occur simultaneously in engineering practice [8].

Thermal conduction is a process where heat energy is transferred interactions between microscopic particles which compose a conducting substance [9]. The direction of such energy flow, i.e., from high to low temperature, is due to the more energetic particles in the higher temperature regions of the substance. Particularly in solids, thermal conduction occurs because of the vibrations of the solid molecules in conjunction with the diffusion of free electrons [10], while in fluids like gases or liquids, it is due to collisions between molecules and the diffusions [7]. Temperature difference is the driving potential in thermal conduction, which intends to achieve a thermal equilibrium in the domain. The parameter to evaluate the material's ability for thermal conduction is the thermal conductivity, which is commonly denoted by k, with SI unit of $\text{Wm}^{-1}\,\text{K}^{-1}$. The defining equation for the conductive heat transfer is called Fourier's Law of heat conduction:

$$\dot{q}'' = -k\nabla T \tag{1}$$

where $\dot{q}''$ is the heat flux, i.e., the amount of heat energy passing through a certain surface/internal boundary in unit time; k is the thermal conductivity which usually is a material property, and ∇T is the temperature gradient in the spatial domain. The negative sign in Eq. (1) on the right-hand side suggests that such heat transfer by thermal conduction is usually in the opposite direction of the temperature gradient, in which temperature decreases.

Thermal convection is a heat transfer process due to the combined effects of the random molecular motion and the movement of fluid [11]. Unlike thermal conduction which usually occurs in the analysis of the solid material system, thermal convection occurs when a fluid (in a state of liquid or gas) carries its heat while 'flowing'. Two general forms of thermal convection include 'natural convection', in which the flow is driven by buoyancy forces due to the fluid density difference possibly caused by the fluid temperature difference, and 'forced convection', in which the flow is driven by an external cause, e.g., the pressure difference caused by the use of a pump. It should be noted that although heat transfer within fluids, either gas or liquid, could involve both mechanisms of conduction and convection, convective heat transfer is usually the dominant mechanism. This is because, within solids, there are relatively close spatial relationships between molecules, whereas in fluids a usual relatively large distance would cause a less effective conduction. When analyzing convective heat transfer between a fluid in motion and a bounding surface, the following equation, normally called Newton's law of cooling, is used to describe this heat transfer process:

$$\dot{q}'' = h\left(T - T_f\right) \tag{2}$$

where h is the heat transfer coefficient, a bulk condition-based parameter; T is the object's surface temperature, and T_f is the fluid temperature. The heat transfer direction in convection is the same as conduction, which is from a high-temperature side to a low temperature side. Note that in Eq. (2), the relationship between the heat flux and the temperature difference could be strongly nonlinear, because with varied flow conditions, the heat transfer coefficient could vary as a function of space and time [8]. Unlike thermal conductivity which is usually a material property, the heat transfer coefficient is strongly dependent on the physical characteristics between the fluid and the object, e.g., the boundary layer flow pattern. Thus, the modeling of the flow field becomes a critical factor in accurately predicting the convective heat transfer. CFD tools are considered to be one of the most important engineering approaches to determine the convective heat transfer coefficient in Eq. (2) and quantify the convective heat transfer.

Thermal radiation, on the other hand, is the transfer of heat energy by means of electromagnetic radiation emitted by the thermal motion of particles in substances, which have a temperature greater than absolute zero [7, 12, 13]. Unlike conduction or convection which requires transferring medium matters, thermal radiation could occur between two objects through a vacuum. For black bodies (with perfect absorptivity at all wavelengths), or so-called ideal radiators, based on Stefan-Boltzmann Law, the maximum radiation heat flux to be emitted from a black body surface is [12].

$$\dot{q}_b{}'' = \sigma T_s^4 \tag{3}$$

where T_s is the surface temperature, σ is the Stefan-Boltzmann constant, 5.670×10^{-8} Wm^{-2} K^{-4}. For a real surface

$$\dot{q}_r{}'' = \sigma \epsilon T_s^4 \tag{4}$$

where ϵ is the emissivity of the surface. The heat radiation from a real surface should be not greater than that from a black body surface, since $0 < \epsilon < 1$. Generally, within radiation acting upon a surface, the energy exchange consists of a portion of absorption, a portion of reflection, and a portion of transmission:

$$q = q_{abs} + q_{ref} + q_{trans} = \alpha q + \rho q + \tau q \tag{5}$$

where α, ρ, and τ are the material surface absorptivity, reflectivity, and transmissivity, separately, which depend on surface temperature, surface roughness, the wavelength of the radiation, etc., and normally could be determined by previous experimental studies, and used as constants in numerical simulations. With Stefan-Boltzmann Law, the net heat transfer rate through radiation from Surface 1 to Surface 2 under the black body assumption could be defined as

$$\dot{Q}_{1\to 2} = \sigma A_1 F_{1\to 2}\left(T_1^4 - T_2^4\right) \tag{6}$$

where σ is the Stefan-Boltzmann constant; T is the temperature; A is the surface area; $F_{1\to 2}$ is the view factor from Surface 1 to Surface 2. For two gray body surfaces (with frequency-independent emissivity other than 'perfect' emissivity as black body), the heat transfer rate is

$$\dot{Q} = \frac{\sigma\left(T_1^4 - T_2^4\right)}{\frac{1-\epsilon_1}{A_1\epsilon_1} + \frac{1}{A_1 F_{1\to 2}} + \frac{1-\epsilon_2}{A_2\epsilon_2}} \tag{7}$$

In Eq. (7), the emissivity is related to the material and its surface condition, which could be critical to the quantity of the radiation heat transfer. The emissivity varies with different materials, different surface conditions (e.g., polished or oxidized), and if the surface is colored or painted. In industry, certain types of coating are also commonly used to change surface conditions, so that the heat transfer through thermal radiation could be then enhanced or mitigated [14]. **Table 1** gives some examples of emissivity values. Since the Stefan-Boltzmann constant is of small value, in a large amount of heat transfer analysis, the energy exchange by thermal radiation is not as significant as thermal conduction and convection and usually can be neglected. However, in applications with extremely high temperatures (or large temperature differences), the heat transfer through thermal radiation has to be considered because the surface temperature in Eq. (7) is in a quadral-order term.

Materials and surface conditions	Emissivity
Polished aluminum	0.05
Alumina	0.78
Polished copper	0.03
Oxidized copper	0.78
Polished cast iron	0.21
Oxidized cast iron	0.57
Polished brass	0.04
Oxidized brass	0.60
Coating: matte white	0.91
Coating: matte black	0.88
Coating: gloss black	0.90

Table 1.
Emissivity of different solid materials and surface conditions [14].

3. Thermal network modeling

The analysis of the heat transfer process in a computational domain could be simplified into a 'nodes and edges' representation, or so-called 'thermal network' modeling. A thermal network consists of mathematical models that can be used to analyze and optimize flow and heat transfer in complex systems. It breaks down a continuous system into smaller, interconnected components and represents the energy transfer between these components as equations that describe the heat energy flow. Thermal network modeling, which can be developed in both steady-state and transient scenarios, could reduce the system's complexity and focus on energy balance and heat transfer. This simplification reduces the number of variables to be solved in thermal analysis, resulting in manageable calculations allowing more efficient solutions. Generally, nodes represent mass substance bodies or parts of the mass substance of interest, and edges represent heat transfer paths between two nodes (mass substances).

Figure 1 gives an example of the simplest thermal network, with two substances and one heat transfer path. The edge indicates a heat transfer process from the mass node 1 (heat source) to the mass node 2 (heat sink). The heat transfer process now could be determined and quantified based on the governing equations mentioned in the previous section, as the energy flows through such edges, among the defined nodes within the computational domain.

Thermal energy flows into this system through the 'heat source' nodes and out of the system through the 'heat sink' nodes. The heat transfer process is

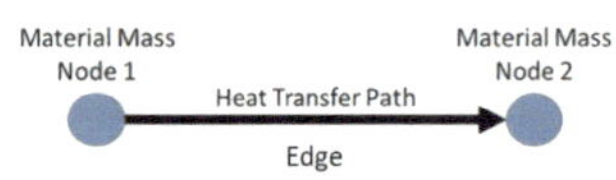

Figure 1.
Node and edge representation of the heat transfer process.

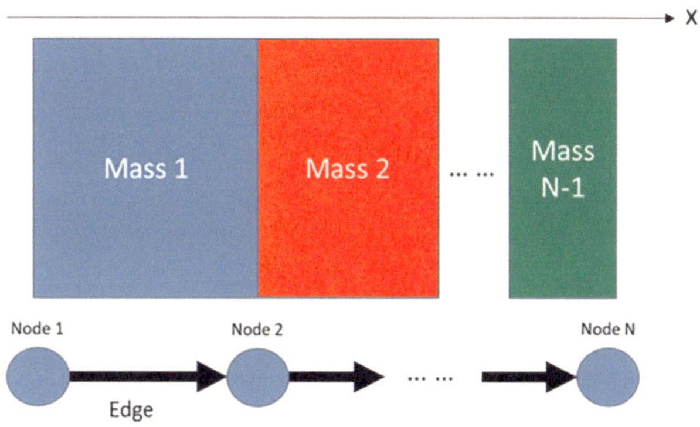

Figure 2.
One-dimensional serial system (1-D solid thermal conduction).

analyzed in the form of the edge individually. Several different scenarios of thermal analysis are shown as follows:

3.1 One-dimensional serial system

A common example of this type of heat transfer is the 1-D solid conduction through multi-layers of the solid bodies (**Figure 2**).

The assumption has been made that temperature gradient only occurs in the x direction, thus the heat energy only transfers along x directions, and uniform temperature distribution occurs in any cross-sectional planes perpendicular to the *x-axis*. So Fourier's Law in Eq. (1) could be written as

$$\dot{q}'' = -k\frac{dT}{dx} \tag{8}$$

In such a physical system, the nodes in the simplified model represent the boundaries between different solid bodies, provided each solid body has separate magnitudes of thermal conductivity. Between two adjacent nodes (solid body boundaries), the amount of heat energy transferred is

$$\dot{q}'' = k\frac{\Delta T}{L} \tag{9}$$

where ΔT is the temperature difference between two adjacent nodes (solid body boundaries) or through a solid body, L is the distance between such two nodes in the x direction (length of the solid body), and k is the thermal conductivity for the material between these two nodes, treated as a constant for each solid body material. And Eq. (9) could also be written as

$$\Delta T = \dot{q}''\frac{L}{k} \tag{10}$$

For the whole system based on a steady-state analysis, if no extra heat source or sink and the system has already reached thermal equilibrium, the heat flux remains constant. Thus the total temperature difference of the system could be

$$\Delta T_{total} = \dot{q}'' \sum_{1}^{N-1} \frac{L_i}{k_i} \tag{11}$$

This apparently indicates the energy transfer in a 'serial' mode. A term 'thermal resistance' could be defined for each individual path:

$$R_i = \frac{\Delta T}{\dot{q}} = \frac{\Delta T}{A\dot{q}''} = \frac{L_i}{Ak_i} \tag{12}$$

where $\dot{q}$ is the heat transfer rate. And like the analogy of Ohm's law in electrical analysis, a serial mode of heat transfer (conduction) could be represented by

$$\Delta T_{total} = \dot{q} \sum_{1}^{N-1} R_i \tag{13}$$

Note the overall thermal resistance of this system with serial mode heat transfer would be the sum of the thermal resistance of each heat transfer edge. For the serial mode of general heat transfer

$$\Delta T_{total} = \dot{q} \sum_{1}^{N-1} \frac{1}{A_i h_i} \tag{14}$$

And thermal resistance could be represented using the heat transfer coefficient:

$$R_i = \frac{1}{A_i h_i} \tag{15}$$

3.2 Parallel system

Another common scenario is a simplified parallel mode heat transfer, with one heat source and one heat sink, as shown in **Figure 3**. Various edges could be involved in this heat transfer process from the heat source to the heat sink, i.e., in one or more heat transfer mechanisms, which could be effective simultaneously. In **Figure 3**, the heat will be generated and brought to this two-node system at Node 1, i.e., the heat source, which normally has a higher temperature. Then there are M edges involved and the thermal energy is transferred from Node 1 to Node 2, i.e., heat sink, which normally has a lower temperature. Unlike the serial mode heat transfer, the heat transfer edges could be by multiple means and each edge could be considered

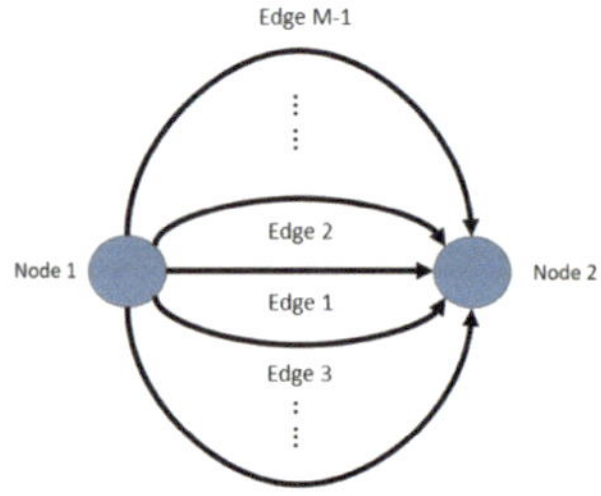

Figure 3.
Parallel system of heat transfer.

individually based on its own thermal resistance, and apparently transfer different amounts of heat energy rate. For edge i, the heat flux could be calculated as

$$T_1 - T_2 = \dot{q}_i'' A_i R_i \tag{16}$$

where R_i is the thermal resistance through edge i, and could be calculated with Eq. (15). The total heat transfer rate from Node 1 to Node 2 should be the sum of the heat transfer rates of all the applicable edges:

$$\dot{q} = \sum_1^M \dot{q}_i = \sum_1^M \dot{q}_i'' A_i \tag{17}$$

where A_i is the heat transfer interface area for the heat transfer edge i. Eq. (16) could be written as

$$\dot{q}_{total} = \sum_1^M \dot{q}_i = \sum_1^M \frac{T_1 - T_2}{R_i} \tag{18}$$

And the 'overall' thermal resistance could be

$$R_{overall} = \frac{1}{\sum_1^M \frac{1}{R_i}} \tag{19}$$

Note that this is consistence with the parallel electric circuit analysis, which brings a similarity in multi-physics systems.

3.3 Complicated system and thermal network

In engineering practice, the heat transfer system could be much more complicated than a simple serial or parallel system. For example, multiple heat sources or heat sinks could exist in the system at the same time. Therefore, multiple nodes and multiple edges are expected to be in such a "complicated system", as shown in **Figure 4** as an example.

In a more realistic analysis of the electric motor, there could be serval heat sources in the system: the copper heat generation from the winding conductors, the iron heat

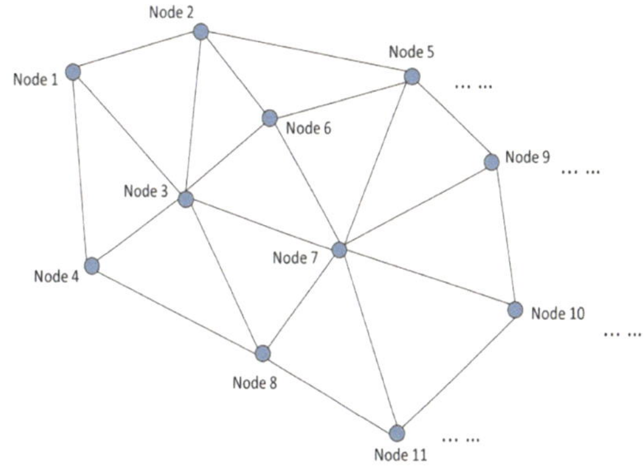

Figure 4.
Nodes and edges in a complicated system.

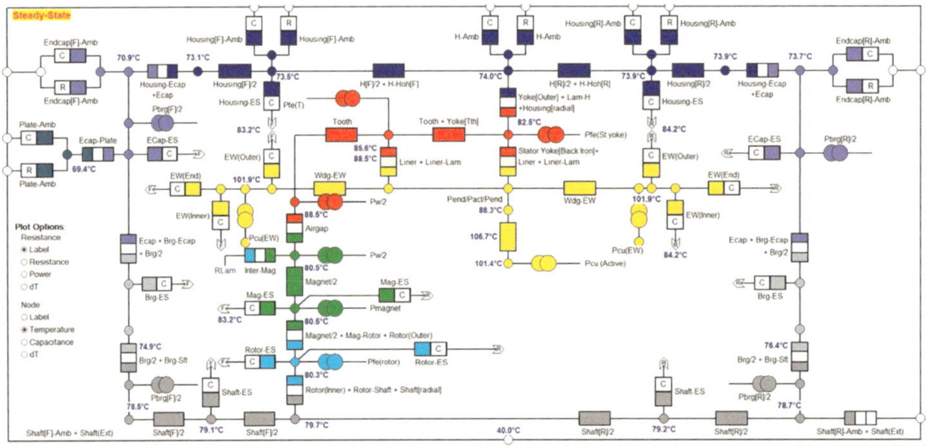

Figure 5.
Thermal network of the electric motor thermal analysis.

loss by eddy current from the steel laminations of the motor body, and the hysteresis heat loss from the iron or the permanent magnets, etc. For a single heat source in the system, one single node may not be enough in the thermal analysis because the temperature difference also occurs within the body of this heat source, which results in a non-uniform temperature distribution within. On the other hand, multiple heat sinks could exist simultaneously, such as the ambient air, the ground, the supporting frames, and the coolant from a built-in cooling system. When we analyze the heat transfer edges, it could be a web-like system of edges involved, or so-called 'thermal network'. **Figure 5** provides an example of the electric motor and its thermal network in thermal analysis, with multiple motor parts and its temperature distributions and heat power transfer during steady-state operations.

When analyzing engineering problems, the thermal network analysis could still have limitations. A most common drawback is that thermal network modeling is effective for relatively simple geometries, and the accuracy of this analysis may be limited for more complex systems. Furthermore, as a lumped approximation, thermal networks can provide less details compared to a CFD model, which solves each system element's flow and heat transfer, by discretizing the continuous system into many more nodes and edges. On the other hand, it should be noted that the thermal resistance for each edge in a thermal network could be difficult to determine sometimes. For example, determining the thermal resistance of the convective heat transfer from the motor body to the coolant flow in the cooling system could be challenging, because the heat transfer coefficient could depend on the temperature and the velocity boundary layers for the internal flow of coolant, and be a variable along the cooling pipeline with non-linearities.

Therefore, in engineering practice, complex geometries or irregular shapes of the parts in the system domain could significantly increase the need for a large number of nodes represented in thermal networks, which could make the typical thermal network models very challenging to get accurate solutions. Considering such difficulties, it will be necessary to additionally perform the approach of discretization in numerical methodologies and simulation to find the solutions to such heat transfer processes.

4. Numerical simulation in thermal analysis

As a scientific analytical method, thermal analysis focuses on the thermal states or behaviors of a substance, either in a steady (static) state or a transient (dynamic) condition, with heat transfer evaluated as a function of temperature. CFD tools have a long history employed to determine the predictive results of temperature distributions, heat fluxes and rates, and locations of potential hotspots, or their time-dependent profiles (if transient), which could be important in research areas such as aerodynamic components, electrical systems, power generation systems, or cooling systems, etc.

During such a design process, a designer is supposed to first build a geometry model of the system to be evaluated. Such a model could be in the level of one-dimensional, two-dimensional, or even three-dimensional analysis based on the requirements of the system to be designed, but the final stage of an engineering design most likely will involve a three-dimensional model of the prototype. The purpose of CFD tools is to achieve modeling of the thermal analysis under various conditions, with predictions on the temperature field results. The model should consider the thermal dependency of the substance materials, material properties, boundary or initial conditions within the system, contact conditions for different components, and other modeling features. Assumptions could be necessary as well in this modeling build-up, to simplify the model complexity and reduce simulation time without jeopardizing the accuracy of simulation.

Once a computational model has been built, basic computational techniques are required to solve thermal problems. Specifically based on features of the system, governing equations and appropriate boundary conditions should be selected, then two main stages of a typical procedure will be established to seek the computational solution of the thermal problems, as shown in **Figure 6**.

The first stage is 'Discretization', possibly with different methods, such as finite difference, finite element, and finite volume methods. The goal of this stage is to convert the continuous state of the governing equations into a discrete system of algebraic equations. Such discretization process could refer to both 'spatial', which is in the form of mesh or grid generation, and 'temporal', which is in the form of appropriate time steps during a transient simulation.

To convert the governing partial differential equations (PDEs) to a discrete system of algebraic equations, there are different discretization methods that could be employed for both time derivatives and spatial derivatives [4]. In engineering practice, time derivatives are normally discretized by the finite difference method, while the spatial derivatives are discretized using finite difference, finite element, or finite volume method based on the features of the computational domain, and requirements on solution accuracy or convergence.

The second stage of the solution process is an equation solver, which provides the solution to the system of algebraic equations. This stage could also introduce errors

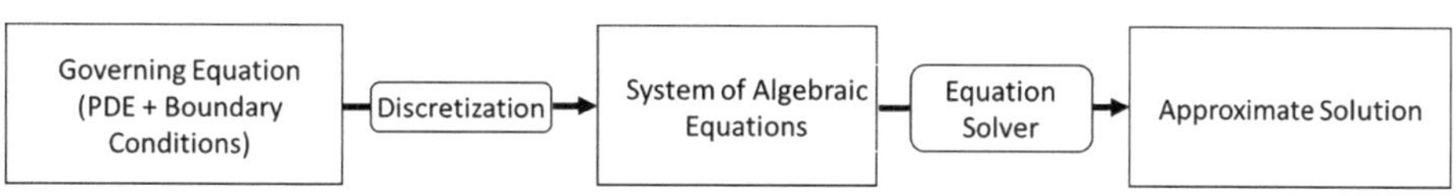

Figure 6.
Computational solution procedure.

but with appropriate handling in simulation, such errors could be neglected compared to the numerical error during the discretization stage [2]. This solver is highly dependent on the CFD code and model, and significantly varied either a commercial CFD software is used, or a self-written code is applied for the solution. So, it is recommended to refer to the CFD software manual for additional descriptions and details of such CFD solver employed.

Hereby an example is provided for the discretization method used in thermal analysis to show the general steps for the discretization and related error analysis. This process covers derivatives of a 1-D transient heat conduction problem described in continuous PDE to discrete algebraic expressions. For a one-dimensional, transient heat conduction problem, the dynamic temperature field could be calculated through

$$\frac{\partial \bar{T}}{\partial t} = \alpha \frac{\partial^2 \bar{T}}{\partial x} \tag{20}$$

where $\bar{T}$ is the transient temperature profile, with the overbar denoting the exact solution, α is the thermal diffusivity, which is a material property defined as

$$\alpha = \frac{k}{\rho C_p} \tag{21}$$

where k is the thermal conductivity; ρ is the material density; C_p is the specific heat capacity. In this case, the thermal diffusivity is treated as a constant of the material and independent of its temperature. Additional initial and boundary conditions could be provided here to specify this thermal problem:

$$\bar{T}(x, 0) = T_0(x), 0 \le x \le 1 \tag{22}$$

$$\bar{T}(0, t) = b, \bar{T}(1, t) = d \tag{23}$$

Eq. (23) suggests a Dirichlet boundary condition, with values that the solution takes along the boundary of the domain are fixed. As mentioned previously, the goal of the discretization is to replace such partial differential derivatives in Eq. (20) with equivalent numerical expressions. Here, we employ the finite difference method. Thus

$$\frac{T_j^{n+1} - T_j^n}{\Delta t} = \frac{\alpha\left(T_{j-1}^n - 2T_j^n + T_{j+1}^n\right)}{\Delta x^2} \tag{24}$$

Note that the discretization has been defined here in both temporal and spatial derivatives, providing the discretized step sizes of Δt and Δx separately. Thus this 1-D transient heat conduction problem could convert to a discrete solution problem as shown in **Figure 7**.

The thermal analysis problem of finding the exact solution of the dynamic temperature field $\bar{T}(x, t)$ defined in the governing equation now has been converted to find out the discrete solution of the dynamic temperature field T_j^n, with a temporal step of n and a spatial step of j. It should be noted that in this discretizing process so far, there are two types of error introduced, which are the truncation error and the solution error. Also, generally, the solution would be expected to vary smoothly

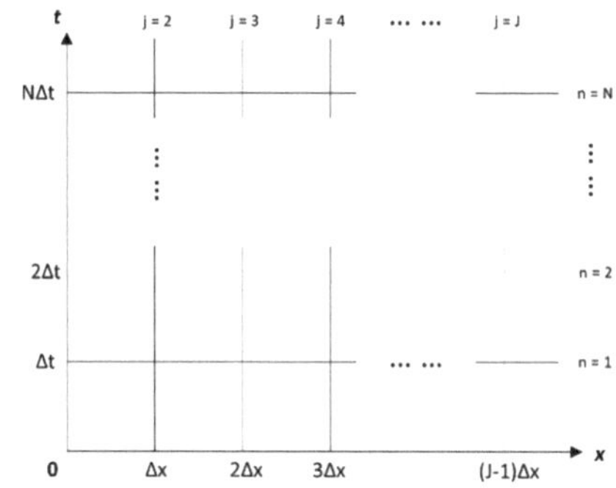

Figure 7.
Discretization of the exact solution $\bar{T}$ to discrete solution T_j^n.

between nodal points in **Figure 7**. For the algebraic equation, it can be given an algorithm for the unknown value of T_j^{n+1}, in terms of the known value of T_j^n:

$$T_j^{n+1} = T_j^n + \frac{\alpha \Delta t}{\Delta x^2}\left(T_{j-1}^n - 2T_j^n + T_{j+1}^n\right) \tag{25}$$

To provide all complete numerical solution at time level (n + 1), the values are needed for all the nodes j = 2, ..., J-1, through the Dirichlet boundary condition defined value T_1^{n+1} and T_J^{n+1}.

For the time derivatives, it is noted that in the continuous PDE, the term $\partial\bar{T}/\partial t$ is now replaced by $\left(T_j^{n+1} - T_j^n\right)/\Delta t$, which only involves the time levels n and n + 1. The finite difference representation in the spatial derivative $\partial^2\bar{T}/\partial x^2$ has been evaluated at the time level of n. So, at the time level n + 1,

$$-\alpha\frac{\Delta t}{\Delta x^2}T_{j-1}^{n+1} + \left(1 + 2\alpha\frac{\Delta t}{\Delta x^2}\right)T_j^{n+1} - \alpha\frac{\Delta t}{\Delta x^2}T_{j+1}^{n+1} = T_j^n \tag{26}$$

Alternatively, a central difference formula $\left(T_j^{n+1} - T_j^{n-1}\right)/2\Delta t$ could be selected here to replace the time derivative:

$$T_j^{n+1} = T_j^{n-1} + 2\alpha\frac{\Delta t}{\Delta x^2}\left(T_{j-1}^n - 2T_j^n + T_{j+1}^n\right) \tag{27}$$

Such an explicit algorithm with a central difference formula is generally more accurate compared to Eq. (25) with fewer potential numerical errors, but apparently more complicated since three levels of time discrete (n-1, n, and n + 1) are used. It usually should be noted that the problem of this 1-D explicit heat conduction with central difference formula in time discretization could include an unstable feature from such an algorithm [2, 15].

After the algebraic formulas are generated, a Taylor series expansion would then be inspected to find out the specific error levels. The spatial and temporal derivatives of the temperature field $\bar{T}$ at the discretized node (j, n) could be expressed with Taylor series expansions:

$$\bar{T}_{j+1}^{n} = \sum_{m=0}^{\infty} \frac{\Delta x^m}{m!} \left[\frac{\partial^m \bar{T}}{\partial x^m}\right]_j^n \tag{28}$$

$$\bar{T}_{j}^{n+1} = \sum_{m=0}^{\infty} \frac{\Delta t^m}{m!} \left[\frac{\partial^m \bar{T}}{\partial t^m}\right]_j^n \tag{29}$$

Such series could be truncated after any number of terms, and the truncation error could be dominated by the next term in the expansion, when $\Delta x \ll 1$ or $\Delta t \ll 1$ in spatial and temporal discretization. Therefore,

$$\bar{T}_{j+1}^{n} = \bar{T}_j^n + \Delta x \left[\frac{\partial \bar{T}}{\partial x}\right]_j^n + \frac{\Delta x^2}{2} \left[\frac{\partial^2 \bar{T}}{\partial x^2}\right]_j^n + o\left(\Delta x^3\right) \tag{30}$$

Apparently, the error term involved in such approximation would rapidly reduce in magnitude as the size of the spatial step (Δx) decreases. In such a scenario, using the finite difference replacement as described before

$$\left[\frac{\partial \bar{T}}{\partial x}\right]_j^n \approx \frac{T_{j+1}^n - T_j^n}{\Delta x} \tag{31}$$

Eq. (31) is considered to have the truncation error term of $o(\Delta x)$. Similarly, such process could be applied to the time derivatives, and an error term is suggested to be $o(\Delta t)$:

$$\left[\frac{\partial \bar{T}}{\partial t}\right]_j^n \approx \frac{T_j^{n+1} - T_j^n}{\Delta t} \tag{32}$$

To sum up, for the finite difference method, the steps to solve a problem are now presented as shown in **Figure 8**.

From the error analysis using Eqs. (31) and (32), the finite difference method (FDM) seems to still have a relatively high error, in orders of Δx and Δt, separately. So, it is necessary to increase the element order to get a higher-order accuracy when

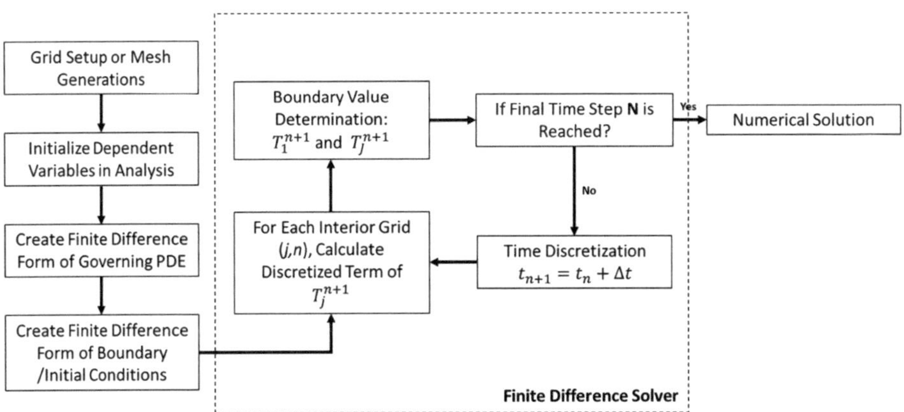

Figure 8.
Finite difference solution process for a 1-D problem.

the FDM is implemented. And for a more complex 2-D or 3-D system, using FDM might be less flexible to find solutions. Therefore, two other commonly used discretization methods are considered as alternatives, which are the finite element method (FEM) and the finite volume method (FVM). FEM is widely used in solid-body-related or multi-physics problems, which is a numerical technique for finding approximate solutions to boundary value problems for PDEs. It uses the subdivision of a whole problem domain into simpler parts, called finite elements, and solves the problem by minimizing an associated error function.

FVM, on the other hand, is mostly used in fluid-related problems. "Finite volume" refers to the small volume surrounding each node point on a mesh. In FVM, volume integrals in a partial differential equation that contain a divergence term are converted to surface integrals, using the divergence theorem. These terms are then evaluated as fluxes through the surfaces of each finite volume. Because FVM treats the volume integrals to have conservative terms, the flux entering a given volume is identical to that leaving the adjacent volume. FVM is easily formulated for unstructured meshes. Due to the page limitation of this chapter, hereby no additional details will be discussed here for FEM and FVM. Interested readers are encouraged to read more in the references [2, 4, 16, 17].

For a 2-D or 3-D problem, the importance of getting an accurate enough discretization solution is to generate an appropriate mesh of the computational domain. Generally speaking, mesh terminology could include components such as cell, cell centroid, node (vertex), and face, as shown in **Figure 9**. 2-D structured (typically quadrilaterals) and unstructured (typically triangles) mesh could be used based on the natures and requirements of the computational domains: the former is efficient in providing good resolution and convergence, but may not provide accurate results in complex geometries; the latter could be less efficient in solution and cause complex algorithms but is able to accommodate arbitrary geometries or fit in certain domain curvatures.

While 3-D mesh could have different shapes, tetrahedral and hexahedral are the two most commonly used types (**Figure 10**). Different features of solving problems might require different mesh types, in order to find the best balance among the accuracy of the simulation, computational time, convergence rate, and complexity in producing numerical modeling with certain mesh techniques. Tetrahedral mesh is commonly constructed as triangular surfaces. They are four-faced polyhedrons where each face is a triangle, optimally equilateral, normally used in geometric scenarios with circular curvature, otherwise could be constructed as isosceles tetrahedral if asymmetry occurs. Thus, it usually captures the complex geometry accurately, with less time in generation. A hexahedral mesh is a six-faced polyhedron where each face

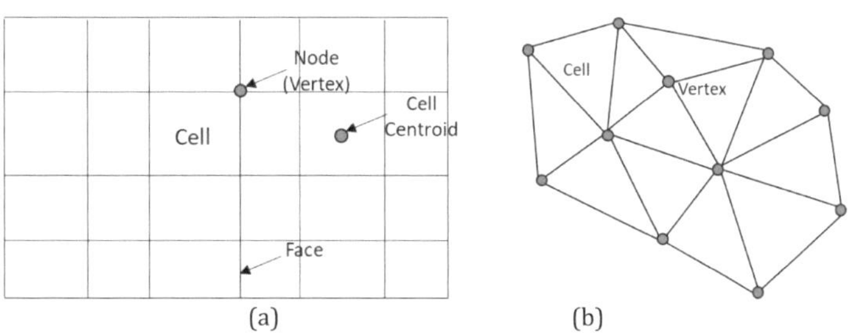

Figure 9.
Example of 2-D mesh cells. (a) Structured, (b) Unstructured.

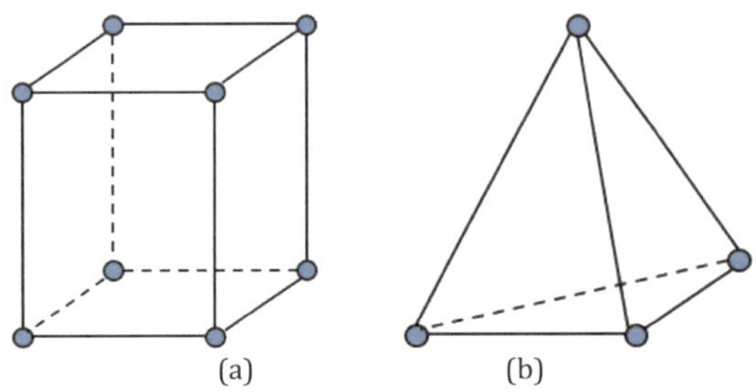

Figure 10.
Example of 3-D mesh cells. (a) Hexahedral, (b) Tetrahedral.

is a quadrilateral. Unlike the four nodes included in a tetrahedral mesh, a hexahedral mesh contains eight nodes. Hexahedral cells generally provide better mesh quality and numerical accuracy compared to tetrahedral elements, especially for problems involving structural mechanics or fluid dynamics [16, 17].

Considering its nature to capture complex geometries, the tetrahedral mesh is normally employed in applications that require accurate discretization, such as fluid-structure interaction studies. For certain types of flows with small curvature geometries, using hexahedral mesh along the curved region, the node density can be increased to capture flows along that particular surface with higher accuracy. Also, the total number of cells generated using hexahedral mesh in the system domain could be fewer than the case using tetrahedral mesh, which reduces the computational time. Therefore, the hexahedral mesh is preferred in applications where computational efficiency and accuracy are crucial, such as structural analysis, heat transfer simulations, and flow simulations.

Additionally, some CFD software also provides the generation of polyhedral cells, which consist of polygons in each face. Polyhedral mesh is proven to have better performance in gradient calculations because of more faces of each cell thus more neighbors, and also better in linear system convergence because of more off-diagonal terms. It is also reported that the polyhedral mesh could result in a fewer iteration to reach numerical convergence, less run time, and a more accurate solution [17], therefore it becomes a better alternative in 3-D mesh generations. Furthermore, some specific types of volume mesh could be used in specific scenarios of thermal analysis. For example, the prism mesh (or boundary layer mesh) is usually used in the flow or temperature boundary layers in analysis, in order to enhance the accuracy in predicting high gradients of velocities or temperatures in boundary layers.

In CFD simulation of thermal analysis, procedures could be used in a solver analyzing different heat transfer processes. In a CFD-based model, heat transfer can be calculated within a fluid (single or multi-species), between different fluid streams, between a fluid and a solid, and within solids [16, 17]. For conductive heat transfer, the major objective is to determine the temperature field in a medium that satisfies the boundary conditions. The First Law of Thermodynamics and Fourier's law (Eq. 1) can be used to derive the heat diffusion equation based on a discretized domain, as discussed in the previous 1-D example, from which the temperature field can be calculated [17].

For convective heat transfer, natural convection and forced convection may have some differences in their analysis. In a gravitational field, temperature differences in fluids could affect their density, and thus the relative buoyancy: components with lower temperature and usually higher density may intend to 'fall', while components with higher temperature and usually lower density (less dense) may intend to 'rise'. Since for gases a temperature difference could potentially cause a larger deviation in its density, natural convection could be more dominant in gases. In CFD models, to quantify when natural convection is possible, the dimensionless Rayleigh number is defined and can be viewed as the ratio of buoyant and viscous forces times the ratio of momentum and thermal diffusivities. Natural convection becomes possibly significant when the Rayleigh number exceeds 100,000 [11]. In forced convection, fluid movement results from external sources and is typically considered a means to increase the heat transfer rate at a surface in cooling or heating applications. In cases of mixed convection, normally magnitudes of the nondimensional parameters, such as Rayleigh (Ra), Nusselt (Nu, representing the enhancement of heat transfer through a fluid as a result of convection relative to conduction across the same fluid layer), Prandtl (Pr, representing relative thickness of the velocity and thermal boundary layer) and Reynolds (Re, representing ratio of inertia forces to viscous forces in the fluid) numbers could indicate which form of convection dominates, and also employed in CFD models to quantify the heat transfer rate of the thermal convection, i.e., to determine the heat transfer coefficient in Eq. (2).

For thermal radiation, the surface radiation exchange models in CFD usually allow to analysis thermal radiative heat transfer process between surfaces of arbitrary complexity that form enclosed spaces without the participation of the medium [12, 13]. The amount of radiation heat that a surface receives or emits is modeled and the net energy exchange is determined through the radiation properties and boundary conditions. In CFD analysis [16, 17], the commonly used surface-to-surface radiation model applies the surfaces of interest into contiguous, non-overlapping 'surface patches', which could be selected based on discretized mesh or grids, and their energy exchange through radiation is enforced between any two of these patches. From the center of each surface patch, a specified number of 'beams' are emitted over the enclosing hemisphere with solid angles that are discretized using an angular quadrature. Each of such beams is traced through the computational domain until it gets intercepted by an opposing patch, creating a radiative heat transfer path between these two selected patches, with the amount of energy exchange calculated from the radiation transport equation, such as Eq. (7). Eq. (7) also suggests that the amount of radiation mainly depends on the emissions from the surfaces and the position and orientation of the surfaces relative to each other. Considering transmission and reflection, the amount of radiation also depends on the presence of any specularly transmissive or reflective surfaces that can alter the transfer between the selected surfaces [13]. The latter two influences are accounted for by the view factors which, by definition, represent the "fraction of uniform diffuse radiation leaving a surface that directly reaches another surface" [17].

5. Example of industrial application in thermal analysis

In this section, an engineering design example will be provided for thermal analysis using CFD tools. The project is for a thermal design and analysis of the permanent magnet electric motor, with a certain target in design to avoid extremely high-temperature hotspots in the motor body. In the meanwhile, a less efficient design with additional heat generation could cause extra energy losses and compromised efficiency. For such reasons, the determination of heat generation and losses would be one of the most important factors in motor design and operations. Unless sufficient heat could be removed from the motor assembly or limited current magnitude would be conducted for necessary cooling to suffice, overheating or melting would happen in the motor and cause system failure, malfunction of operation, or even bring safety issues.

In modern industry and motor design business, it is critical to resolve the problem of the above-the-limit temperatures, or hotspots, to achieve a design target in improving a motor's efficiency and output performance. The relevant thermal analysis in the motor design would initially contain tasks such as investigation of the electric-magnetic field and its resultant energy losses. The next task is to analyze heat transfer processes in which the heat power generated would be transferred to the lower temperature region. In the second task, the design process focuses on the improvement of the heat dissipation to ambient or cooling fluids. The temperature rises during a motor operation could cause changes in the magnet's properties, and if this motor has long durations exposed to an above-the-limit temperature could cause a failure in windings or insulations as well. During motor operation, normally a rated evaluation would be conducted with a continuously operated load until a thermal equilibrium state. In addition to these rated evaluations, in some design cases, the dynamic thermal evaluations are necessary to add a thermal cycle with intermittent overloads as well.

To carry out an accurate thermal design of motors, it is necessary to first correctly understand the heat generation amount and its location, i.e., heat source, and from there to carry out a thermal analysis using the resulting loss distribution. In this example, it will be addressed how to evaluate a motor's temperature distribution, by creating a thermal analysis model that can investigate heat losses and temperature distributions and then analyzing the elevated temperature in form of hotspots. Then the design targets should be determined based on the requirements of such design, as thresholds for a numerical evaluation of alternative cases. For instance, the design of an electric system normally requires a limitation of the maximum temperature within the embodiment, to avoid burning out of electronic components; a constantly high winding temperature would cause a meltdown of the insulation and cause a short circuit; if the magnets exceed their temperature thresholds, they instantly demagnetized and the entire component must be replaced; while in a cooling system, there are always targets of the outlet temperatures of the working fluids, to meet the performance requirements of cooling.

In a permanent magnet (PM) motor, the heat losses should be quantified, and the heat transfer process needs to be studied during the design stages. The following steps are considered in a thermal analysis of CFD motor design:

First, the 3-D model for the motor embodiment is to be developed. A general PM motor consists of different bodies with various types of materials. Therefore, in a thermal analysis, the following regions need to be built: (a) winding coils and conductors; (b) stator slot epoxy and thermal insulation; (c) stator iron laminations; (d) iron body in rotor including the shaft; (e) permanent magnets; (f) air gap region; (g) cooling system (if applicable); and (h) ambient environment including air and ground.

With design purpose, the specifications of all these parts, material properties, and the motor operating conditions such as current, voltage, and frequency (for an AC motor) of the electric power supply, should be considered and a baseline design is to be developed for later analysis of design alternatives. Normally, thermal-related material properties could include the thermal conductivity or diffusivity in view of conduction, the fluid density, and viscosity to simulate the flow boundary layer in view of convection, and the parts material surface emissivity, absorptivity, reflectivity, and transmissivity for analysis on radiation. Necessary assumptions could be made at this stage to simplify the motor geometry, but note that less assumptions should be involved in the last round of design analysis to avoid potential design bias before the prototype manufacturing. Also, sometimes upper bounds of scalar values, such as temperature, could be set up in a CFD simulation to avoid numerical divergence or 'unreal' solution in iterations.

Secondly, pre-simulation analysis is performed to better understand this heat transfer problem and determine the required governing equations of thermal analysis. In motor operation, there could be multiple heat sources where heat energy is generated. There are two main sources for the heat generation while a PM motor machine operates:

Copper losses, which are generated within the copper windings because of its carried electric current, are proportional to the winding resistance and the square of the current magnitude. The equation used to quantify this part of heat generation is Joule's First Law:

$$P_{copper} = I^2 R \tag{33}$$

where P_{copper} is the heat generation power known as motor copper losses; I is the current flowing in the winding conductor (root-mean-square value if the alternating current is used); and R is the resistance of the winding conductor. As a result, the calculated heat power will be treated as a heat source in the computational domain of the designed motor, in the body of copper windings to be specific.

Iron losses, which generally occur in the iron or magnet parts of the motor, include eddy current loss and hysteresis loss. The eddy current loss is the result of Faraday's law as well, it occurs when any magnetic changes happen in conductors which would cause an induced voltage and small current within. In a PM motor, the eddy current is in the iron body produced by relative movement between 'conductors' (stator iron core) and magnetic flux lines due to the rotation of the rotor. The hysteresis loss is due to the reversal of magnetization of the stator iron core whenever it is subjected to changing magnetic forces. The iron losses could depend on variables such as the thickness of the stator laminations, the magnitude of the flux density on a transient basis, the stator iron material resistivity, and most importantly, the electrical frequency at which the motor operates. Modeling of P_{iron}, the iron loss power per unit volume of the stator material, was performed as

$$P_{iron} = k_e B_p^2 f^2 + k_{ex} B_p^{1.5} f^{1.5} + k_h B_p^2 f \tag{34}$$

where k_e, k_{ex}, and k_h are the coefficients of classical eddy current loss, excess eddy current loss, and hysteresis loss, respectively. The variable B_p is the peak flux density in the 3-D stator domain, which needs to be determined through a magnetic field analysis, and f is the motor supply frequency. The coefficients here for iron losses are related to the given stator material properties from manufacturers. Apparently, iron losses depend on the time-dependent magnetic flux profiles within the motor iron body, which could be quantified through semi-analytical modeling or numerical simulation. Since the prediction of the magnetic flux profile in a motor case could be complicated and not in the original scope of this chapter, the readers are recommended to refer to related publications or citations for more details of such analysis [18–20].

Some other heat sources within a motor operation, such as friction losses in a bearing, could be included as well with related empirical equations or models based on the design requirement. Once the heat generation power is determined, CFD tools could consider such values (even on a transient or time-dependent basis) as heat sources of the system of thermal analysis.

Thirdly, the discretization should be performed based on the given motor specs and operating conditions. A 3-D model has been built, to represent different regions of a motor. The energy terms indicating the heat generation, as modeled in Eqs. (33) and (34), will be properly introduced in the regions of the copper windings, iron laminations, and magnets. As illustrated in the previous section, different heat transfer mechanisms could be involved, such as the thermal conduction within the solid body of the motor (even within different material parts: magnets, copper windings, steel laminations, etc.), the thermal convection at the external surface of the motor (to ambient air) and at the surface of the cooling system (to the flowing coolant), and the thermal radiation to the environment. Note that different motor cooling approaches might lead to different heat transfer mechanisms in simulation, for example, housing design types of the motor may have natural convection as the main heat transfer for cooling purposes, while an air-cooling (external fan) or water-cooling industrial motor may consider the forced convection dominant, but with the different medium in convection flows (air or coolant). This pre-simulation analysis would be helpful in identifying the locations with potential high-temperature gradient, which would require high resolution in mesh cells when discretization is performed.

Both the solid parts of this designed PM motor assembly as well as the ambient air surroundings/cooling system should be included in the computational domain unless an approximated heat transfer coefficient is used for modeling simplification at the motor's external boundary. Transient solution is conducted with proper time step length, and to solve the time-dependent temperature distributions for changes in motor operations (e.g., the start of the motor).

For this example, an alternative current (AC) surface permanent magnet (SPM) motor [20], with 10 poles, and 12 slots, is modeled for thermal analysis through the CFD tool ANSYS Motor-CAD. **Figure 11** shows a 2-D cross-sectional view of the motor embodiment, including the stator and the rotor parts. **Figure 12** shows the copper windings configuration within the stator slot. Since the electric heat

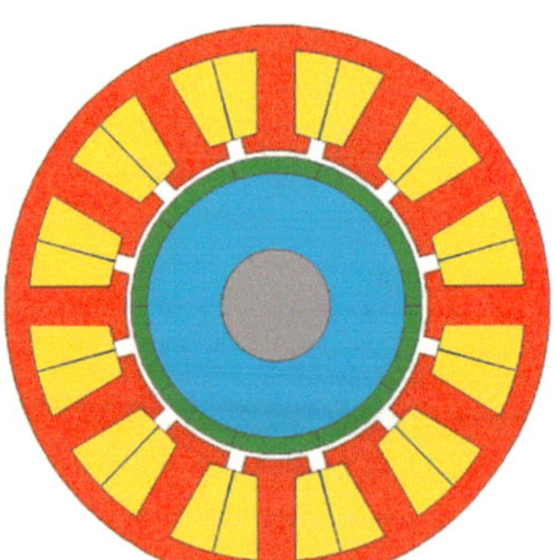

Figure 11.
SPM motor example. Parts: Red-Stator Yoke, Yellow-Stator Slots with Windings, White-Airgap and Slot Opening, Green-Permanent Magnets, Cyan-Rotor Yoke, Gray-Shaft.

DOI: http://dx.doi.org/10.5772/intechopen.1006819

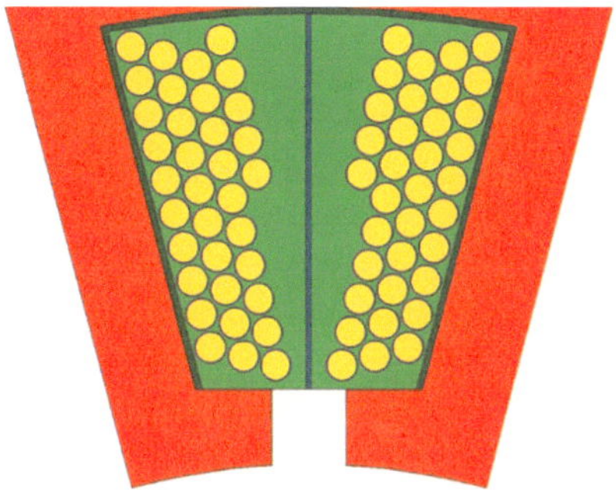

Figure 12.
Motor stator slot example. Parts: Red-Stator Yoke, Yellow-Copper Windings, Green-Slot Epoxy and Thermal Insulation.

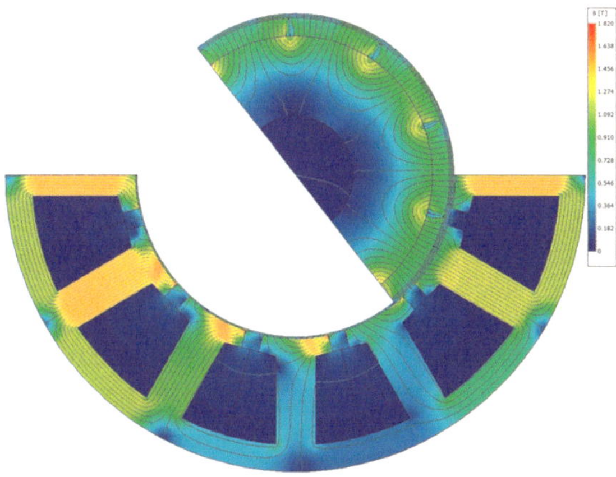

Figure 13.
CFD-based electric-magnetic flux field modeling results.

generations within the copper windings is usually one of the most significant heat generation in a motor, a detailed model of the copper conductors would be needed, with thermal analysis on the slot epoxy and the insulation of windings. Such heat generation within the copper windings could then be calculated using Eq. (33), with a given AC current value. In this simulation, the electric-magnetic field within the motor body needs to be determined first, as an interaction of the current armature in stationary windings and the rotating magnetic field from the rotor magnets. **Figure 13** gives a solution of such an armature field in the motor operation, with given operating conditions such as supply electric current and frequency. Note that as the rotor rotates, this electric-magnetic field is also a transient one with time-dependent magnetic flux densities, which will later be used in the calculation of the motor iron loss as shown in Eq. (34).

Once the heat generation source of this motor has been quantified and modeled, the heat transfer process will be performed to determine the temperature distributions. The heat sinks in this system could be the ambient air, ground, and the cooling flow (if a water-cooled machine). The boundary conditions will be defined for the computational domain of systems. Based on the engineering assumptions of the analysis, the boundary conditions could be selected with (1) specified heat flux,

(2) specified temperature, or (3) specified heat transfer coefficient, etc. Generally speaking, the specified heat flux boundary condition would require a clear understanding of the heat transfer of the system and an assumption about the heat energy flow distributions; the specified temperature boundary condition is normally used for the ambient, which has a fixed temperature at 'free' boundaries or enough distance away from heat sources, or if the coolant flow used in a 'perfectly efficient' cooling system; and the boundary condition with specified heat transfer coefficient is usually used if the heat transfer on boundaries, in convection for example, has a relatively stable condition and an empirical estimation to simplified the modeling process. Note that one commonly used boundary condition, the adiabatic boundary, is actually a type of specific heat flux when the flux is fixed to be zero.

In this case, a thermal network could be generated first to specify the heat transfer process, as shown in **Figure 5**. Then based on the discretized model, numerical simulation will be performed. The solid material properties for the copper, iron, and magnet will be used to specify the thermal conduction, and the intermedium and external radiation could be modeled as well with a given emissivity on the external surface of the motor (Eqs. (4), (7), and **Table 1**). The convective heat transfer could occur on the motor surface to the ambient air (natural convection for a normal case, or forced convection if it is an air-cooling motor), or in the heat exchangers with the coolant flow (forced convection if it is water-cooling). Flow simulation will be conducted to solve the velocity boundary layers of the airflow or the coolant flow, in order to determine the heat transfer coefficient for this convective heat transfer (Eq. (2)).

As a result of such simulation, the temperature distribution of the system could be obtained. For a steady-state scenario, the temperature field is solved for a constant motor operating condition (e.g., fixed rotor speed or power), as shown in **Figure 14**. For a transient scenario, e.g., the starting process of the SPM motor (**Figure 15**), the time-dependent temperature profiles for each motor part could be identified. The designer could easily evaluate such thermal analysis solution for each design iteration about its potential risks due to high-temperature hotspots. The maximum temperature within each categorized region in the motor can be simulated, as shown

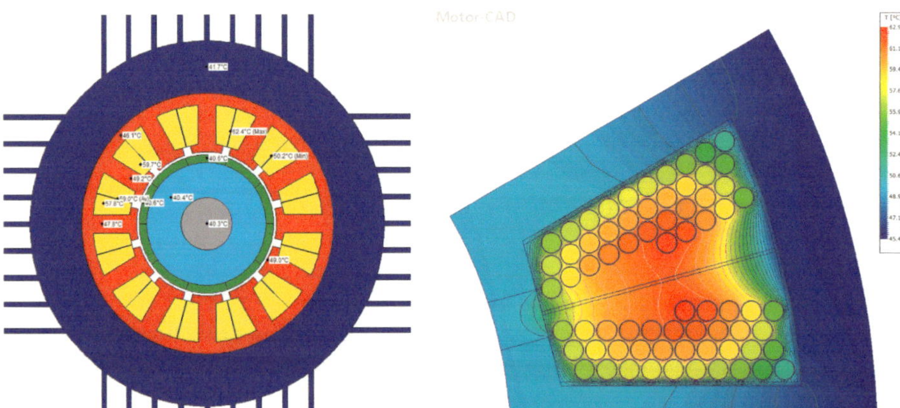

(a) (b)

Figure 14.
Steady-state temperature field of the SPM motor with water-cooling system: (a) Maximum temperature for each categorized region, and (b) the stator slot temperature distributions.

DOI: http://dx.doi.org/10.5772/intechopen.1006819

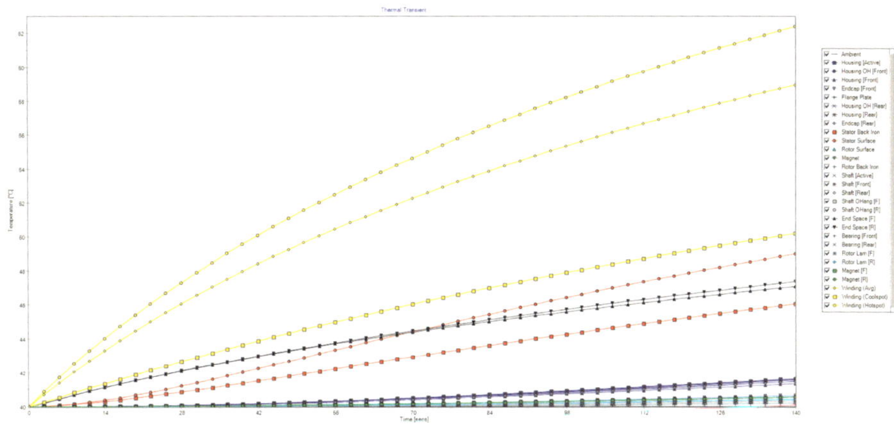

Figure 15.
Transient temperature profiles of a starting SPM motor.

in **Figure 14(a)**, and evaluated if any of those triggers an above-the-limit temperature. In most cases, the maximum temperature of a motor while it operates occurs near the motor windings, because the copper heat losses are usually the most dominant heat source in a motor. Locally in such key areas, the temperature gradient (**Figure 14(b)**) could also be critical to analyze. If high-temperature hotspots are observed in the CFD-based thermal analysis, the designer can choose to either re-identify motor input conditions for less heat generation or modify designs based on certain constraints to enhance the heat transfer process, e.g., to increase the thermal conduction through winding insulation and stator slot epoxy, redesign the cooling system specs by increasing the coolant flow rates or increasing contacting area by introducing fins, etc. Other design interests could include the temperature of permanent magnets, bearings, and other thermal-sensitive electric control equipment to avoid overheating or an intense temperature increase when the motor is started. For instance, the magnet temperature during a motor's rated operation could be simulated using CFD-based thermal analysis, and then compared with the threshold temperature of such magnet material to avoid potential demagnetization. Then the thermal analysis on motor design alternatives could be conducted with the same procedure discussed in this section until a model with better thermal performance is found and ready for prototype testing.

6. Remarks and conclusions

Heat energy and its resultant temperature field both play important roles in systems of engineering design such as electronic devices, aerodynamic components, and energy/power systems. A thermal network methodology could be used to analyze the heat generation and transfer process, but in a system with complex geometries or complicated multi-physical conditions, the analytical method could be challenging to reach a solution of the continuous field, while it is practical to obtain a numerical (discretized) solution as an approximation. By using CFD tools, the discretized solution is obtained from a developed model of the system to be evaluated, considering the temperature dependency of the material, properties of the solids and fluids, contact conditions, and with a discretized form of the governing equations to solve the non-linearity of energy transfer. As a result, modern CFD thermal analysis simulations are capable of acquiring information to accurately predict heat generations, flow fields, heat transfer rates, and temperature distributions with hot spots. These profiles could then be employed to design a better thermal management solution and ensure system thermal safety in engineering prototypes.

In design alternatives, different thermal and material attributes and system dimensions can be assigned to the developed CFD thermal analysis models for optimization, which significantly reduce the design cost both in time and expense. The iterations of design alternatives could be done completely in a CFD-based environment, and only the last design needs to be validated in performance and efficiency in prototype testing. With boosted computational powers, such numerical thermal analysis could be much more efficient compared to an experimental way without compromising solution accuracy. With proper modeling, the numerical simulation could be performed to avoid serious maintenance costs because of future potential thermal failures and design reworks in prohibitive prototype thermal testing. Therefore, this comprehensive CFD thermal analysis remains a practical and efficient way in the early stages of engineering designs.

Author details

Yingjie Tang
University of Houston, Houston, TX, United States

*Address all correspondence to: tangyingjie823@gmail.com

References

[1] McLean D. Understanding Aerodynamics: Arguing from the Real Physics. Chichester: John Wiley & Sons Ltd; 2012

[2] Fletcher C. Computational Techniques of Fluid Dynamics. Berlin: Springer; 1991

[3] Kundu PK, Cohen IM, Dowling D. Fluid Mechanics. Waltham: Academic Press – Elsevier; 2015

[4] Anderson JD, Wendt J. Computational Fluid Dynamics. New York: McGraw-hill; 1995

[5] Chung T. Computational Fluid Dynamics. Cambridge: Cambridge University Press; 2002

[6] Lienhard JH IV, Lienhard JH V. A Heat Transfer Textbook. Cambridge, MA: Phlogiston Press; 2001

[7] Mills A. Heat Transfer. Homewood, IL: Richard D. Irwin, Inc. – CRC Press; 1992

[8] Nellis G, Klein S. Heat Transfer. New York: Cambridge University Press; 2008

[9] Carslaw H. Introduction to the Mathematical Theory of the Conduction of Heat in Solids. London: Macmillan and Company, Limited; 1906

[10] Ezekoye O. Conduction of heat in solids. In: SFPE Handbook of Fire Protection Engineering. New York: Springer; 2016

[11] Bejan A. Convection Heat Transfer. Hoboken, NJ: John Wiley & Sons; 2013

[12] Siegel R, Howell JR. Thermal Radiation Heat Transfer. London: Hemisphere Publishing Co.; 1992

[13] Modest M. Radiative Heat Transfer. London: Academic Press – Elsevier; 2003

[14] Thulukkanam K. Heat Exchanger Design Handbook. New York: CRC Press – Taylor & Francis; 2000

[15] Gear C. Numerical Initial Value Problems in Ordinary Differential Equations. Englewood Cliffs, NJ: Prentice-Hall; 1971

[16] ANSYS. ANSYS FLUENT 12.0 Theory Guide. Canonsburg, PA, USA: Software Guide Documentation, ANSYS, Inc.; 2009

[17] Siemens PLM Software. Simcenter STAR-CCM+ Manual. Plano, TX, USA: Software Guide Documentation, Siemens Digital Industries Software; 2020

[18] Wu LJ, Zhu ZQ, Staton D, Popescu M, Hawkins D. An improved subdomain model for predicting magnetic field of surface-mounted permanent magnet machines accounting for tooth-tips. IEEE Transactions on Magnetics. 2011;**47**(6):1693-1704

[19] Tiang TL, Ishak D, Lim CP, Jamil MK. A comprehensive analytical subdomain model and its field solutions for surface-mounted permanent magnet machines. IEEE Transactions on Magnetics. 2014;**51**(4):1-14

[20] Mallek M, Tang Y, Lee J, Wassar T, Franchek M, Pickett J. An analytical subdomain model of torque dense Halbach array motors. Energies. 2018; **11**(12):3254

Chapter 6

Plasmonic Excitations in Carbon Nanotubes: PIC Simulations vs Hydrodynamic Model

Pablo Martín-Luna, Alexandre Bonatto, Cristian Bontoiu, Guoxing Xia and Javier Resta-López

Abstract

Charged particles traveling along a carbon nanotube (CNT) may produce the collective oscillation of the free electrons within the cylindrical graphene shell that makes up the nanotube wall. The associated electromagnetic modes (called plasmonic modes) are a potential candidate to achieve ultra-high accelerating gradients for particle acceleration. The plasmonic excitations can be studied by particle simulations and with analytical models. In this chapter, we firstly review different works that employ particle-in-cell (PIC) codes to simulate plasmonic excitations in carbon nanostructures. Then, the linearized hydrodynamic model is presented to analytically describe the plasmonic modes excited by a localized point-like charge propagating along a single-walled nanotube. In this model, the free electron gas at the nanotube wall is treated as a plasma, which satisfies the linearized continuity and momentum equations with specific solid-state properties. Finally, we compare the plasmonic excitations obtained using the hydrodynamic model with those from Fourier-Bessel PIC (FBPIC) simulations. A comprehensive analysis is conducted to examine similarities, differences, and limitations of both methods. This research offers an insightful viewpoint on the potential use of CNTs to enhance particle acceleration techniques, paving the way for future progress in high-energy physics and related fields.

Keywords: carbon nanotubes (CNTs), particle-in-cell (PIC) simulations, linearized hydrodynamic model, wakefield excitation, plasmonics, particle acceleration

1. Introduction

Particle accelerators play a crucial role in various domains, such as scientific research, industry, and national security. They have enabled cutting-edge research in high-energy physics for over half a century and are essential in synchrotron radiation and free electron lasers. Furthermore, they are widely used for medical applications, for example, in radiotherapy and hadrontherapy for cancer treatments. However, the conventional accelerators based on the radio frequency (RF) technology are limited to gradients in the order of 100 MV/m due to surface RF breakdown [1, 2]. For this reason, several alternative solutions are currently explored to overcome the limitation, such as dielectric laser or wakefield accelerators [3–6] and plasma-based accelerators [7–10].

In the 1980s and 1990s, Tajima *et al.* [11–13] proposed solid-state wakefield acceleration using the ion lattice of crystals as a potential technique to achieve acceleration gradients of several TV/m. In this technique, X-rays or charged particle bunches may be able to excite longitudinal wakefields while traveling along the crystal. However, the angstrom-sized channels in natural crystals impose limitations on beam intensity acceptance and dechanneling rates. These challenges need to be addressed to fully realize the potential of solid-state wakefield acceleration using crystals.

In this context, carbon nanotubes (CNTs) represent a potential candidate for obtaining ultra-high gradients. CNTs can display either metallic or semiconductor properties based on their rolling pattern and possess remarkable thermo-mechanical and electronic characteristics, along with dimensional flexibility. Since their discovery by Iijima in 1991 [14], buckling and bending [15–17], vibrational [18–20], and electronic and optical properties [21–25] of CNTs have been extensively studied in both theoretical and experimental aspects. Because of their unique hollow structure, CNTs may be used for channeling and steering charged particles (e.g., ions [26] and electrons [27]), akin to crystal channeling. Thus, CNTs offer several advantages over natural crystals when it comes to channeling and steering charged particles, such as wider channels in two dimensions and longer dechanneling lengths [28, 29], becoming a robust alternative for TV/m acceleration due to their exceptional electronic properties, greater flexibility, and thermo-mechanical strength. Consequently, carbon nanostructures (CNTs or even graphene layers) are currently being studied for wakefield acceleration [30–34].

Wakefields in CNTs can arise due to the collective oscillation of the free electron gas confined to the surfaces of the nanotubes, which is triggered by the driving bunch. Therefore, the CNT surface can be understood as a plasma composed of free carbon ions and free electrons, whose dynamics is highly susceptible to external electromagnetic fields. In particular, this collective oscillation of electrons is called plasmons because of their similarities with plasmas. On the one hand, plasmons can be theoretically studied by solving the Maxwell's equations in a medium characterized by the electric permittivity. In this sense, several models have been employed, for example, a dielectric theory [35], kinetic models [36, 37], and hydrodynamic models [38–40]. On the other hand, plasmon excitations can also be described by particle simulations, which take into account the motion of the particles. In this context, particle-in-cell (PIC) codes have emerged as a potent instrument to simulate these excitations, although with some limitations [41].

The motivation of the present work is to compare the excited wakefields predicted by the linearized hydrodynamic model [31] with PIC simulations performed with Fourier-Bessel PIC (FBPIC), analyzing the commonalities and discrepancies between both approximations.

This chapter is organized as follows. Section 2 describes the capabilities and limitations of PIC codes, mentioning different recent studies of carbon nanostructures that have been made using such simulations. Section 3 reviews the linearized hydrodynamic, providing general expressions for the longitudinal wakefield excited by the interaction of a charged particle that is traveling paraxially along a CNT. In Section 4, we describe the hollow plasma channel that we consider for modeling the plasmonic excitations in a CNT. Then, we compare the results obtained from the simulations with the hydrodynamic model in Section 5, discussing the similarities and limitations in Section 6. Finally, a summary of the main findings of this study is presented in Section 7.

2. PIC codes

PIC simulations are primarily designed for modeling the behavior of plasmas, where the focus is on the self-consistent dynamics of charged particles and electromagnetic fields. In PIC simulations, the medium is usually modeled as a free electron gas and a neutralizing ionic background. However, most modern PIC codes now incorporate some form of ionization model, such as the Ammosov-Delone-Krainov (ADK) model [42], which allows for the simulation of ionizable atoms. This addition enables a more dynamic and realistic representation of the medium, particularly in scenarios where ionization processes are significant, such as in high-intensity laser-plasma interactions [43]. In these simulations, the electrons are initially randomly distributed according to the density. Then, in each temporal step, the electromagnetic fields (calculated by solving the Maxwell's equations) are used to update the position of each electron, solving the Newton-Lorentz equation. It is worth mentioning that the movement of the electrons produces a local current density, which will affect the electromagnetic fields calculated using Maxwell's equations in the next temporal step. Thus, PIC codes are capable of obtaining the position of each electron as a function of time. PIC simulations can be very time consuming, and for this reason, macroparticles (representing many electrons or positive ions) are normally used. As far as we know, traditional PIC methods are not well-suited for incorporating solid-state effects, such as interactions with the lattice or inter-band transitions, and others are normally neglected due to very time consuming, such as collisions [41]. Nevertheless, PIC codes are especially effective for solving problems that are extremely rapid ($\sim$ sub-fs) and occur at small scales ($\sim$ nm) with a large assembly of charged particles. Consequently, they can be used to study the interaction of laser and beams with plasmas in the context of ultra-high-gradient particle acceleration and plasmon excitation.

Several PIC codes, for example, EPOCH, FBPIC, OSIRIS, PIConGPU, SLIPs, Smilei, and WarpX, have been used in the last decade to simulate laser wakefield acceleration (LWFA) and plasma wakefield acceleration (PWFA) in carbon nanostructures. In the following subsections, we will briefly review these codes and some recent studies conducted using them.

2.1 EPOCH

EPOCH [44] is an open-source PIC code capable to simulate laser-plasma interactions. As with many PIC codes, EPOCH utilizes a staggered Yee grid to resolve the fields and employs a Boris-type pusher [45, 46] to simulate the particle dynamics. Zhang *et al.* [47] performed two-dimensional simulations with EPOCH, showing that a laser propagating inside a nanotube can excite gradients in the order of TV/cm and the emission of high-energy photons in the order of 10–100 MeV. Furthermore, Sangwan *et al.* [48] studied the interaction of a 10 PW laser incident on a fully ionized planar target of carbon. It was shown that carbon ions can be accelerated to energies $\sim$GeV for a target thickness $\sim$ 120 nm, demonstrating that this setup can be used to obtain C^{6+} ions for hadrontherapy treatments.

2.2 FBPIC

Fourier-Bessel PIC (FBPIC) [49] is a code whose solver uses 2D radial grids to calculate azimuthal modes within cylindrical hollow plasma channels. Consequently, this code allows us to reduce the high computational cost in geometries with cylindrical symmetry compared to 3D PIC codes.

FBPIC has been used to simulate CNTs and bundles of CNTs, modeled as hollow cylindrical channels composed of a pre-ionized cold plasma of two species (electrons and carbon ions). In particular, Bonatto *et al.* [30] have studied the wakefields excited by an external beam traveling along carbon nanostructures, obtaining their dependence on the tube aperture and the wall thickness.

2.3 OSIRIS

OSIRIS [50] is a massively parallel and fully relativistic open-source PIC code for modeling plasma-based accelerators. The PIC algorithm can take into account ionization, classical radiation damping, and quantum electrodynamics (QED) effects. Furthermore, this code implements different geometries (1D, 2D, and 3D) and quasi-3D with azimuthal expansion for cylindrical problems. The OSIRIS code has been employed to study ion acceleration in the radiation pressure regime with 100 TW ultrashort (<50 fs) pulse lasers interacting with nanometer-scale carbon targets [51].

2.4 PIConGPU

PIConGPU [52] is a computational code designed for laser-plasma simulations that permits the free movement of both electrons and carbon ions. This code is capable to enhance performance proportionally with the number of graphics cards provided. Additionally, it offers a comprehensive array of technical functionalities, including the initialization of macroparticles, various ionization processes, and field-solving techniques, among others. Sharma and Kamperidis [53] used the PIConGPU code to study the interaction of ultrashort, ultra-intense (2 PW, 20 fs) laser pulses with near-critical-density partially ionized plasmas (hydrogen and carbon species), demonstrating the creation of high-energy monochromatic proton micro-bunches of high quality. More recently, Bontoiu *et al.* [34] have utilized this code to simulate electron laser-driven acceleration in ionized graphene layers, revealing a new collective and self-injection mechanism (which was called *catapult acceleration*) based on edge plasma oscillations where gradients of several TV/m can be obtained.

2.5 SLIPs

The spin-resolved laser interaction with plasma simulation code (SLIPs) is capable of performing spin-resolved quantum electrodynamic PIC simulations in laser-matter interactions [54, 55]. In particular, its Monte-Carlo algorithm implements nonlinear Compton scattering and the nonlinear Breit-Wheeler process in the local constant field approximation, vacuum birefringence processes, spin dynamics, and high-energy bremsstrahlung.

Dou *et al.* [56] have recently employed the SLIPs code to investigate the dynamics of positrons in a novel polarized positron acceleration method that combines transition radiation along with electrostatic fields to attain ultra-high gradients. In this method, an ultrarelativistic high-density electron beam traverses any aperture of multi-layer microhole array carbon films, ionizing the film around the aperture that induces a plasma return current. Then, this current produces a net charge accumulation on the film surface, which excites transverse and longitudinal electrostatic fields within and behind the hole that can be utilized to accelerate witness particles.

2.6 Smilei

Smilei [57] is an open-source, highly parallel code that includes additional modules to take into account field ionization and collisions in plasmas. Heppe and Kumar [58] have employed Smilei to demonstrate that laser energy can be efficiently converted into high-energy, brilliant photon beams (~ 300 MeV, $\gtrsim 10^{20}\,\mathrm{s}^{-1}\mathrm{mm}^{-2}\mathrm{mrad}^{-2}(0.1\%\mathrm{BW})^{-1}$) in carbon plasmas. In 2024, on the one hand, Azamoum *et al.* [59] have shown that nanometer-thin diamond-like carbon foils may experience a transition from solid to plasma during the laser rising edge employing one-dimensional Smilei simulations; and, on the other hand, Pan *et al.* [60] have demonstrated that electrons can be accelerated through direct laser acceleration in a near-critical density plasma channel composed of carbon nanotube foams.

2.7 WarpX

WarpX is probably the most complete PIC code available today, as it includes the main capabilities of other codes, while also offering dynamic load balancing for CPUs and GPUs, as well as groundbreaking mesh refinement capabilities [61]. Maffini *et al.* [62] have employed WarpX to study laser-driven particle acceleration in a completely ionized homogeneous carbon foam attached to a completely ionized aluminum foil with a 0.05-μm-thick ionized hydrogen layer added on the rear surface of the substrate to simulate contaminants. They demonstrated that MeV protons could be used to conduct particle-induced X-ray emission with comparable performances to conventional sources.

3. Linearized hydrodynamic model

In this work, the excitations caused by a single-charged particle Q on the surface of a CNT are modeled using the linearized hydrodynamic model [31, 40]. This approach conceptualizes a single-walled CNT as a cylinder that is both infinitesimally thin and infinitely long, with a radius a. The delocalized electrons of carbon ions form a two-dimensional Fermi gas (i.e., a free electron gas) initially uniformly distributed (with the surface density n_0) over the CNT surface. Thus, the scenario under consideration involves a point charge moving at a uniform velocity v along the axis of the CNT (the z-axis), as illustrated in **Figure 1**. The position of the driver charge over time t is represented in cylindrical coordinates as $\mathbf{r}_0(t) = (r_0, \varphi_0, vt)$ if we assume that it does not lose energy because of the interaction with the surrounding medium. As a result, the electron gas experiences a perturbation, which is characterized by a velocity field $\mathbf{u}(\mathbf{r_a}, t)$ and a surface density $n(\mathbf{r_a}, t) = n_0 + n_1(\mathbf{r_a}, t)$, where $n_1\,(\mathbf{r_a}, t)$ is the perturbed density per unit area. Here, $\mathbf{r_a} = (a, \varphi, z)$ specifies the cylindrical coordinates of the CNT surface. In this theory, it is assumed that the electron gas motion is restricted to the CNT surface (i.e., the radial component of $\mathbf{u}$ is zero), and both quantities $\mathbf{u}$ and n_1 are considered to be small perturbations. In the study of wakefield dynamics, it is acceptable to overlook the movement of ions because their motion occurs on a time scale that is significantly slower than the motion of electrons [63, 64]. This is due to the fact that carbon ions have a much larger mass than electrons, resulting in a significantly slower response time.

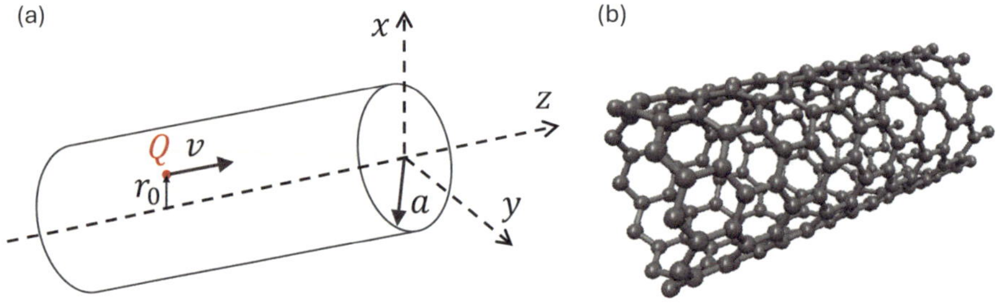

Figure 1.
(a) Schematic of a charge Q moving parallel to the z-axis inside a tube. (b) Schematic of the hexagonal lattice of a CNT.

In this model, the electronic excitations on the CNT wall are characterized by (i) the continuity equation

$$\frac{\partial n_1(\mathbf{r_a},t)}{\partial t} + n_0 \nabla_{\parallel} \cdot \mathbf{u}(\mathbf{r_a},t) = 0, \tag{1}$$

(ii) the Poisson's equation

$$\nabla^2 \Phi(\mathbf{r},t) = \frac{1}{\varepsilon_0}[en_1(\mathbf{r_a},t)\delta(r-a) - Q\delta(\mathbf{r}-\mathbf{r}_0)], \tag{2}$$

and (iii) the momentum-balance equation

$$\frac{\partial \mathbf{u}(\mathbf{r_a},t)}{\partial t} = \frac{e}{m_e}\nabla_{\parallel}\Phi(\mathbf{r_a},t) - \frac{\alpha}{n_0}\nabla_{\parallel}n_1(\mathbf{r},t) + \frac{\beta}{n_0}\nabla_{\parallel}\left[\nabla_{\parallel}^2 n_1(\mathbf{r_a},t)\right] - \gamma\mathbf{u}(\mathbf{r_a},t). \tag{3}$$

where we have kept terms up to the first order in $\mathbf{u}(\mathbf{r_a},t)$ and $n_1(\mathbf{r_a},t)$. In the previous equations, Φ is the electric scalar potential, e is the elementary charge, m_e is the electron mass, ε_0 is the vacuum electric permittivity, δ is the Dirac-delta function, $\nabla = \hat{\mathbf{r}}\frac{\partial}{\partial r} + \hat{\boldsymbol{\varphi}}\frac{1}{r}\frac{\partial}{\partial \varphi} + \hat{\mathbf{z}}\frac{\partial}{\partial z}$ is the nabla operator in cylindrical coordinates, and $\nabla_{\parallel} = \hat{\boldsymbol{\varphi}}\frac{1}{a}\frac{\partial}{\partial \varphi} + \hat{\mathbf{z}}\frac{\partial}{\partial z}$ only differentiates tangentially to the tube surface. Eq. (3) illustrates the sum of four different elements. The initial term on the right-hand side denotes the force acting upon electrons at the nanotube's surface due to the tangential component of the electric field. The second term accounts for the possible interaction with acoustic modes, establishing the parameter $\alpha = v_F^2/2$ with $v_F = \hbar(2\pi n_0)^{1/2}/m_e$ ($\hbar$ is the reduced Planck constant), representing the Fermi velocity in the 2D electron gas. The third term introduces a quantum correction, which details the single-electron excitations within the electron gas, defining the parameter $\beta = \frac{1}{4}\left(\frac{\hbar}{m_e}\right)^2$. The final term describes a damping force acting on electrons due to their scattering with the ionic-lattice background, denoted by the frictional parameter γ.

The electric potential satisfies the boundary conditions: (i) It vanishes at $r \to \infty$ and (ii) is finite at the origin $r = 0$. Thus, the electric potential can be expanded employing a Fourier-Bessel expansion, that is, using the modified Bessel functions $I_m(x)$ and $K_m(x)$ of integer order m. We will denote the total electric potential inside the nanotube ($r < a$) as Φ_{in} and the total electric potential outside the nanotube ($r > a$) as Φ_{out}. The electric potential inside the nanotube can be expressed as $\Phi_{in} = \Phi_0 + \Phi_{ind}$

with the Coulomb potential Φ_0 generated by the driving charge and the induced potential Φ_{ind} produced by the perturbation of the electron fluid on the CNT surface. The Fourier-Bessel expansion of these magnitudes at a general point $\mathbf{r} = (r, \varphi, z)$ is given by:

$$\Phi_0(\mathbf{r}, t) = \frac{1}{4\pi\varepsilon_0} \frac{Q}{\|\mathbf{r} - \mathbf{r_0}\|} = \frac{Q}{4\pi^2\varepsilon_0} \sum_{m=-\infty}^{+\infty} \int_{-\infty}^{+\infty} dk \, e^{ik\zeta + im(\varphi-\varphi_0)} I_m(|k|r_{\min}) K_m(|k|r_{\max}), \tag{4}$$

$$\Phi_{ind}(\mathbf{r}, t) = \frac{Q}{4\pi^2\varepsilon_0} \sum_{m=-\infty}^{+\infty} \int_{-\infty}^{+\infty} dk \, e^{ik\zeta + im(\varphi-\varphi_0)} I_m(|k|r_0) I_m(|k|r) A_m(k), \tag{5}$$

$$\Phi_{out}(\mathbf{r}, t) = \frac{Q}{4\pi^2\varepsilon_0} \sum_{m=-\infty}^{+\infty} \int_{-\infty}^{+\infty} dk \, e^{ik\zeta + im(\varphi-\varphi_0)} I_m(|k|r_0) K_m(|k|r) B_m(k), \tag{6}$$

where k is the wavenumber, $\zeta = z - vt$ is a comoving coordinate and $r_{\min} = \min(r, r_0), r_{\max} = \max(r, r_0)$. Imposing the boundary conditions: (i) continuity of the electric potential at the nanotube surface

$$\Phi_{in}(\mathbf{r_a}, t) = \Phi_{out}(\mathbf{r_a}, t), \tag{7}$$

and (ii) discontinuity of the radial component of the electric field due to perturbed density n_1 of the electron fluid

$$\left.\frac{\partial \Phi_{out}(\mathbf{r}, t)}{\partial r}\right|_{r=a} - \left.\frac{\partial \Phi_{in}(\mathbf{r}, t)}{\partial r}\right|_{r=a} = \frac{e n_1(\mathbf{r_a}, t)}{\varepsilon_0}, \tag{8}$$

the coefficients $A_m(k)$ and $B_m(k)$ can be easily calculated, obtaining

$$A_m(k) = \frac{\Omega_p^2 a^2 (k^2 + m^2/a^2) K_m^2(|k|a)}{kv(kv + i\gamma) - \omega_m^2(k)}, \tag{9}$$

$$B_m(k) = \frac{kv(kv + i\gamma) - \alpha(k^2 + m^2/a^2) - \beta(k^2 + m^2/a^2)^2}{kv(kv + i\gamma) - \omega_m^2(k)}, \tag{10}$$

where $\Omega_p = \sqrt{\frac{e^2 n_0}{\varepsilon_0 m_e a}}$ is the two-dimensional plasma frequency and $\omega_m(k)$ the dispersion relations:

$$\omega_m^2(k) = \alpha\left(k^2 + \frac{m^2}{a^2}\right) + \beta\left(k^2 + \frac{m^2}{a^2}\right)^2 + \Omega_p^2 a^2 \left(k^2 + \frac{m^2}{a^2}\right) K_m(|k|a) I_m(|k|a). \tag{11}$$

The induced longitudinal electric wakefield inside the tube is given by

$$W_{z,\text{ind}} = -\frac{\partial \Phi_{\text{ind}}}{\partial z} = \frac{Q}{4\pi^2\varepsilon_0} \sum_{m=-\infty}^{+\infty} \int_{-\infty}^{+\infty} dk \, k e^{im(\varphi-\varphi_0)} I_m(|k|r_0) I_m(|k|r) \times [\text{Re}[A_m(k)] \sin(k\zeta) + \text{Im}[A_m(k)] \cos(k\zeta)] = W_{z,\text{Re}} + W_{z,\text{Im}}, \tag{12}$$

which has been separated into two different terms which come from the real and imaginary parts of $A_m(k)$, respectively. To minimize computational time and avoid artificial numerical errors, a cutoff for large wavenumbers k has to be implemented in the numerical integration. Nevertheless, when the frictional parameter is negligible ($\gamma \to 0^+$), the term $W_{z,\mathrm{Im}}$ can be analytically integrated (see Appendix B of [31]), obtaining

$$W_{z,\mathrm{Im}}(\gamma \to 0^+) = \frac{-Q}{2\pi\varepsilon_0} \sum_{m=-\infty}^{+\infty} e^{im(\varphi-\varphi_0)} k_m I_m(k_m r_0) I_m(k_m r) \times \Omega_p^2 a^2 \left(k_m^2 + \frac{m^2}{a^2}\right) K_m^2(k_m a) \left| \frac{\partial Z_m}{\partial k} \right|_{k=k_m}^{-1} \cos(k_m \zeta), \tag{13}$$

where k_m are the positive roots of $Z_m(k) = (kv)^2 - \omega_m^2(k)$. Furthermore, it is satisfied that $W_{z,\mathrm{Re}} \approx W_{z,\mathrm{Im}}$ [31]. Thus, the longitudinal wakefield can be approximated by:

$$W_z(\mathbf{r}, t) \approx \frac{-Q}{\pi\varepsilon_0} \sum_{m=-\infty}^{+\infty} e^{im(\varphi-\varphi_0)} k_m I_m(k_m r_0) I_m(k_m r) \times \Omega_p^2 a^2 \left(k_m^2 + \frac{m^2}{a^2}\right) K_m^2(k_m a) \left| \frac{\partial Z_m}{\partial k} \right|_{k=k_m}^{-1} \cos(k_m \zeta). \tag{14}$$

In particular, if the charged particle travels on the z-axis (i.e., $r_0 = 0$), Eq. (14) can be simplified because the only non-vanishing mode is $m = 0$:

$$W_z \approx W_z^{max} \cos(k_0 \zeta), \tag{15}$$

$$W_z^{max} = -\frac{Q}{\pi\varepsilon_0} k_0^3 I_0(k_0 r_0) I_0(k_0 r) \Omega_p^2 a^2 K_0^2(k_0 a) \left| \frac{\partial Z_0}{\partial k} \right|_{k=k_0}^{-1}. \tag{16}$$

It is worth noting that Eq. (16) gives the amplitude of the excited longitudinal wakefield.

4. PIC simulations

Carbon nanotubes can be modeled as hollow plasma tubes consisting of layers filled with a uniformly distributed, pre-ionized cold plasma of carbon ions and electrons. We will define a hollow plasma channel model with an inner radius r_{in} and wall thickness w_{t} with a volumetric density $n_{\mathrm{V}} = 10^{28}\,\mathrm{m}^{-3}$ of free electrons within this region, as shown in **Figure 2**. Because of the high computational cost of 3D PIC simulations and the cylindrical symmetry of the geometry, the FBPIC code is chosen to perform the simulations. We will consider a beam driver with a bi-Gaussian distribution

$$n_b(\zeta, r) = n_b e^{-\zeta^2/\left(2\sigma_\zeta^2\right)} e^{-r^2/\left(2\sigma_r^2\right)}, \tag{17}$$

where $n_b = (Q_b/e)/\left[(2\pi)^{3/2}\sigma_\zeta\sigma_r^2\right]$ is the peak beam density, $Q_b = -44$ fC is the beam charge, and σ_ζ and σ_r are the beam longitudinal and radial root mean square (RMS) sizes, respectively. It is worth mentioning that, for the sake of comparison, we choose $\sigma_\zeta = \sigma_r = 3.33\,\mathrm{nm} \ll r_{\mathrm{in}}$ to ensure that the beam driver can be approximated by

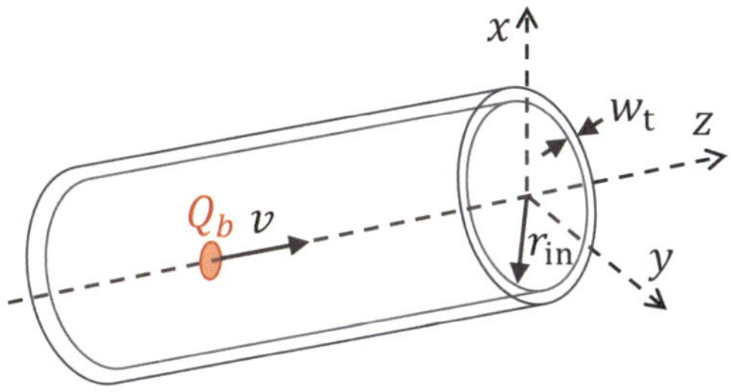

Figure 2.
(a) Scheme considered for PIC simulations: a bi-Gaussian beam driver with charge Q_b moving along the z-axis inside a hollow plasma channel with inner radius r_{in}, wall thickness w_t, and uniform volumetric density n_V.

a point-like charge as in the hydrodynamic model. To minimize changes in the beam distribution along the propagation, all PIC simulations utilized a high-energy beam of 1 GeV (Gaussian distribution, with a standard deviation of 0.005 GeV). The simulation domain was longitudinally set to approximately ~920 nm, accommodating around 2 to 3 complete wakefield oscillations in the wake of the driver. The maximum radius was defined as the sum of the internal radius (r_{in}), wall thickness (w_{t}), plus an additional empty region of half of the plasma wavelength ($0.5\lambda_p$) to allow observations of potential oscillations on the external face of the CNT wall. The spatial resolution in z and r coordinates was set to 1/15th of σ_ζ and σ_r, respectively, with a total of 20 particles per cell (8 along z, 8 along r, and 4 along the azimuthal angle θ).

5. Results

The main difference between the PIC simulations and the hydrodynamic model is the definition of a 3D region with free electrons compared to the infinitesimally thin cylindrical surface which is assumed in the hydrodynamic model. Consequently, to compare the results obtained from the simulations and the hydrodynamic model (Eq. (16)), we have to relate the volumetric and surface densities, that is, n_{V} and n_0. Thus, we will assume that the number of free electrons within the cylindrical surface of radius $a = r_{\text{in}}$ is equal to the number of free electrons in the wall thickness w_{t}. This relation is given by the expression:

$$n_0 = \frac{n_{\text{V}} w_{\text{t}}}{2r_{\text{in}}}(2r_{\text{in}} + w_{\text{t}}). \tag{18}$$

Figure 3 shows the amplitude of the longitudinal wakefield as a function of the CNT radius $a = r_{\text{in}}$ for different wall thicknesses. It can be observed that there is a good qualitative agreement between the PIC simulations and the hydrodynamic model.

However, the results obtained with the PIC simulations are lower than the wakefields obtained with the hydrodynamic model for small values of the CNT radius, in particular, for larger wall thicknesses. This behavior is produced because not all free electrons of the wall thickness excite the wakefield effectively; that is, the electrons that are closer to the inner radius have a stronger contribution than the external electrons. For this reason, it seems more reasonable to consider an effective density or an effective wall thickness. In the effective density approach, we consider that the surface density of the linearized hydrodynamic model is given by

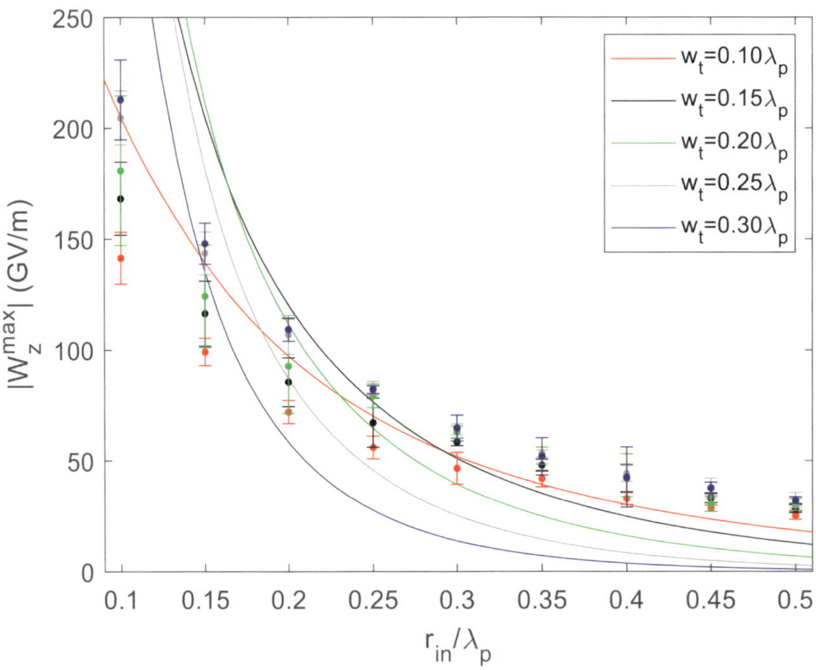

Figure 3.
Amplitude of the longitudinal wakefield on the z-axis (excited by a driver traveling on the axis) as a function of the CNT radius for different wall thicknesses. Solid lines indicate the linearized hydrodynamic model, and the points display results from the PIC simulations. Error bars are estimated from the simulations for different propagation times. The parameters used are: $Q = Q_b = -44\,fC$, $n_V = 10^{28}\ m^{-3}$ *and* $v \to c$*. The lengths are expressed in units of the plasma wavelength* $\lambda_p = 2\pi c/\omega_p$*, where* $\omega_p = \sqrt{e^2 n_V/(m_e \varepsilon_0)}$ *is the plasma frequency.*

$$n_0 = \kappa \frac{n_V w_t}{2r_{in}} (2r_{in} + w_t), \tag{19}$$

which is Eq. (18), but introducing the parameter $\kappa \in (0, 1]$ to take into account the effectiveness of the electrons. In the effective wall thickness approach, the surface density is given by

$$n_0 = \frac{n_V w_t^{eff}}{2r_{in}} \left(2r_{in} + w_t^{eff}\right), \tag{20}$$

where the effective wall thickness $w_t^{eff} \le w_t$ is introduced compared to Eq. (18). Thus, **Figures 4** and **5** depict the amplitude of the excited longitudinal wakefield employing the effective density and effective wall thickness approaches, respectively, using the numerical values collected in **Table 1** and plotted in **Figure 6**. We can see a better agreement between these approximations and the PIC simulations. As it is intuitively expected, the parameter κ decreases as a function of the wall thickness, while the effective wall thickness w_t^{eff} increases.

6. Discussion

The discrepancies observed between the hydrodynamic model and the PIC simulations can be attributed to the distinctions in both approaches. In the hydrodynamic

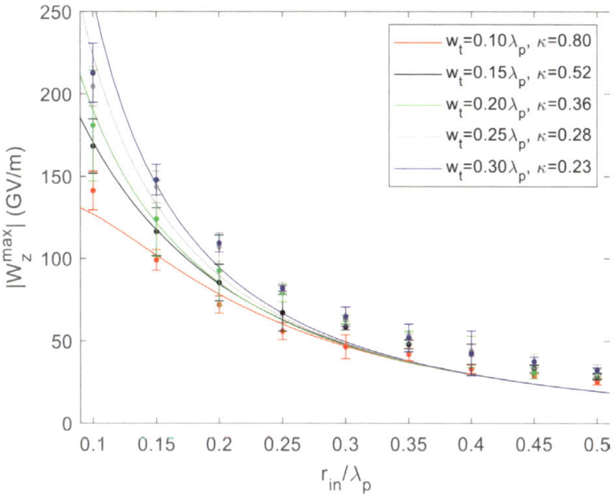

Figure 4.
Amplitude of the longitudinal wakefield on the z-axis (excited by a driver traveling on the axis) as a function of the CNT radius for different wall thicknesses. Solid lines indicate the linearized hydrodynamic model using the effective density approach with the values shown in ***Table 1****, and the points display results from the PIC simulations. Error bars are estimated from the simulations for different propagation times. We use the same parameters as in* ***Figure 3****:* $Q = Q_b = -44\,fC$, $n_V = 10^{28}\ m^{-3}$ *and* $v \to c$.

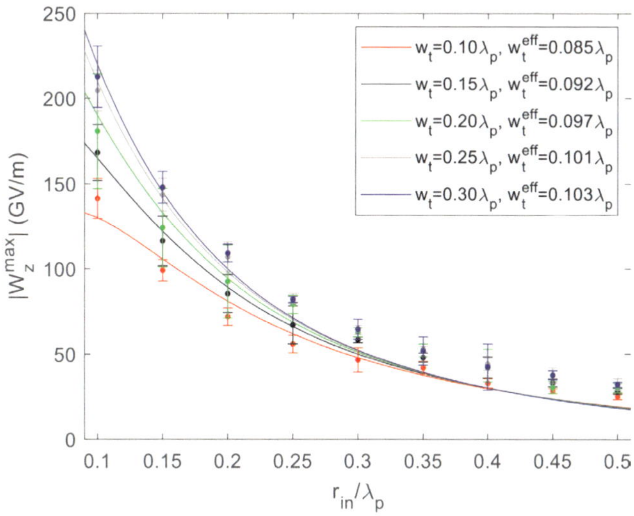

Figure 5.
Amplitude of the longitudinal wakefield on the z-axis (excited by a driver traveling on the axis) as a function of the CNT radius for different wall thicknesses. Solid lines indicate the linearized hydrodynamic model using the effective wall thickness approach with the values shown in ***Table 1*** *and the points display results from the PIC simulations. Error bars are estimated from the simulations for different propagation times. We use the same parameters as in* ***Figure 3****:* $Q = Q_b = -44\,fC$, $n_V = 10^{28}\ m^{-3}$ *and* $v \to c$.

model, the driver is a point-like particle, with dynamics decoupled from the surrounding medium. Once defined, its properties (charge, radial position, and longitudinal velocity) remain unchanged. In contrast, in PIC simulations, the driver is modeled as a bi-Gaussian beam consisting of a large number of particles distributed within a finite volume. Moreover, since PIC simulations self-consistently resolve the

Wall thickness: w_t	Effective density approach: κ	Effective wall thickness approach: w_t^{eff}
$0.10\lambda_p$	0.80	$0.085\lambda_p$
$0.15\lambda_p$	0.52	$0.092\lambda_p$
$0.20\lambda_p$	0.36	$0.097\lambda_p$
$0.25\lambda_p$	0.28	$0.101\lambda_p$
$0.30\lambda_p$	0.23	$0.103\lambda_p$

Table 1.
Summary of the numerical values of the parameters κ and w_t^{eff} employed in the effective density and effective wall thickness approaches.

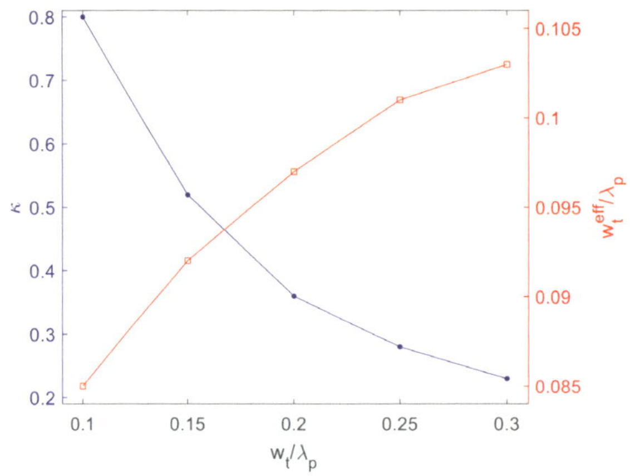

Figure 6.
Parameter κ (defined in Eq. (19)) and effective wall thickness w_t^{eff} as a function of the wall thickness w_t.

dynamics of all charged particles, the driver and the surrounding medium are both affected by mutual interactions and self-interactions.

For a proper comparison with the hydrodynamic model, the beam size in a PIC simulation must be smaller than the tube aperture. This requires a higher spatial resolution, which in turn increases the simulation time. Additionally, it results in a high-density beam, making it susceptible to stronger space-charge effects. These effects can alter the beam distribution and modify the properties of the driven wakefields. For highly relativistic beams ($v \sim c$), the driver evolution along short propagation distances is negligible. However, lower-energy beams are susceptible to significant changes in their spatial distribution and velocity reduction due to energy loss caused by interactions with their self-driven fields.

Hence, while PIC simulations with highly relativistic beams behave more similar to the hydrodynamic model, the latter was developed within a nonrelativistic framework. This discrepancy introduces additional challenges in establishing a direct comparison, but there are strategies to mitigate these issues. For example, one could artificially increase the mass of the beam particles and eliminate the energy spread.

Another source of discrepancy arises from the absence of solid-state properties in PIC simulations. These characteristics, modeled in the hydrodynamic model using parameters α, β, and γ, describe the collective behavior of electrons within CNTs. In CNTs, electron motion along the surface is facilitated, allowing for easier longitudinal movement while remaining constrained. However, in PIC simulations, electrons can move equally in any direction. This is because there are no mechanisms to specifically constrain them to the CNT surface. Even if collisions or interactions are modeled using algorithms like Monte-Carlo Collisions (MCC) or Fokker-Planck operators, conventional PIC solvers cannot differentiate between radial and longitudinal movements, nor do they restrict electron motion to the CNT surface. In contrast, the hydrodynamic model incorporates these interactions and restrictions through the friction parameter γ.

Furthermore, we are contrasting a three-dimensional region populated with free electrons (initially a thick cylinder with inner radius r_{in} and wall thickness w_{t}) in the PIC simulations against a two-dimensional cylindrical surface with radius a in the hydrodynamic model. While the electrons and carbon ions that make up the carbon nanotube are free to move in three dimensions in the simulations, in the hydrodynamic model they are considered to be restricted to the surface.

Thus, taking into account the differences between both approaches, the agreement between the results is satisfactory and consequently, the linearized hydrodynamic model (with the effective approaches) becomes an interesting tool to estimate the excited longitudinal wakefields in CNTs, avoiding time-consuming PIC simulations.

7. Conclusions and outlook

In this chapter, we have described the main characteristics and limitations of PIC codes. Furthermore, we have reviewed some works that have analyzed the wakefield excitation in carbon nanostructures with different PIC codes. Also, the wakefield excited by a driving charge moving through a carbon nanotube has been determined making use of the linearized hydrodynamic model and compared with PIC simulations. The results demonstrate a relatively good agreement of the wakefield amplitude when comparing the theoretical model with simulations. The quantitative agreement is improved if we consider an effective density or an effective wall thickness to take into account an average response of the electrons within the wall thickness in the PIC simulations. We have also discussed the main similarities and differences between the theoretical model and PIC simulations. Thus, it has been demonstrated that the hydrodynamic model may offer a fast method for estimating the excited wakefield in hollow plasmas with thin walls, becoming an alternative for bypassing the need for time-consuming PIC simulations.

Acknowledgements

This work has been supported by Ministerio de Universidades (Gobierno de España) under grant agreement FPU20/04958 and the Generalitat Valenciana under grant agreement CIDEGENT/2019/058.

Abbreviations

ADK	Ammosov-Delone-Krainov
CNT	carbon nanotube

EPOCH	extendable particle-in-cell open collaboration
FBPIC	Fourier-Bessel particle-in-cell
LWFA	laser wakefield acceleration
MCC	Monte Carlo Collisions
PIC	particle-in-cell
PIConGPU	particle-in-cell on graphics processing unit
PWFA	plasma wakefield acceleration
QED	quantum electrodynamics
RF	radio frequency
RMS	root mean square
SLIPs	spin-resolved laser interaction with plasma simulation

Author details

Pablo Martín-Luna[1*], Alexandre Bonatto[2], Cristian Bontoiu[3,4], Guoxing Xia[4,5] and Javier Resta-López[6*]

1 Instituto de Física Corpuscular (IFIC), Universitat de València - Consejo Superior de Investigaciones Científicas, Paterna, Spain

2 Graduate Program in Information Technology and Healthcare Management, and the Beam Physics Group, Federal University of Health Sciences of Porto Alegre, Porto Alegre, RS, Brazil

3 Department of Physics, The University of Liverpool, Liverpool, United Kingdom

4 The Cockcroft Institute, Sci-Tech Daresbury, Warrington, United Kingdom

5 Department of Physics and Astronomy, The University of Manchester, Manchester, United Kingdom

6 Instituto de Ciencia de los Materiales (ICMUV), Universitat de València, Valencia, Spain

*Address all correspondence to: pablo.martin@uv.es and javier2.resta@uv.es

References

[1] Abe T, Kageyama T, Sakai H, Takeuchi Y, Yoshino K. Breakdown study based on direct in situ observation of inner surfaces of an rf accelerating cavity during a high-gradient test. Physical Review Accelerators and Beams. 2016;**19**:102001. DOI: 10.1103/PhysRevAccelBeams.19.102001

[2] Shao J. Investigations on rf Breakdown Phenomenon in High Gradient Accelerating Structures. Singapore, Singapore: Springer; 2018. DOI: 10.1007/978-981-10-7926-9

[3] Thompson MC, Badakov H, Cook AM, Rosenzweig JB, Tikhoplav R, Travish G, et al. Breakdown limits on gigavolt-per-meter electron-beam-driven wakefields in dielectric structures. Physical Review Letters. 2008;**100**:214801. DOI: 10.1103/PhysRevLett.100.214801

[4] Breuer J, Hommelhoff P. Laser-based acceleration of nonrelativistic electrons at a dielectric structure. Physical Review Letters. 2013;**111**:134803. DOI: 10.1103/PhysRevLett.111.134803

[5] England RJ, Noble RJ, Bane K, Dowell DH, Ng CK, Spencer JE, et al. Dielectric laser accelerators. Reviews of Modern Physics. 2014;**86**:1337-1389. DOI: 10.1103/RevModPhys.86.1337

[6] O'Shea B, Andonian G, Barber S, Fitzmorris K, Hakimi S, Harrison J, et al. Observation of acceleration and deceleration in gigaelectron-volt-per-metre gradient dielectric wakefield accelerators. Nature Communications. 2016;**7**(1):1-7. DOI: 10.1038/ncomms12763

[7] Tajima T, Dawson JM. Laser electron accelerator. Physical Review Letters. 1979;**43**:267-270. DOI: 10.1103/PhysRevLett.43.267

[8] Chen P, Dawson JM, Huff RW, Katsouleas T. Acceleration of electrons by the interaction of a bunched electron beam with a plasma. Physical Review Letters. 1985;**54**:693-696. DOI: 10.1103/PhysRevLett.54.693

[9] Leemans WP, Nagler B, Gonsalves AJ, Tóth C, Nakamura K, Geddes CG, et al. GeV electron beams from a centimetre-scale accelerator. Nature Physics. 2006;**2**(10):696-699. DOI: 10.1038/nphys418

[10] Litos M, Adli E, An W, Clarke C, Clayton C, Corde S, et al. High-efficiency acceleration of an electron beam in a plasma wakefield accelerator. Nature. 2014;**515**(7525):92-95. DOI: 10.1038/nature13882

[11] Tajima T, Cavenago M. Crystal x-ray accelerator. Physical Review Letters. 1987;**59**:1440-1443. DOI: 10.1103/PhysRevLett.59.1440

[12] Chen P, Noble RJ. A solid state accelerator. In: AIP Conf. Proc. Vol. 156. New York: AIP; 1987. pp. 222-227. DOI: 10.1063/1.36458

[13] Chen P, Noble RJ. Crystal channel collider: Ultra-high energy and luminosity in the next century. In: AIP Conf. Proc. Vol. 398. New York: AIP; 1997. pp. 273-285. DOI: 10.1063/1.53055

[14] Iijima S. Helical microtubules of graphitic carbon. Nature. 1991;**354**(6348):56-58. DOI: 10.1038/354056a0

[15] Wong EW, Sheehan PE, Lieber CM. Nanobeam mechanics: Elasticity, strength, and toughness of nanorods and nanotubes. Science. 1997;**277**(5334):1971-1975. DOI: 10.1126/science.277.5334.1971

[16] Salvetat JP, Briggs GAD, Bonard JM, Bacsa RR, Kulik AJ, Stöckli T, et al. Elastic and shear moduli of single-walled carbon nanotube ropes. Physical Review Letters. 1999;**82**:944-947. DOI: 10.1103/PhysRevLett.82.944

[17] Yakobson BI, Brabec CJ, Bernholc J. Nanomechanics of carbon tubes: Instabilities beyond linear response. Physical Review Letters. 1996;**76**: 2511-2514. DOI: 10.1103/PhysRevLett.76.2511

[18] Yoon J, Ru CQ, Mioduchowski A. Vibration of an embedded multiwall carbon nanotube. Composites Science and Technology. 2003;**63**(11):1533-1542. DOI: 10.1016/S0266-3538(03)00058-7

[19] Wang CY, Ru CQ, Mioduchowski A. Free vibration of multiwall carbon nanotubes. Journal of Applied Physics. 2005;**97**(11):114323. DOI: 10.1063/1.1898445

[20] Ghavanloo E, Fazelzadeh SA. Vibration characteristics of single-walled carbon nanotubes based on an anisotropic elastic shell model including chirality effect. Applied Mathematical Modelling. 2012;**36**(10):4988-5000. DOI: 10.1016/j.apm.2011.12.036

[21] Bursill LA, Stadelmann PA, Peng JL, Prawer S. Surface plasmon observed for carbon nanotubes. Physical Review B. 1994;**49**:2882-2887. DOI: 10.1103/PhysRevB.49.2882

[22] Lin MF, Chuu DS, Huang CS, Lin YK, Shung KWK. Collective excitations in a single-layer carbon nanotube. Physical Review B. 1996;**53**: 15493-15496. DOI: 10.1103/PhysRevB.53.15493

[23] Östling D, Tománek D, Rosén A. Electronic structure of single-wall, multiwall, and filled carbon nanotubes. Physical Review B. 1997;**55**:13980-13988. DOI: 10.1103/PhysRevB.55.13980

[24] Marinopoulos AG, Reining L, Rubio A, Vast N. Optical and loss spectra of carbon nanotubes: Depolarization effects and Intertube interactions. Physical Review Letters. 2003;**91**: 046402. DOI: 10.1103/PhysRevLett.91.046402

[25] Dmitrović S, Vuković T, Nikolić B, Damnjanović M, Milošević I. Plasmon excitations of single-wall carbon nanotubes. Physical Review B. 2008;**77**: 245415. DOI: 10.1103/PhysRevB.77.245415

[26] Zhu Z, Zhu D, Lu R, Xu Z, Zhang W, Xia H. The experimental progress in studying of channeling of charged particles along nanostructure. In: Dabagov SB, editor. International Conference on Charged and Neutral Particles Channeling Phenomena, Society of Photo-Optical Instrumentation Engineers (SPIE) Conference Series. Vol. 5974. Bellingham, WA: SPIE; 2006. pp. 382-389. DOI: 10.1117/12.640101

[27] Chai G, Heinrich H, Chow L, Schenkel T. Electron transport through single carbon nanotubes. Applied Physics Letters. 2007;**91**(10):103101. DOI: 10.1063/1.2778551

[28] Biryukov VM, Bellucci S. Nanotube diameter optimal for channeling of high-energy particle beam. Physics Letters B. 2002;**542**(1):111-115. DOI: 10.1016/S0370-2693(02)02276-1

[29] Bellucci S, Biryukov VM, Cordelli A. Channeling of high-energy particles in a multi-wall nanotube. Physics Letters B. 2005;**608**(1):53-58. DOI: 10.1016/j.physletb.2005.01.003

[30] Bonatto A, Xia G, Apsimon O, Bontoiu C, Kukstas E, Rodin V, et al.

Exploring ultra-high-intensity wakefields in carbon nanotube arrays: An effective plasma-density approach. Physics of Plasmas. 2023;**30**(3):033105. DOI: 10.1063/5.0134960

[31] Martín-Luna P, Bonatto A, Bontoiu C, Xia G, Resta-López J. Excitation of wakefields in carbon nanotubes: A hydrodynamic model approach. New Journal of Physics. 2023; **25**(12):123029. DOI: 10.1088/1367-2630/ad127c

[32] Martín-Luna P, Bonatto A, Bontoiu C, Xia G, Resta-López J. Plasmonic excitations in double-walled carbon nanotubes. Results in Physics. 2024;**60**:107698. DOI: 10.1016/j.rinp.2024.107698

[33] Martín-Luna P, Resta-López J. Excitation of plasmonic wakefields in multi-walled carbon nanotubes: A hydrodynamic approach. In: Kuznetsov DA, editor. Carbon Nanotubes - Recent Advances, Perspectives and Applications. Rijeka: IntechOpen; 2024. DOI: 10.5772/intechopen.114270

[34] Bonţoiu C, Apsimon O, Kukstas E, Rodin V, Yadav M, Welsch C, et al. TeV/m catapult acceleration of electrons in graphene layers. Scientific Reports. 2023; **13**:1330. DOI: 10.1038/s41598-023-28617-w

[35] Wang YN, Mišković ZL. Energy loss of charged particles moving in cylindrical tubules. Physical Review A. 2002;**66**:042904. DOI: 10.1103/PhysRevA.66.042904

[36] Song YH, Zhao D, Wang YN. Kinetic study on self-energy and stopping power of charged particles moving in metallic carbon nanotubes. Physical Review A. 2008;**78**:012901. DOI: 10.1103/PhysRevA.78.012901

[37] Zhang YY, Sun JZ, Song YH, Mišković ZL, Wang YN. Channeling of protons in single-walled carbon nanotubes based on kinetic and molecular-dynamics treatment. Carbon. 2014;**71**:196-205. DOI: 10.1016/j.carbon.2014.01.030

[38] Stöckli T, Bonard JM, Châtelain A, Wang ZL, Stadelmann P. Collective oscillations in a single-wall carbon nanotube excited by fast electrons. Physical Review B. 2001;**64**:115424. DOI: 10.1103/PhysRevB.64.115424

[39] Mowbray DJ, Mišković ZL, Goodman FO, Wang YN. Wake effect in interactions of fast ions with carbon nanotubes. Physics Letters A. 2004; **329**(1):94-99. DOI: 10.1016/j.physleta.2004.06.090

[40] Wang YN, Mišković ZL. Interactions of fast ions with carbon nanotubes: Self-energy and stopping power. Physical Review A. 2004;**69**: 022901. DOI: 10.1103/PhysRevA.69.022901

[41] Ding WJ, Lim JZJ, Do HTB, Xiong X, Mahfoud Z, Png CE, et al. Particle simulation of plasmons. Nano. 2020; **9**(10):3303-3313. DOI: 10.1515/nanoph-2020-0067

[42] Ammosov MV, Delone NB, Krainov VP. Tunnel ionization of complex atoms and of atomic ions in an alternating electromagnetic field. Soviet Journal of Experimental and Theoretical Physics. 1986;**64**(6):1191. DOI: 10.1117/12.938695

[43] Chen M, Cormier-Michel E, Geddes CGR, Bruhwiler DL, Yu LL, Esarey E, et al. Numerical modeling of laser tunneling ionization in explicit particle-in-cell codes. Journal of Computational Physics. 2013;**236**: 220-228. DOI: 10.1016/j.jcp.2012.11.029

[44] Arber TD, Bennett K, Brady CS, Lawrence-Douglas A, Ramsay MG, Sircombe NJ, et al. Contemporary particle-in-cell approach to laser-plasma modelling. Plasma Physics and Controlled Fusion. 2015;**57**(11):113001. DOI: 10.1088/0741-3335/57/11/113001

[45] Boris JP. Relativistic plasma simulation-optimization of a hybrid code. In: Proc. Fourth Conf. Numerical Simulations of Plasmas. Washington, D. C.: Naval Research Laboratory; 1970. pp. 3-67 Available from: https://apps.dtic.mil/sti/citations/ADA023511

[46] Ripperda B, Bacchini F, Teunissen J, Xia C, Porth O, Sironi L, et al. A comprehensive comparison of relativistic particle integrators. The Astrophysical Journal Supplement Series. 2018;**235**(1):21. DOI: 10.3847/1538-4365/aab114

[47] Zhang X, Tajima T, Farinella D, Shin Y, Mourou G, Wheeler J, et al. Particle-in-cell simulation of x-ray Wakefield acceleration and betatron radiation in nanotubes. Physical Review Accelerators and Beams. 2016;**19**:101004. DOI: 10.1103/PhysRevAccelBeams.19.101004

[48] Sangwan D, Culfa O, Ridgers CP, Aogaki S, Stutman D, Diaconescu B. Simulations of carbon ion acceleration by 10 PW laser pulses on ELI-NP. Laser and Particle Beams. 2019;**37**(4):346-353. DOI: 10.1017/S0263034619000648

[49] Lehe R, Kirchen M, Andriyash IA, Godfrey BB, Vay JL. A spectral, quasi-cylindrical and dispersion-free particle-in-cell algorithm. Computer Physics Communications. 2016;**203**:66-82. DOI: 10.1016/j.cpc.2016.02.007

[50] Fonseca RA, Silva LO, Tsung FS, Decyk VK, Lu W, Ren C, et al. OSIRIS: A three-dimensional, fully relativistic particle in cell code for Modeling plasma based accelerators. In: PMA S, Hoekstra AG, CJK T, Dongarra JJ, editors. Computational Science — ICCS 2002. Berlin, Heidelberg: Springer Berlin Heidelberg; 2002. pp. 342-351. DOI: 10.1007/3-540-47789-6_36

[51] Dover NP, Najmudin Z. Ion acceleration in the radiation pressure regime with ultrashort pulse lasers. High Energy Density Physics. 2012;**8**(2):170-174. DOI: 10.1016/j.hedp.2012.02.002

[52] Bussmann M, Burau H, Cowan TE, Debus A, Huebl A, Juckeland G, et al. Radiative signatures of the relativistic kelvin-Helmholtz instability. In: Proceedings of the International Conference on High Performance Computing, Networking, Storage and Analysis. SC'13. Association for Computing Machinery: New York, NY, USA; 2013. DOI: 10.1145/2503210.2504564

[53] Sharma A, Kamperidis C. High energy proton micro-bunches from a laser plasma accelerator. Scientific Reports. 2019;**9**:13840. DOI: 10.1038/s41598-019-50348-0

[54] Wan F, Lv C, Xue K, Dou ZK, Zhao Q, Ababekri M, et al. Simulations of spin/polarization-resolved laser–plasma interactions in the nonlinear QED regime. Matter and Radiation at Extremes. 2023;**8**(6):064002. DOI: 10.1063/5.0163929

[55] Xue K, Sun T, Wei KJ, Li ZP, Zhao Q, Wan F, et al. Generation of high-density high-polarization positrons via single-shot strong laser-foil interaction. Physical Review Letters. 2023;**131**:175101. DOI: 10.1103/PhysRevLett.131.175101

[56] Dou ZK, Lv C, Salamin YI, Zhang N, Wan F, Xu ZF, et al. Compact

spin-polarized positron acceleration in multi-layer microhole array films. ARXIV. 2024. Available from: https://arxiv.org/abs/2405.11186

[57] Derouillat J, Beck A, Pérez F, Vinci T, Chiaramello M, Grassi A, et al. Smilei: A collaborative, open-source, multi-purpose particle-in-cell code for plasma simulation. Computer Physics Communications. 2018;**222**:351-373. DOI: 10.1016/j.cpc.2017.09.024

[58] Heppe C, Kumar N. High brilliance γ-ray generation from the laser interaction in a carbon plasma channel. Frontiers in Physics. 2022:**10**:987830. DOI: 10.3389/fphy.2022.987830

[59] Azamoum Y, Becker GA, Keppler S, Duchateau G, Skupin S, Grech M, et al. Optical probing of ultrafast laser-induced solid-to-overdense-plasma transitions. Light: Science Applications. 2024;**13**:109. DOI: 10.1038/s41377-024-01444-y

[60] Pan Z, Liu J, Wang P, Mei Z, Cao Z, Kong D, et al. Electron acceleration and x-ray generation from near-critical-density carbon nanotube foams driven by moderately relativistic lasers. Physics of Plasmas. 2024;**31**(4):043108. DOI: 10.1063/5.0202843

[61] Fedeli L, Huebl A, Boillod-Cerneux F, Clark T, Gott K, Hillairet C, et al. Pushing the Frontier in the design of laser-based electron accelerators with groundbreaking mesh-refined particle-in-cell simulations on exascale-class supercomputers. In: SC22: International Conference for High Performance Computing, Networking, Storage and Analysis. Dallas, Texas, USA: IEEE; 2022. pp. 1-12. DOI: 10.1109/SC41404.2022.00008

[62] Maffini A, Mirani F, Galbiati M, Ambrogioni K, Gatti F, Galli De Magistris MS, et al. Towards compact laser-driven accelerators: Exploring the potential of advanced double-layer targets. EPJ Techniques and Instrumentation. 2023;**10**:15. DOI: 10.1140/epjti/s40485-023-00102-8

[63] Hakimi S, Nguyen T, Farinella D, Lau CK, Wang HY, Taborek P, et al. Wakefield in solid state plasma with the ionic lattice force. Physics of Plasmas. 2018;**25**(2):023112. DOI: 10.1063/1.5016445

[64] Hakimi S, Zhang X, Lau C, Taborek P, Dollar F, Tajima T. X-ray laser Wakefield acceleration in a nanotube. International Journal of Modern Physics A. 2019;**34**(34):1943011. DOI: 10.1142/S0217751X19430115

Chapter 7

Influence of Working Fluid on the Mean Flow Vortex Structure in Industrial Flow Pipelines of Arbitrary Bend Angle via RANS Simulation

Keagan Church-Forbes, Mohammad Zamir A. Koshuriyan and Ahmed A. Abuhatira

Abstract

Numerical simulations of flow through industrial pipelines is important for design purposes in commercial and domestic transport of oil and gas. In this paper we use steady Computational Fluid Dynamics (CFD) to investigate the effect of different working fluids and streamwise bend angle on the spatial distribution, wall shear stress and turbulence kinetic energy of the fully-developed Dean vortex. Using Ansys Fluent, steady Reynolds-Averaged Navier Stokes (RANS) simulations are performed in a parametric study for various gas flows (air, methane and carbon dioxide) and pipe elbow bend angles $(90°, 60°, 45°, 30°, 0°)$. Our results show the flow properties (velocity streamlines, turbulence kinetic energy and induced wall shear stress) of the Dean vortex for inflow velocities ranging from 5 to 15 m/s at Dean numbers of $O(10^4)$ corresponding to working fluids having the same inflow mass flux. Reducing the pipe bend angle was found to minimize the wall shear stress and subsequent vortex magnitude for all gas flows types. We discuss these results quantitatively in terms of turbulent kinetic energy/the degree of secondary flow comparing to experimental literature. Reference to pipeline longevity is made.

Keywords: steady CFD, ANSYS fluent, secondary flow, dean vortices, wall shear stress

1. Introduction

Industrial flow pipelines will carry different fluids depending on their usage. For example, air is used in domestic heating, ventilation, and air conditioning (HVAC) systems; carbon dioxide (CO_2) is used in carbon capture, and storage applications, and methane in gas pipelines used to heat homes and buildings across the globe. It is often important to understand how these common gas flows behave when it comes to the design of pipelines used to transport them from one location to another and how the different flow patterns within the fluids can affect both the structural integrity and overall life cycle of pipelines.

CO_2 is becoming of greater importance as we continue on the road to net zero where it is planned to be pumped underground in depleted wells across the North Sea for example. This will involve the transportation of CO_2 in pipelines with possibly multiple 90° bends when a change of flow direction is required. In such case, maintenance of pipeline systems requires understanding of the flow patterns along the pipe and particularly at the pipe elbow where the shear stress is largest. Other natural gases such as methane is an example of a natural gas that is used in household flows for cooking and heating. It is obviously crucial to understand how this gas behaves within bends in pipelines as it travels on its way to its destination. In this paper we assess the effect of pipe working fluid and bend angle on the mean flow vortex structure at the location of maximum flow shear stress (i.e. at the elbow along the inner pipe curve referred to as the *intrados* location). We accomplish this using Computational Fluid Dynamics (CFD) via steady Reynolds averaged Navier Stokes (RANS) simulations for pipe flow that is fully developed at inlet to the pipe elbow.

Pipe elbows are not only required for a change of flow direction in pipelines, they are also used to overcome man-made or geological obstacles along the pathway of the piping system. The pipe turning angle and radius of curvature are the two parameters that define the elbow. Although there may be some variety in their selection, the most common elbow joints being 45° and 90° but some can be greater or less than this depending on the requirements of the system or route for transportation of hydrocarbons, water, or other chemicals in a pipeline.

The impacts and effects of elbow joints are of great interest to researchers and have been explored both experimentally [1] and numerically in many different studies; see for example, refs. [2–11]. Sudo et al. [1] reported the secondary flow found immediately after the elbow location which was induced by the centrifugal force due to the elbow geometry and the primary flow was forced towards the outer wall. The experimental results produced by Sudo et al. have been replicated and validated by many authors, each with a slightly different aim. For example, similar secondary flow fluid mechanics were found in the numerical simulations published by Ganesh et al. [11]. The secondary flow results in swirling eddies close to the inner wall at the elbow outlet due to recirculation of flow [2, 7, 8, 11–13]. These eddies, known as *Dean vortices*, are counter rotating motions induced by pipe wall curvature and their numerical simulation is our focus here. That is, we use a commercial CFD code to parametrically investigate the wall shear stress and turbulence kinetic energy contours in the elbow region where the vortex motion persists. The main parameters in our simulations are the working gas fluid the pipe flow carries and the pipe bend angle. For self-consistency in the numerical analysis we consider fixed mass flow rate (equal to the baseline case of air in a 90° bend used by Church-Forbes et al. [16] who validated the experiments of Sudo et al. [1]) such that the flow coefficient of the pipe depends only on the specific gravity (and therefore the working gas flow under consideration) when the pressure drop is fixed. Note that the latter (pipe pressure drop) is set in CFD pre-processing as an input to the flow solver and is kept at the Church-Forbes et al./ Sudo et al. value.

2. CFD simulation of dean vortices in industrial pipelines

Dean vortices, visible near the inner (left hand side) wall location on the 90° location, are physically caused by the existence of a centrifugal force at the curved inner elbow which creates a secondary flow in the pipe due to the primary (streamwise) flow being forced towards the outside wall [12]. They are described as a pair of counter-rotating vortices [14] located in the flow and can be seen to move further towards the middle of the pipe with less degree of swirl at reducing angle locations. Studies by Dutta et al. [2], Zhang et al. [13], Hufnagel et al. [8], and Röhrig et al. [5] also observed this swirling eddy flow at the inner wall at the 90° location in the elbow. To give an idea of the spatial morphology of these vortices, the streamlines produced by Dutta et al. [2] at 90° azimuthal location and at Reynolds number of 1×10^5 are shown in **Figure 1a**. The spatial distribution and maximum intensity of the Dean vortices found by Dutta et al. [2] were in good agreement with the CFD study by Wang et al. [12].

Turning our attention to the latter, Wang et al. [12] investigated the effects of pipe flow Reynolds number on the flow separation and secondary flow experienced in a 90° elbow with large curvature ratio (typically, $R_c/D > 0.5$) at various angles in the elbow from $(30° - 90°)$ using air as the fluid (see Dean number (De) definition below). Wang [12] found that the degree of boundary layer separation at the inner wall decreased with increasing Reynolds number (and as a consequence, increasing Dean number (De) at fixed curvature ratio) such that the Dean vortices displayed a higher degree of swirling due to turbulence increasing in the flow. Here, $De = Re\sqrt{D/2R_c}$, represents the ratio of viscous to centrifugal force acting in a curved pipe of Reynolds number, $Re = \rho UD/\mu$ based on pipe diameter, D, inflow pipe velocity U, where μ is the working fluid viscosity and R_c the radius of curvature of the inner pipe wall.

Kim et al. [3] attempted to replicate the results published by Sudo et al. [1] (which is used herein to establish a baseline test case) using Open Foam CFD software in order to further understand the secondary flow experienced at the elbow outlet. They found the swirl intensity of the eddies at the inner wall to be a strong function of the radius of curvature of the elbow and a weak function of Reynolds number. They also performed a comparative assessment to find the optimum k-ϵ turbulence model to use and found an RNG model provided sufficiently accurate results in comparison to

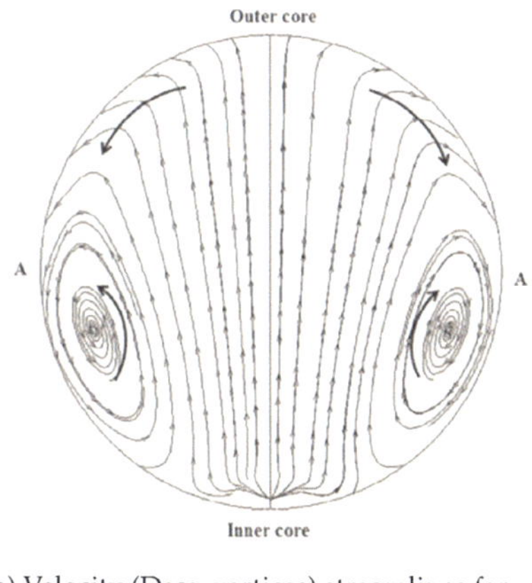

(a) Velocity (Dean vortices) streamlines for 90° elbow [1].

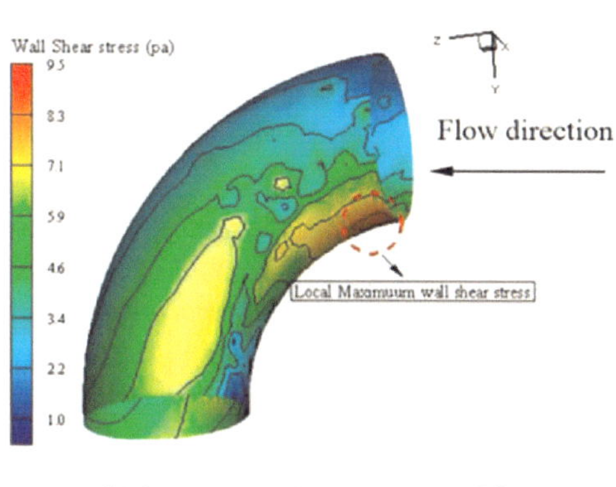

(b) Shear stress along inner wall [2].

Figure 1.
Vortex structure and shear stress distributions along the intrados of typical pipe elbows (from Refs. [2, 15]).

other published research. This study was further developed by Abuhatira et al. [4] who investigated the 90° elbow with normalized radius of curvature of $R_c/D = 2$. Their goal was to determine an optimum turbulence model by selecting a 'balanced' value for Y^+ in order to compromise between computational time on the one hand versus accuracy of results compared to experiment on the other. Abuhatira et al performed a comparative assessment looking at four types of RANS models: k-ε, SST k-Ω, Reynolds Stress Model (RSM) and Spalart-Allmaras to determine which most closely predicted the experimental results published by Sudo et al. [1]. They found RSM is able to predict the steady mean flow spatial structure accurately, but these simulations did not rule out (at least in numerical simulation terms) whether flow unsteadiness plays any role in pipe vortex development. The latter was fully addressed in ref. [16] using Unsteady-RANS simulations and time-stationary nature of vortex flow in the Sudo et al. experiments is discussed at length in this paper.

Notwithstanding the above discussion into the physics of pipe flow, two pertinent issues remain unclear. These are: how the change in working fluid and/or pipe elbow angle has on the spatial development of the vortex. The second is (presumably) the more important of the two. In this paper we, therefore, address both issues. We compare contours of the wall shear stress (WSS) and turbulence kinetic energy (TKE) at bend angles ranging from $(0° - 90°)$ for three types of working fluids commonly used in industrial pipeline flows. That is, air, methane and carbon dioxide. The vortex flow structure obtained in the experiments and simulations (discussed in Wang et al. [12], Abuhatira et al. [4] and more recently in the U-RANS simulations of Church-Forbes et al. [16]) confirm that the maximum shear stress is at the location pipe elbow where the flow turns and vorticity magnitude is greatest. Hence, we focus on the structure of Dean vortices for three fluid types and all bend angles at the elbow location within the pipeline. The simulations were conducted using the Ansys Fluent workbench for pipe flows of fixed geometry that is the same as the airflow experiments conducted in a 90° pipe bend by [1]. The Sudo et al. experiments were validated using steady RANS by [4] and more recently via U-RANS simulation in Church-Forbes et al. [16]. Since the latter indicate that Dean vortices are largely time-invariant, we assess the impact of pipe bend angle using a steady RANS calculation. For the non-air gas flows cases we keep the mass flow rate $(\dot{m})$ fixed and determine the inlet velocity and Reynolds number using the thermodynamic properties of methane and carbon dioxide at standard room temperature. The pipe inlet velocity, elbow angle, Reynolds and Dean numbers are summarized in (**Table 1**).

Our results show that the wall shear stress increases with elbow angle with Dean vortex being fully developed at the intrados location for the 90° bend angle. Although the spatial structure of the Dean vortex is not unaltered by the use of alternative pipe working gas (at fixed bend angle, θ), the magnitude of shear stress and turbulence kinetic energy is effected wing to the difference in inlet flow velocities for pipe flow simulations at fixed mass flow rate. The main physical change that occurs when both working fluid and bend angle are altered in the flow simulations is the intensification of vortex energy (i.e. turbulence kinetic energy associated with the streamwise flow) is greater in magnitude as $\theta \to 90°$ when the inflow velocity is greater. We discuss these results in comparison to the baseline case of airflow at the 90° bend angle corresponding to a Dean vortex with maximum vorticity magnitude. In §.3 we describe the pre-processing required for the Fluent CFD methodology. In Section 4, we present the main results of the paper with a full description of the fluid mechanics of the Dean vortex in terms of streamwise mean velocity distribution, wall shear stress

Case	Gas	Elbow angle	Inlet velocity (ms^{-1})	Re. No.	Dean No.
1	Air	90°	8.427	60,000	30,000
2	Air	60°	8.427	60,000	30,000
3	Air	45°	8.427	60,000	30,000
4	Air	30°	8.427	60,000	30,000
5	Air	0°	8.427	60,000	30,000
6	Methane	90°	15.421	99,385.50	49692.75
7	Methane	60°	15.421	99,385.50	49692.75
8	Methane	45°	15.421	99,385.50	49692.75
9	Methane	30°	15.421	99,385.50	49692.75
10	Methane	0°	15.421	99,385.50	49692.75
11	Carbon Dioxide	90°	5.762	78,199.68	39099.84
12	Carbon Dioxide	60°	5.762	78,199.68	39099.84
13	Carbon Dioxide	45°	5.762	78,199.68	39099.84
14	Carbon Dioxide	30°	5.762	78,199.68	39099.84
15	Carbon Dioxide	0°	5.762	78,199.68	39099.84

Table 1.
Test cases investigated.

distributions and turbulence kinetic energy as a function of pipe bend angle and inflow working fluid.

3. CFD methodology of Church-Forbes et al.

We follow the CFD methodology of Church-Forbes et al. [16], which we remind the reader is based on the Abuhatira et al. [4] mesh that itself is a reproduction of the Sudo et al. [1] experiment. With this in mind, consider a fully developed turbulent flow at inlet to the pipe (**Figure 2**). The flow turns through a bend of angle θ with radius of curvature, $R_c/D = 2$. Adopting a cylindrical coordinate system given (x, r, ϕ), to represent the flow in the pipe, where the velocity $\boldsymbol{v}(\boldsymbol{x}, t) = (u, v_r, v_\phi)(\boldsymbol{x}, t)$ enters at inflow with a streamwise component, $\boldsymbol{v}(\boldsymbol{x}, t) = \boldsymbol{e}_1 U_0$ where $u(x, t) = U_0$ is the constant mean flow.

The pipe geometry is created using the SolidWorks 3D modeling software mainly due to its ease of use but also because of its ability to export the CAD model as an IGES file into Fluent's meshing software. The pipe elbow is modeled according to the dimensions set in the Abuhatira et al. [4] paper, in which a round pipe with a 90° elbow bend downwards has dimensions upstream of the elbow set as $10D$ and downstream at, $20D$. The normalized radius of curvature, $R_c/D = 2$ is kept constant at throughout to keep the geometry in line with Sudo et al. [1], Abuhatira et al. [4], and Kim et al. [3] papers. Abuhatira et al. measure the pipe diameter of the Sudo et al. [1] experimental setup as 0.104 m. But note that the experimental setup in [1] included a longer upstream section but this was reduced by Abuhatira et al. due to the flow being fully developed within a shorter, $10D$ length from inlet [4, 16], which we follow in our

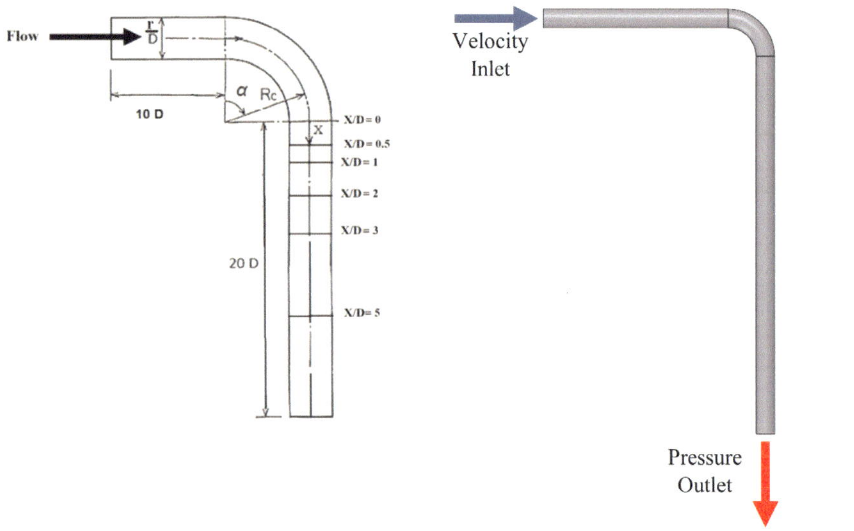

Figure 2.
90-degree pipe elbow schematic reprinted from Abuhatira et al. [4] paper and SolidWorks generated 3D pipe elbow geometry outlining direction of flow.

work here. The mesh for varying elbow angles (**Table 1**) can be easily created by adjusting the elbow curvature (denoted by α in the Abuhatira et al paper reproduced in **Figure 2** and measured from the outlet section to the straight line extending parallel to the inlet) directly in SolidWorks–note that this is a more efficient method than creating new geometries each time. The SolidWorks CAD model can be seen alongside the schematic of the pipe elbow [4] in **Figures 2** and **3**.

The meshing of the pipe elbow is carried out using the inbuilt Ansys meshing tool (after inputting the Solid Works CAD rendering). The Y^+ value at the wall is required to remain less than 3.0 throughout the pipe wall to ensure that the viscous sublayer is adequately resolved following the turbulence model optimization guidelines of

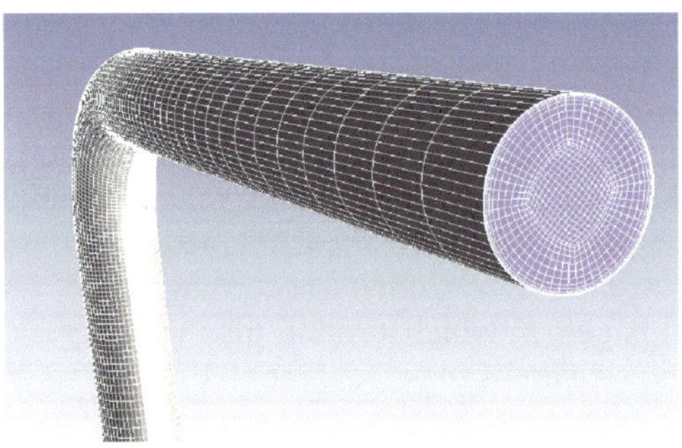

(a) Medium mesh of pipe elbow showing inflation layers on inlet to pipe and structured mesh across body of pipe

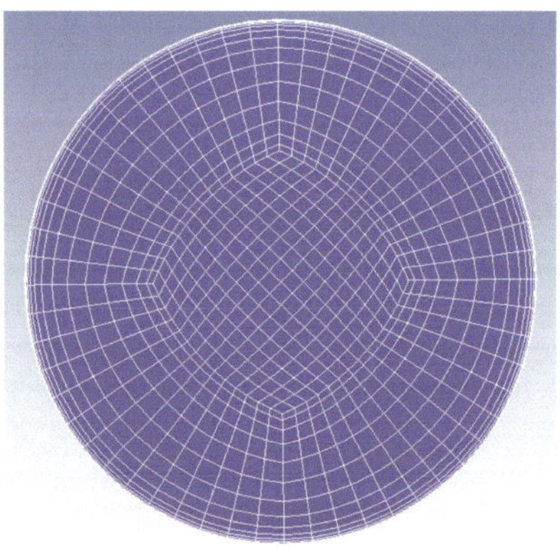

(b) Zoomed section at inlet showing inflation layers

Figure 3.
Meshing of pipe elbow using Ansys Fluent built-in meshing.

Abuhatira et al. (see recommendations in [4]). Although, the Y+ value varies along the wall, as noted in the above papers, ensuring the maximum value stays below a certain size is a good way to control it. Using the meshing rules in an inflation layer is added to the inlet and outlet faces, with first cell size set at 1.000756×10^{-4}m designed to remain consistent with the upper bound for Y^{+}. The remaining geometry can now be meshed with a standard cylindrical mesh having elements of length 1.0×10^{-2}m. The total mesh volume is then 159, 422 elements, which from the grid resolution study in ref. [16] was deemed good enough to achieve converged outlet velocity. The mesh for different bend angles follow similarly but with the finer mesh using a lower value of element size near the elbow section. The skewness of the mesh is kept below 0.95 in all cases (see methodology for the 90° bend angle reported by ref. [16]).

The turbulence model and material settings are fixed in the setup tab in Fluent at the Abuhatira et al. values (i.e., the standard k-ϵ with enhanced wall function with model constants were left at default; and aluminum being chosen for as pipe solid material consistent with the Sudo et al. experiment). The boundary conditions were setup with the gauge pressure of 0 MPa at outlet and the inlet velocity set at 8.427ms^{-1} corresponded to a Reynolds number of Re 60,000 used in [1]. Flow initialization can then begin at the uniform flow velocity at inlet consistent with the experimental setup in Sudo et al. [1]. Spatial discretization was set to second order to ensure the accuracy for a maximum of 2000 iterations which is found to be adequate to achieve residuals of $O(10^{-10})$. Further details regarding the optimum choice on turbulence model, grid convergence study, post-processing of flow data and a detailed comparison with the Sudo et al. experiment and Abuhatira et al.'s study can be found in Church-Forbes et al. [16]. The test cases used in this paper are summarized in **Table 1** and the mesh is depicted in **Figure 3**.

4. Effect of pipe elbow bend angle and working fluid on vortex energy spatial distribution

We consider the two basic parameters of the wall shear stress, $\tau_{\mathrm{w}} = \mu(\partial U(\boldsymbol{x}))/\partial r)|_{r=0}$, and the turbulence kinetic energy (k), to assess the strength of the vortex at the pipe elbow. These results are given in **Figures 4** and **5**. We also compare the velocity streamlines pattern in **Figure 6** to discern the localization of vortex lobes within the pipe as θ increases. All contours are taken at the inlet or exit to the elbow (see caption in **Figures 5–6**). For example **Figure 6a,b** and **c** displays how the velocity streamlines at the elbow exit vary between elbow angles for each fluid type. The legend shown in the bottom of each figures is used to show where the highest velocities occurs in each elbow exit, and this is observed to occur near the middle of the pipe for each elbow configuration. At 90° (on far left of each row) there is a high degree of separation in the flow and swirling occurs in the form of eddies in the contours at the inside pipe wall. This data correlates well with the velocity profiles found in literature (see **Figure 7** and further discussion in §. 5). At 60°, separation starts to decrease to a lesser degree and moves towards the walls perpendicular to the inside wall. The 30° contours appears to be the most structured of the four elbow angles, with the eddies occurring only towards the wall perpendicular to the inside and outside walls. The limit of this trend is consistent with the flow features of Wang's results (reproduced in **Figure 7**) for the 0° bend angle (straight pipe) test case.

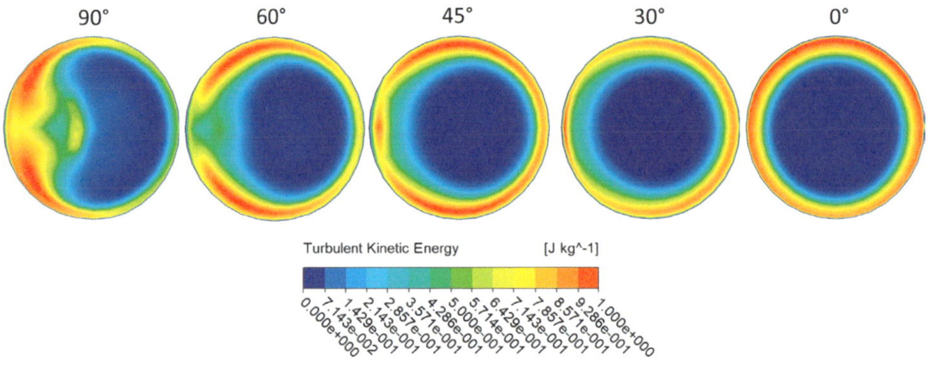

(a) Air.

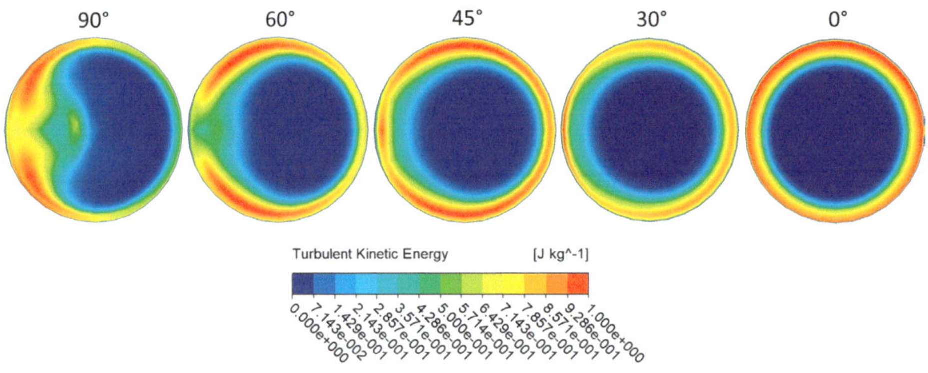

(b) Methane.

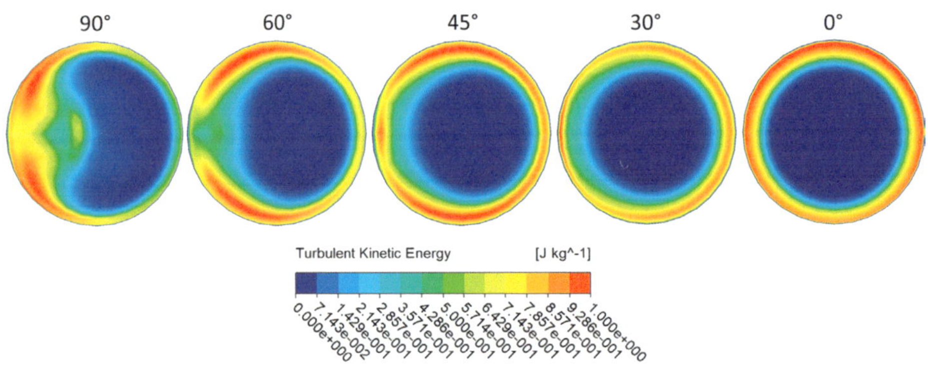

(c) Carbon dioxide.

Figure 4.
Contours of turbulent kinetic energy $k(\mathbf{x})$ *at elbow inlet for* $(90°, 60°, 45°, 30°)$ *pipe elbow bend angles under different working fluid.*

4.1 Variation of $\max \tau_w(x)$ and $\max k(x)$ with working fluid at fixed θ

Figure 8 shows the change in maximum value of the wall shear stress at the different elbow angles in **Table 1** (ranging from $30° - 90°$). As expected, τ_w increases more or less proportionally with θ for each working fluid. Taking air between bend angles between $30° - 45°$ for example, τ_w increases by 2.85% from 0.424 Pa to

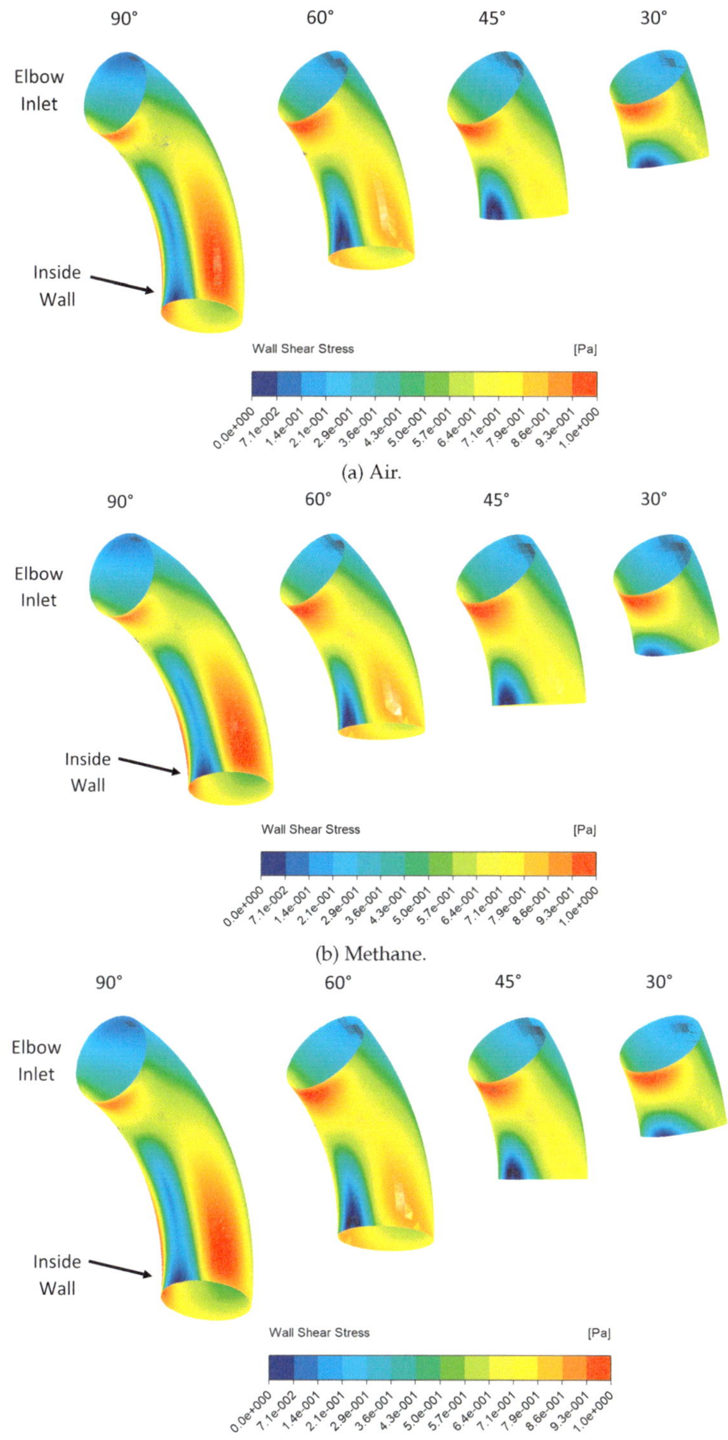

Figure 5.
Wall shear stress (τ_w) *distribution along the wall of pipe at* $(90°, 60°, 45°, 30°)$ *elbow bend angles under different working fluids.*

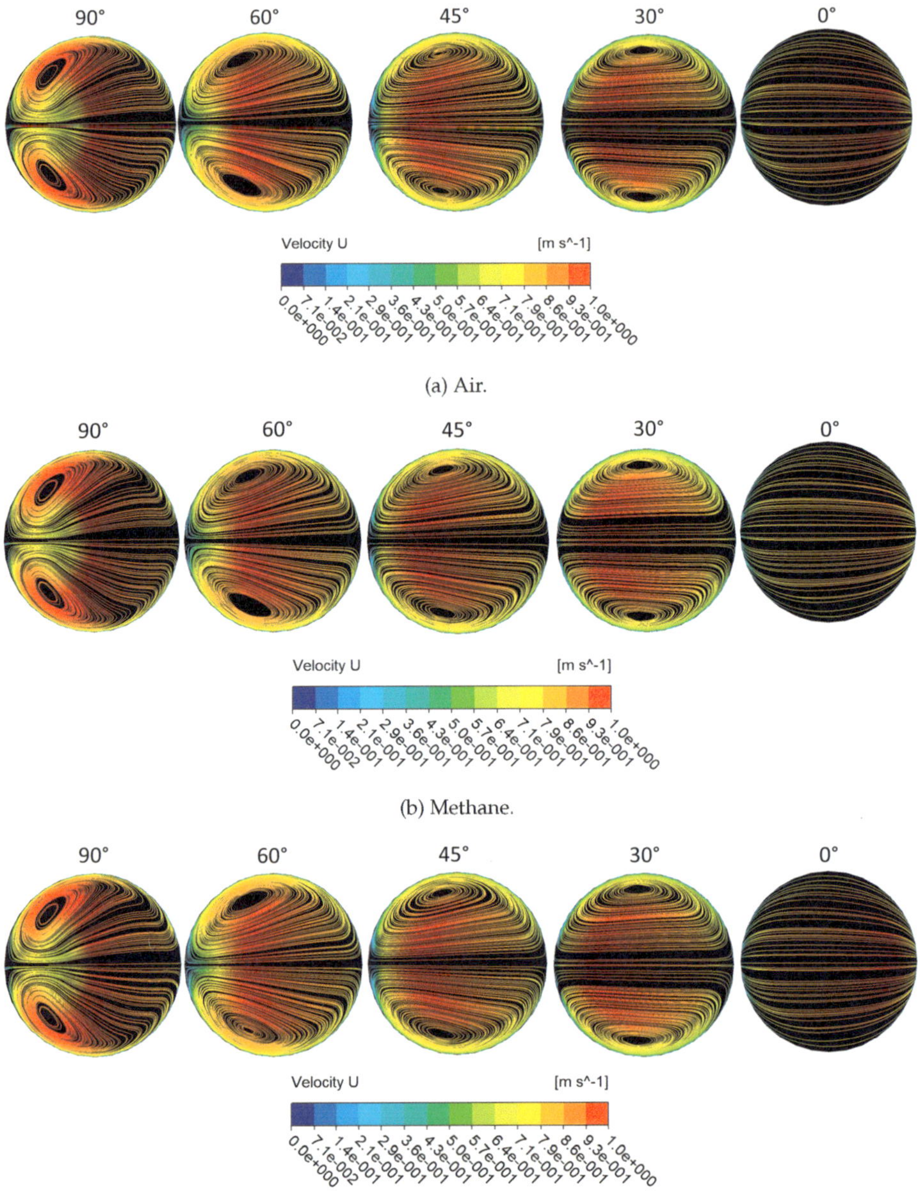

(a) Air.

(b) Methane.

(c) Carbon dioxide.

Figure 6.
Velocity streamlines $\overline{u}(\boldsymbol{x};\theta)$ at elbow exit for $(90°, 60°, 45°, 30°)$ pipe elbow bend angles under different working fluid.

0.436 Pa (3 d.p.). Interestingly, between elbow angles of 45° and 60°, there is little to no change in τ_w, with numerical variation of 0.436 Pa to 0.439 Pa, which correspond to about 0.65% difference between the two cases. Not surprisingly, the largest difference is between the angles of 60° and 90° which results in a 6.2% increase in τ_w. The individual wall shear stress surface contours are displayed in **Figure 5a–c**.

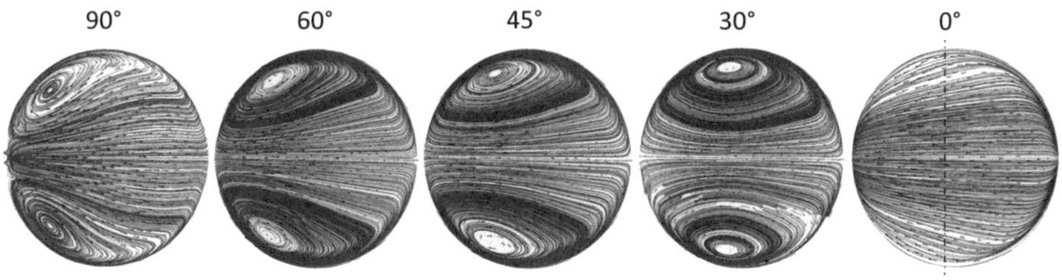

Figure 7.
Velocity streamlines (Wang et al. [12]) at elbow locations of 90°, 60°, 45° and 30° with 0° at the elbow inlet.

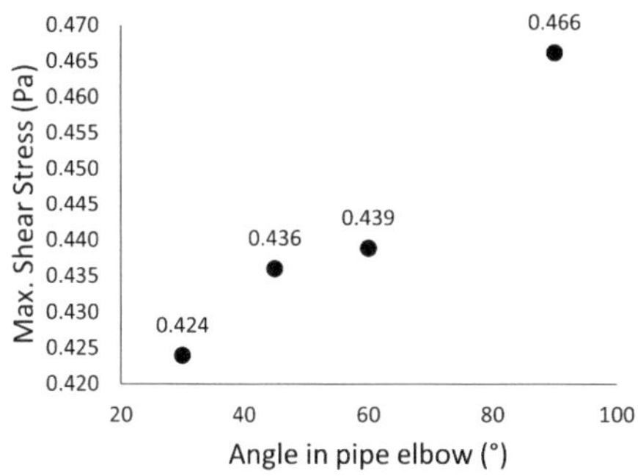

(a) Air.

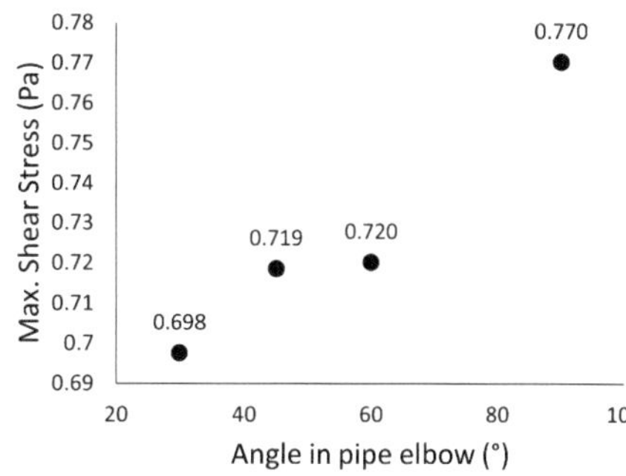

(b) Methane.

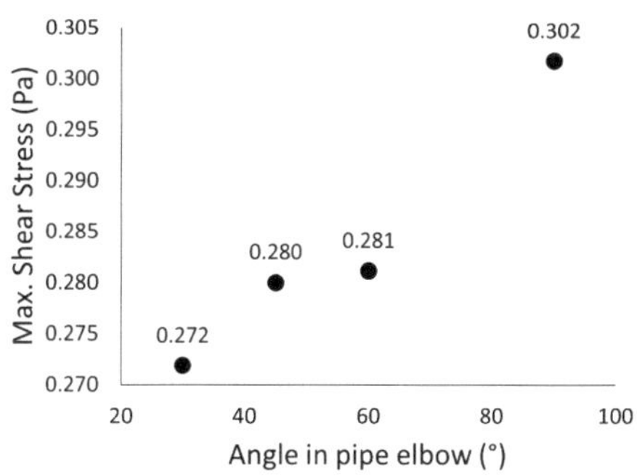

(c) Carbon dioxide.

Figure 8.
*Wall shear stress (τ_w) at the elbow exit plane in **Figure** 5 as a function of bend angle θ.*

The 90° (**Figure 5a** for air) elbow exhibits the highest shear stress at inlet to the bend. The high τ_w at the perpendicular walls in the 90° elbow is consistent with the high velocity values apparent at these locations in the pipe (see **Figure 6a** for air and cf. **Figure 6b** and **c**).

The main differences in τ_w distribution over the pipe surface when using methane/carbon dioxide as working fluid is the higher/lower values for magnitude of shear stress. This is direct consequence of the fixed inlet mass flow rate through the pipe and increased/decreased inlet velocity (**Table 1**). While the spatial distributions for wall shear stress in **Figure 5** is virtually the same for all working fluids (e.g. compare **Figure 4a-b** and similarly compare 7a to 7b). But, nonetheless, the higher/lower

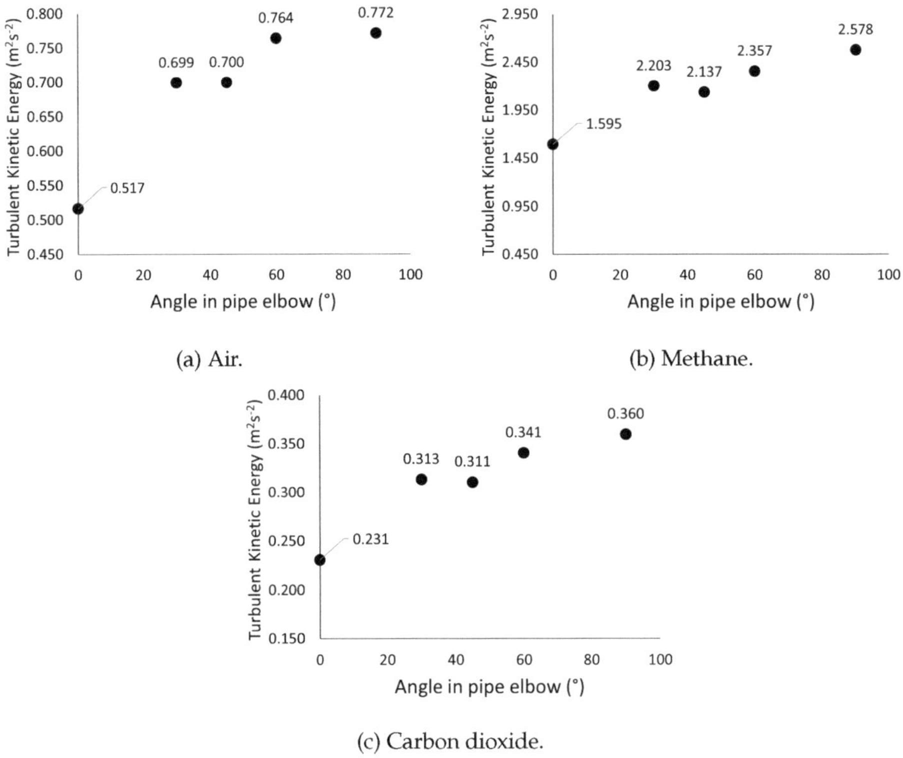

Figure 9.
*max $k(\boldsymbol{x})$ at the elbow exit plane in **Figure 4** as a function of bend angle θ.*

inflow velocity for methane/carbon dioxide compared to air, will result is higher/lower numerical values for the maximum wall shear stress (WSS) at the intrados location. This can be easily observed in the distribution of $\max \tau_w(\boldsymbol{x})$ values with θ in **Figure 8a–c**. At any fixed value of bend angle θ, while almost doubling when air and methane (**Figure 8a** vs. **b**), will almost halve in value when carbon dioxide is used (**Figure 8c**). Note, that for all three working fluids $\max \tau_w(\boldsymbol{x}) < 1$.

The turbulence kinetic energy defined in cylindrical polar velocity components is given by $2k(\boldsymbol{x}) = \left(\overline{u'^2 + v_r'^2 + v_\phi'^2}\right)(\boldsymbol{x})$ where the velocity perturbation $\boldsymbol{v}'(\boldsymbol{x}, t) = \{u', v_r', v_\phi'\}(\boldsymbol{x}, t) = \boldsymbol{v}(\boldsymbol{x}, t) - \overline{\boldsymbol{v}}(\boldsymbol{x})$ is defined about the time averaged mean flow:

$$\overline{\boldsymbol{v}}(\boldsymbol{x}) = \lim_{T \to \infty} \frac{1}{2T} \int_{-T}^{T} \boldsymbol{v}(\boldsymbol{x}, t)\, dt \tag{1}$$

where $\overline{\boldsymbol{v}}(\boldsymbol{x}) = \left(\overline{u}, \overline{v}_r, \overline{v}_\phi\right)(\boldsymbol{x})$ is given directly by the Fluent CFD calculation in which the average flow quantities are determined over the simulation time, T. In **Figure 4** we show the $k(\boldsymbol{x})$ distribution for all working fluids across bend angles, $\theta = (0° - 90°)$. As expected, the maximum in the TKE distribution at elbow exit is proportional to the the inflow pipe velocity. The methane results in **Figure 9b** are higher than both air and carbon dioxide in **Figure 9a** and **c**. The distribution of max $k(\boldsymbol{x})$ values at the elbow exit plane in **Figure 9a** vs. **b** and **c** shows methane as the working fluid will result in an almost three-fold increase in turbulence kinetic energy compared to air. Carbon dioxide, on the other hand, results in almost a halving of max $k(\boldsymbol{x})$ values compared to air (**Figure 9c** vs. **a**).

4.2 Variation of $\max \tau_w(\boldsymbol{x})$ and max $k(\boldsymbol{x})$ with θ at fixed working fluid

The mean flow simulation results in **Figures 4** and **5** show that maximum vortex intensification occurs at at the 90° bend angle. For example, **Figure 8** shows that for $\theta \in [30°, 90°]$, $\max \tau_w(\boldsymbol{x})$ increases from 0.272 Pa to 0.280 Pa at 45° (3% increase) for air. Methane as the working fluid results in an increase of $\max \tau_w(\boldsymbol{x})$ of 0.4% to 0.281 Pa. Again, the largest increase in wall shear stress at the elbow is observed between 60° and 90° with a 7.33% change to a maximum value of 0.302 Pa. While the magnitudes of $\tau_w(\boldsymbol{x})$ are smaller for carbon dioxide compared to air and methane, the changes in $\max \tau_w(\boldsymbol{x})$ as θ increases is similar to the trends above. For example, the largest increase in wall shear stress at the elbow is observed at 90° for cabon dioxide under a 7.33% change to a maximum value of 0.302 Pa (**Figure 8c**).

The change in max $k(\boldsymbol{x})$ with θ is similar for all working fluids and peaks at $\theta = 90°$. For example, comparing numerical values in **Figure 9a**, the RANS simulation shows that flow turning through a 90° bend will experience a 49.25% increase in max $k(\boldsymbol{x})$ compared to $\theta = 0°$. Using methane (**Figure 9b**) or carbon dioxide (**Figure 9c**) increases/decreases this change compared to $\theta = 0°$ to 61.65%/55.80% respectively. Note that the overall trend of $k(\boldsymbol{x})$ increasing with θ is not the case for every elbow angle and gas combination; for example, methane and carbon dioxide experience a smaller decrease between 30° and 45° compared to the percentage change in $k(\boldsymbol{x})$ between 60° to 90°. The reason for this apparent 'intermediate leveling' of $k(\boldsymbol{x})$ with bend angle (see **Figures 8–10** and **Table 2**) is unexplained within our findings and would be an interesting avenue for mathematical modeling.

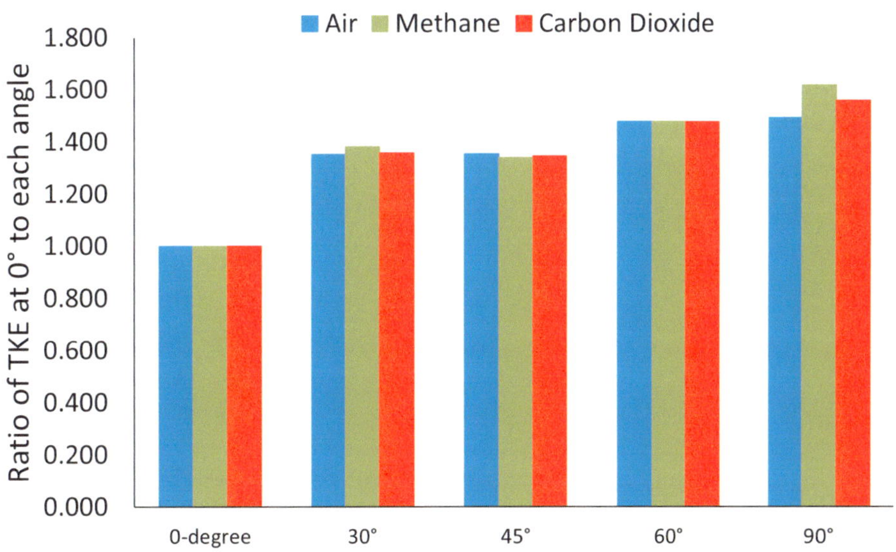

Figure 10.
$\max k(\boldsymbol{\theta}) / \max k(\boldsymbol{\theta} = 0°)$ *with working fluid.*

Gas	Elbow Angle	TKE (m^2s^{-2})
Air	90°	0.772
Air	60°	0.764
Air	45°	0.700
Air	30°	0.699
Air	0°	0.517
Methane	90°	2.578
Methane	60°	2.357
Methane	45°	2.137
Methane	30°	2.203
Methane	0°	1.595
Carbon Dioxide	90°	0.360
Carbon Dioxide	60°	0.341
Carbon Dioxide	45°	0.311
Carbon Dioxide	30°	0.313
Carbon Dioxide	0°	0.231

Table 2.
max $k(\boldsymbol{x})$ at varying bend angle, θ.

5. Discussion: Consistency of results

Flow simulations reported in this paper reveal that at the intrados location near the elbow inlet there will be a higher shear stress (**Figure 5**). $\tau_w(\boldsymbol{x})$ reduces at elbow exit on the intrados location for each bend angle, θ, and for any working fluid. The 90° elbow bend angle exhibits a high wall shear stress at the walls perpendicular to the inside and outside walls. Intense $\tau_w(\boldsymbol{x})$ is also visible on the 60° elbow but to a lesser degree and is not found on the 45° and 30° elbows (cf. contours at each θ in **Figure 5a, b** or **c**). **Figure 4a-c** display $k(\boldsymbol{x})$ contours at elbow inlet for all gases as θ reduces from

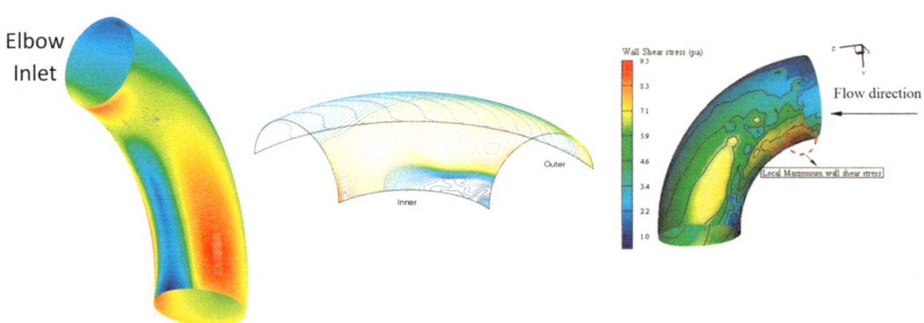

Figure 11.
Comparison between wall shear stress contours our simulations for air at 90° (left) and Homicz et al. (middle) and Liu et al. [15] (right).

90° to 30° compared to 0° (straight pipe). $k(\boldsymbol{x})$ is a maximum at the inner wall (red color in contours) for the 90° bend angle. Indeed, the contour at 0-degrees was found to correlate well with results in [18] which also showed the minimum TKE occur at the middle of the pipe.

Homicz et al. investigated pipe elbows in the nuclear power industry using the Fluent software. The $\tau_w(\boldsymbol{x})$ contours at 90° from our RANS simulations match the Homicz et al. results (see **Figure 11**) with the max $\tau_w(\boldsymbol{x})$ occurring at the intrados of the elbow inlet and outer pipe walls. The wall shear stress contours at 90° were also compared with Liu et al. [15] who investigated max $\tau_w(\boldsymbol{x})$ for a corrosion effect study of water and oil mixture in a steel pipeline.

Work published by Dutta et al. [2] also found Dean Vortices at the 90° location consistent with our results. Wang et al. [12] published velocity streamline data at 0° (elbow inlet) and 90° (elbow exit) and also observed such vortices (see **Figure 7**). In comparing velocity streamlines for all working fluids (**Figure 6a**-**c**) with Wang et al. [12] (**Figure 7**), we find a reasonable degree of spatial similarity between both datasets (theirs and ours) indicating further consistency of present results.

6. Conclusions

In this paper, we conducted a computational fluid dynamics (CFD) study using steady Reynolds-averaged Navier-Stokes (RANS) simulations of flow through an industrial flow pipeline carrying a different working fluid and and elbow angle (θ) measured from the horizontal inflow section. Understanding the magnitude of wall shear stress and maximum turbulence kinetic energy are important parameters in monitoring overall pipe longevity (high shear stress is linked to higher levels of corrosion; see ref. [15]). Our results show that commercial RANS CFD simulations are capable of determining the effect of the elbow angle on the wall shear stress and turbulence kinetic energy effectively. We investigated three working gases (air, methane & carbon dioxide) at fixed inflow mass flow rate for pipe geometry based on the Sudo et al. [1] experiment (also reported in Abuhatira et al. [4]). The elbow angles ranging from $\theta = [0°, 30°, 45°, 60°, 90°]$. Our results show that while working gas does not change the spatial structure of the turbulence kinetic energy and wall shear stress distribution compared to baseline calculation for air flow (as used in refs. [1, 4]), their magnitudes are affected because the inflow velocity is increased/ decreased depending on the working gas density relative to air. The velocity streamlines visible in **Figure 6a-c** can be observed to show the same trend for all three gases. The 90° elbow exhibits the highest degree of recirculation in the flow shown near the inner wall (Dean vortices). These vortices are caused by the centrifugal force and viscous forces being at their maximum at this location within the flow [12] and are weakened as the elbow angle approaches the $\theta = 0°$ straight pipe configuration. Reducing the elbow angle (θ), reduces the wall shear stress. Reducing the elbow angle from 90° to 60° was found to result in the greatest percentage reduction in wall shear stress value (see **Figure 8**) for all working fluids. This could be used for industrial applications when designing pipeline systems with increase system longevity in mind.

The methodology and results of the present paper can be used for industrial flow problems (e.g. as in the Sudo et al. [1] and Abuhatira et al. studies) or medical flow problems e.g., determining surface shear stress on skin grafts [16] that are required to be within certain tolerances for maximum longevity.

Author details

Keagan Church-Forbes[1], Mohammad Zamir A. Koshuriyan[2*] and
Ahmed A. Abuhatira[3]

1 Department of Mechanical and Aerospace Engineering, University of Strathclyde, Glasgow, UK

2 Department of Mathematics, University of York, Heslington York, UK

3 Nuclear, HE Engineering and Science, Lakes College West Cumbria, Workington, UK

*Address all correspondence to: zamir.koshuriyan@york.ac.uk

References

[1] Sudo K, Sumida M, Hibara H. Experimental investigation on turbulent flow in a circular-sectioned 90-degree bend. Experiments in Fluids. 1998;**25**(1): 42-49

[2] Dutta P, Saha S, Nandi N, Pal N. Numerical study on flow separation in 90° pipe bend under high Reynolds number by k- modelling. Engineering Science and Technology, an International Journal. 2016:904-910

[3] Kim J, Yadav M, Kim S. Characteristics of secondary flow induced by 90-degree elbow in turbulent pipe flow. Engineering Applications of Computational Fluid Mechanics. 2014;**8** (2):229-239

[4] Abuhatira AA, Salim SM, Vorstius JB. Numerical simulation of turbulent pipe flow with 90-degree elbow using wall Y+ approach. In: Proceedings of the ASME 2021 International Mechanical Engineering Congress and Exposition IMECE2021. IMECE2021 November 1-5, 2021, Virtual, Online

[5] Röhrig R, Jakirlić S, Tropea C. Comparative computational study of turbulent flow in a 90° pipe elbow Author links open overlay panel. International Journal of Heat and Fluid Flow. 2015;**55**:120-131

[6] Wang S, Ren C, Sun Y, Yang X, Tu J. A study on the instantaneous turbulent flow field in a 90-degree elbow pipe with circular section. Science and Technology of Nuclear Installations. 2016;**2016**: e5265748

[7] Dutta P, Nandi N. Effect of Reynolds number and curvature ratio on single phase turbulent flow in pipe bends. Mechanics and Mechanical Engineering. 2015;**19**:5-16

[8] Hufnagel L, Canton J, Örlü R, Marin O, Merzari E, Schlatter P. The three-dimensional structure of swirl-switching in bent pipe flow. Journal of Fluid Mechanics. 2018;**835**:86-101

[9] Korei Z, Benissaad S. Turbulent forced convection and entropy analysis of a nanofluid through a 3D 90° elbow using a two-phase approach. Heat Transfer. 2021;**50**(8):8173-8203

[10] Mazumder Q, Nallamothu VT, Mazumder F. Comparison of characteristic particle velocities in solid-liquid multiphase flow in elbow. International Journal of Thermofluids. 2020;**5–6**:100032

[11] Ganesh N, Jain P, Choudhury A, Dutta P, Kalita K, Barsocchi P. Random forest regression-based machine learning model for accurate estimation of fluid flow in curved pipes. PRO. 2021;**9**(1111): 2095

[12] Wang Y, Dong Q, Wang P. Numerical investigation on fluid flow in a 90-degree curved pipe with large curvature ratio. Mathematical Problems in Engineering. 2015;**2015**:1-12

[13] Zhang P, Gros Y, Roberts R, Benard A. Modeling of Turbulent Flow with Particle Deposition in Curved Pipes. Tampa, FL USA: 7th International Conference on Multiphase Flow ICMF; 2010; May 30-June 4, 2010

[14] Bernad SI, Totorean A, Bosioc A, Stanciu R, Bernad ES. Numerical investigation of Dean vortices in a curved pipe. AIP Conf. Proc. 2013;**1558**: 172-175

[15] Liu X, Gong C, Zhang L, Jin H, Wang C. Numerical study of the hydrodynamic

parameters influencing internal corrosion in pipelines for different elbow flow configurations. Engineering Applications of Computational Fluid Mechanics. 2020;**14**(1):122-135

[16] Church-Forbes K, Koshuriyan MZA, Abuhatira AA. CFD investigation of unsteady fluid effects through industrial flow pipelines and pipe elbows, Submitted to Proc. IMechE Part C. Journal of Mechanical Engineering Science. 2024

Chapter 8

Analysis of Flow Dynamics and Energy Dissipation in Piano Key and Labyrinth Weirs Using Computational Fluid Dynamics

Haytham F. Isleem, Mohamed K. Elshaarawy and Abdelrahman K. Hamed

Abstract

This chapter investigates the flow dynamics and energy dissipation of Piano Key Weirs (PKWs) and Labyrinth Weirs (LWs) using Computational Fluid Dynamics (CFD) models. PKWs and LWs are nonlinear weirs designed to enhance discharge capacity and energy dissipation in reservoirs and flood control facilities. Our research employs the FLOW-3D and ANSYS models to analyze various geometric parameters and their effects on discharge performance and energy dissipation. For PKWs, the analysis indicates that trapezoidal PKWs outperform rectangular PKWs regarding discharge efficiency due to their larger inlet flow area and improved flow distribution. Additionally, moving the PKW overhangs toward upstream-zone enhances discharge performance, while moving them toward downstream zone increases energy dissipation. For LWs, the analysis demonstrated that LWs with smaller sidewall angles increase crest length, enhancing discharge performance but leading to early submergence, decreasing the energy dissipation. Finally, it concludes that PKWs and trapezoidal LWs have a new function as structures that dissipate energy near the maximum limit. The findings confirmed the ability of the FLOW-3D and ANSYS models to accurately predict the various flow characteristics. It also provides valuable insights for designing and optimizing PKWs and LWs to balance discharge efficiency and energy dissipation, ensuring the safety and resilience of flood control structures.

Keywords: piano key weir, labyrinth weir, computational fluid dynamics (CFD), energy dissipation, discharge performance, FLOW-3D, ANSYS

1. Introduction

The increasing frequency of extreme weather occurrences has significantly heightened concerns over the safety of flood control facilities, particularly dams and reservoirs. Nonlinear weirs, such as Piano Key Weirs (PKWs) and Labyrinth Weirs (LWs), are viable solutions for enhancing the safety and resilience of these facilities [1].

1.1 Overview of nonlinear weirs

Nonlinear weirs, compared to traditional linear weirs such as ogee crest, can enhance the flow discharge capacity in narrow channels without requiring for increased upstream head. LWs are multi-linear weirs folded in plan view to increase the length of their crest. Typically, the crest is symmetric and consists of N-cycles of triangular, trapezoidal, or rectangular shape [2].

PKWs, as introduced by Lempérière and Ouamane [3], have been commonly employed for the same purpose as LWs. The PKW is an advancement of the LWs, featuring cantilevered overhangs on both ends and inclined floors in the inlet and outlet keys. The PKW's cantilevered overhangs create a longer weir crest length than an LW with the same footprint. Therefore, PKWs are especially suitable for use in areas with restricted space, such as on the tops of concrete dams. PKWs have been classified into four types according to the presence of overhangs [4]: which are type-A with symmetrical upstream and downstream overhangs, type-B with only upstream overhangs, type-C with only downstream overhangs, and type-D without any overhangs (**Figure 1**).

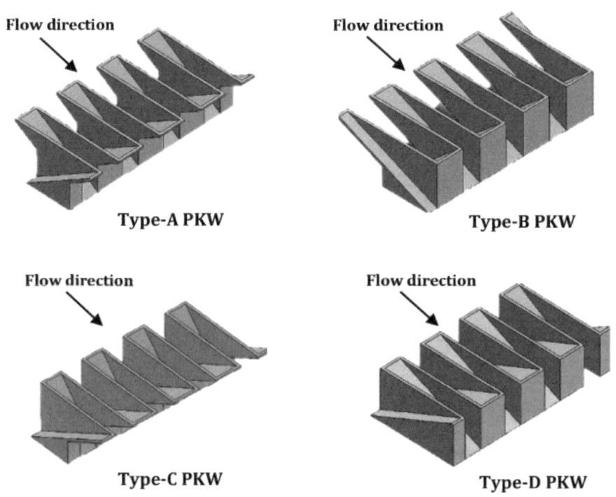

Figure 1.
Different types of PKW.

1.2 Discharge capacity and energy dissipation studies

Numerous current reservoirs require strengthening and enlargement to improve their discharge capacities, thus preventing overflow and the risk of dam failure caused by insufficient discharge capacity. The insufficient discharge capacity significantly increases the energy of the water flowing downstream, leading to higher flow velocity. This, in turn, causes damage to the installations and erosion of the riverbed in the downstream region. Hence, implementing nonlinear weirs can greatly enhance the flow rate and minimize the dissipation of energy downstream. The nonlinear shape of weirs with a zigzag pattern contributes to energy dissipation. Therefore, the dimensions of the stilling basin in the downstream region were decreased.

One of the challenges associated with PKW designs is the large number of geometric parameters. Important geometric parameters for PKW design include the total crest length (L), the weir height (P), the inlet and outlet key widths

(W_i and W_o, respectively), PKW cycle width (W_u), upstream and downstream overhang lengths (B_o and B_i, respectively), total lateral crest length (B), crest thickness (T_s), and dam height (P_d). On the other hand, a small set of geometric parameters for the LW design are the sidewall angle (α) and the outside and inside apex widths (D and A, respectively). The majority of these parameters are shown in **Figure 2**.

Several experimental studies have assessed the effect of geometric parameters for different designs of PKW and LW on enhancing the flow discharge capacity [6–10]. Furthermore, numerous experimental studies have been conducted to estimate energy dissipation rates and evaluate the flow regime downstream of the PKW and LW [11–15]. Recently, Computational Fluid Dynamics (CFD) models, such as FLOW-3D, Ansys-CFX, and Ansys-Fluent, are increasingly utilized to enhance the discharge performance over PKWs and LWs [16].

Savage et al. [17] used the FLOW-3D model to expand the headwater ratio (H_t/P) range, corresponding to the discharge coefficient (C_d) up to 2.0 for a trapezoidal LW (TLW) with a sidewall angle (α) of 15°. Carrillo et al. [18] studied the C_d under free and submerged flow conditions over a 30° TLW using the Ansys-CFX model. The results indicated that the CFD model can agree very well with experimental results, with maximum relative errors (RE) of 1.0 and 3.6% for free and submerged flows, respectively. In addition, the Re-Normalization Group (RNG) turbulence model produced more accurate results than the other turbulence models. Ghaderi et al. [19] used the FLOW-3D model to investigate the effect of sloping the weir crest edge and cutting the weir walls on enhancing the C_d of TLW. The results indicated that modifications significantly increase the C_d up to H_t/P equal to 0.20 and decrease gradually as the head above the weir crest increases.

Meanwhile, for PKWs, Safarzadeh and Noroozi [20] explored the effect of changing the sidewall angle of a rectangular type-A PKW with 6° on the discharge efficiency using the FLOW-3D model. Their findings revealed that trapezoidal PKW was 18% more discharge-efficient than rectangular PKW. That is due to the increase in the inlet flow area, which reduced the approach velocity and gave a better distribution over the inlet key. Mahabadi and Sanayei [21] investigated the effects of outlet key inclination type-A PKW on discharge capacity using the FLOW-3D model. Five inclination angles of 5, 10, 15, 20, and 25° were considered. A PKW with an angle of 25° produced 3.37% more discharge than a standard PKW. Ghanbari and Heidarnejad [22] investigated the effect of a triangular notch beneath type-A PKW overhangs on the C_d using the FLOW-3D model. The discharge coefficient of the PKW with a

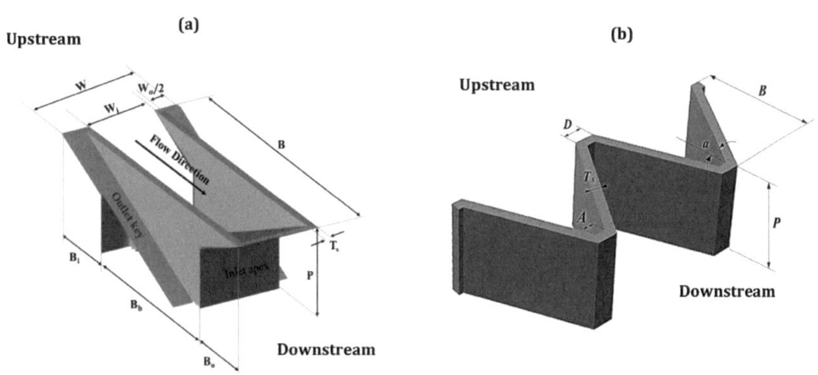

Figure 2.
3D view with the main geometric parameters for (a) PKW and (b) trapezoidal LW [4, 5].

triangle notch was increased by 36%. Besides the experimental study, Bekheet et al. [23] used the Ansys-Fluent software to evaluate the C_d of types A and B PKW. Type B PKW was found to have 2.0 to 4.5% higher discharge than type A PKW.

Finally, a few studies assessed the downstream energy dissipation and its characteristics using CFD models. For example, Ghaderi et al. [24] numerically investigated the effect of geometrical parameters on energy dissipation of the Trapezoidal-Triangular Labyrinth Weirs (TTLW). Three weir heights (P = 8, 10, and 12 cm) and five sidewall angles (α = 10, 15, 20, 25, and 30°) were considered. Their results indicated that the TTLW approximately dissipated the maximum amount of energy due to the collision of nappes in the upstream apexes and the circulating flow in the pool generated behind the nappes. Results also showed a considerable increase in energy dissipation by increasing the α and P. Also, Selim et al. [4] discovered that TLWs with an angle of α greater than 12° exhibited higher energy dissipation than PKWs.

1.3 Chapter objectives

Rapid climate changes over the last two decades have urgently necessitated the expansion of water reservoirs and the enhancement of their discharge capabilities using nonlinear weirs. This chapter will explore the flow dynamics (discharge, free-water surface, and flow characteristics) and energy dissipation capabilities of nonlinear weirs (Piano Key and Labyrinth). Utilizing various CFD commercial software, such as FLOW-3D and ANSYS, the study compares the effectiveness of these weirs in different configurations and assesses their ability to enhance discharge performance while minimizing downstream energy. The analysis aims to provide valuable guidelines for optimal nonlinear designs, assisting designers in deciding on the best configuration weir to offer suitable discharge as well as reasonable energy dissipation. This chapter emphasizes the importance of advanced numerical tools in the design and evaluation of hydraulic structures, contributing to more effective water management practices.

2. Materials and methods

2.1 Dimensional analysis

2.1.1 Discharge performance

The discharge performance of PKWs and LWs can be estimated by discharge coefficient (C_d) according to the common equation for the head-discharge relationship of sharp-crested linear weirs, expressed as [25]:

$$C_d = \frac{1.5\,Q}{L\,\sqrt{2g}\,{H_t}^{1.5}} \tag{1}$$

In Eq. (1), Q is the flowing discharge capacity, g is the gravitational acceleration, and H_t is the total upstream head [H_t = piezometric head (h) + approach velocity head ($V_o^2/2g$)] measured relative to the weir crest elevation, where V_o is the depth-averaged upstream velocity.

The discharge performance, which expresses the discharge flow efficiency of nonlinear weirs, is a function of several dimensionless geometric parameters, fluid parameters, and flow characteristics that can be summarized as follows:

$$C_d = \text{Ø}(g, \upsilon, \sigma, H_t, L, P, B, B_o, B_i, W_o, W_i, w, \alpha, N, A, D) \tag{2}$$

where Ø is a functional symbol, υ is the kinematic viscosity, σ is the surface tension, w is the cycle width, and N is the number of cycles. By using Buckingham (π) theorem, the dimensionless parameters affecting the discharge performance of the nonlinear weirs can be expressed as:

$$C_d = \text{Ø}\left(\frac{H_t}{P}, \frac{L}{W}, \frac{P}{W}, \frac{B_o}{B_i}, \frac{W_i}{W_o}, \frac{w}{P}, \frac{A}{D}, W_e, F_r, R_n, N, \alpha\right) \tag{3}$$

where W_e, F_r, and R_n are the Weber, Froude, and Reynolds numbers, respectively.

2.1.2 Energy dissipation

The ability of nonlinear weirs to dissipate energy can be evaluated by calculating the energy dissipation rate ($\Delta E/E_o$). The $\Delta E/E_o$ can be calculated by using the difference between the total upstream specific energy of weirs (E_o) and the downstream specific energy (E_1) that remained unaffected by the complex flow created by the weirs and hydraulic jumps (if that occurs). The E_o, E_1, and $\Delta E/E_o$ are expressed as:

$$E_o = y_o + \frac{V_o^2}{2g} = y_o + \frac{Q^2}{2g\,(Wy_o)^2} \tag{4}$$

$$E_1 = y_1 + \frac{V_1^2}{2g} = y_1 + \frac{Q^2}{2g(Wy_1)^2} \tag{5}$$

$$\frac{\Delta E}{E_o}(\%) = \left(\frac{E_o - E_1}{E_o}\right) \times 100 \tag{6}$$

where W is the total weir width, (y_o and y_1) are the upstream and downstream flow depths, respectively, and V_1 is the depth-averaged downstream velocity. In accordance with the specific energy curve, it is possible to perform a comparison between the downstream residual energy (E_1) measured in a rectangular channel and minimum residual energy (E_{min}) that develops during critical flow conditions, as well as the maximum limit energy dissipation (E_{max}), which can be computed using the following equations:

$$E_{min} = 1.5\,y_c = 1.5\sqrt[3]{\frac{Q^2}{g\,W^2}} \tag{7}$$

$$E_{max} = E_o - E_{min} = E_o - 1.5y_c \tag{8}$$

where y_c is the critical depth, once dividing both sides of Eq. (8) by (E_o), the following equation is derived for the maximum rate of energy dissipation:

$$\frac{E_{max}}{E_o} = 1 - 1.5\frac{y_c}{E_o} \tag{9}$$

2.2 Weir submergence

Weirs are typically designed to function under free-flow conditions but can become submerged in specific circumstances. Weir submergence happens when the water level downstream rises above the crest elevation. This is more likely to occur in weirs located in rivers or canals with a mild slope, areas of restricted flow, or an abundance of vegetation downstream. Submergence (S) is measured by determining the ratio of the total head downstream of the weir (H_d) to the total head upstream (H_t), as shown in Eq. (10):

$$S = \frac{H_d}{H_t} \tag{10}$$

The free-flow total upstream head is not influenced by the elevated tailwater at low submergence levels, a phenomenon referred to as modular submergence. The potential for upstream flooding is increased when the modular submergence limit is exceeded, resulting in an increase in the H_t for a given weir discharge (Q). The modular submergence in PKW and LW was $S < 0.48$ and $S < 0.50$, as suggested by Dabling and Tullis [26] and Carrillo et al. [18], respectively. **Figure 3** shows the hydraulic features of PKW under the free-flow and submerged conditions.

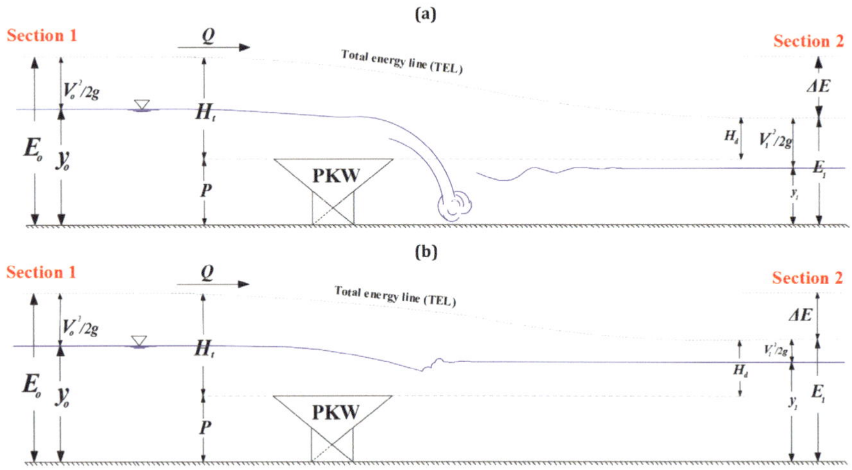

Figure 3.
Hydraulic features of PKW under (a) free-flow and (b) submerged flow conditions.

2.3 CFD modeling

Numerical models are increasingly being employed to analyze the energy dissipation and flow dynamics of nonlinear weirs. They were preferred by numerous researchers due to their ability to accurately simulate a variety of scenarios and complex geometries. In the current investigation, the numerical models of Ansys-Fluent and FLOW-3D were implemented. These models are widely used CFD models that solve the Reynolds-averaged Navier-Stokes (RANS) equation through the finite-volume method (FVM).

Initially, the geometries of the flume and nonlinear weirs are generated using a robust 3D design tool, such as SolidWorks software, in order to perform a run on these models. Subsequently, the mesh is generated by selecting the optimal mesh to reduce the runtime. Following this, the initial depth, the time step size, the general settings, and the boundary conditions are established. Ultimately, the user runs the solution and presents the results in the form of the water surface profile, the total head, and the corresponding flow discharge. The CFD modeling steps are illustrated in **Figure 4**.

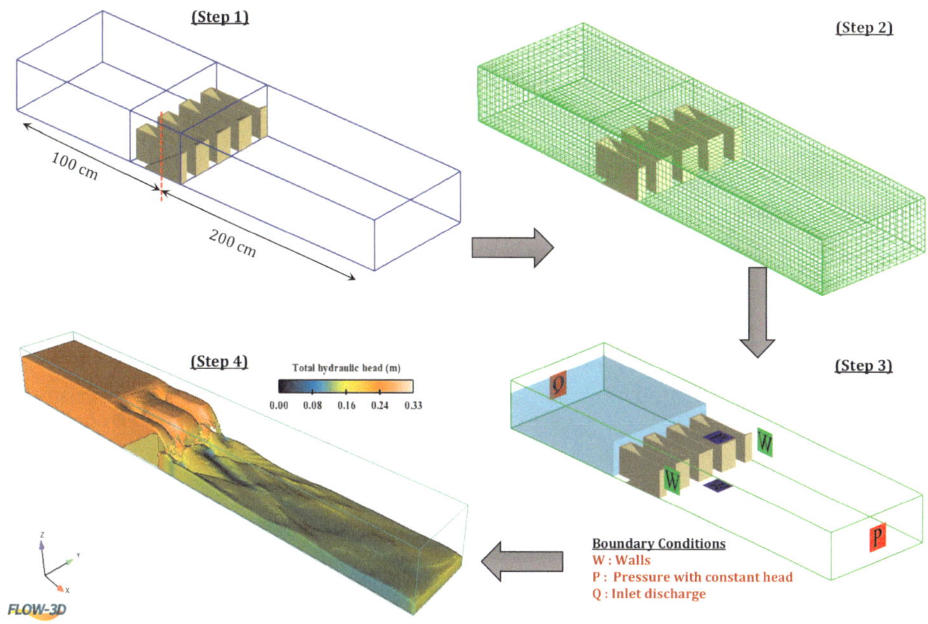

Figure 4.
The CFD modeling steps.

2.3.1 Governing equations

The governing equations of the used CFD models are the continuity and the momentum equation [17], which is expressed in the following vector forms:

$$\frac{\partial}{\partial x}(uA_x) + \frac{\partial}{\partial y}(vA_y) + \frac{\partial}{\partial z}(wA_z) = 0 \tag{11}$$

$$\frac{\partial U_i}{\partial t} + \frac{1}{V_F}\left(U_i A_j \frac{\partial u_i}{\partial x_j}\right) = -\frac{1}{\rho}\frac{\partial p}{\partial x_i} + g_i + f_i \tag{12}$$

where (A_x, A_y, and A_z) represents the cross-sectional area of the cell respective x, y, and z directions, respectively; (u, v, and w) represent the fluid velocity at any point within the cell x, y, and z directions, respectively. In addition, (i and j) are indicated (i and j = 1, 2, 3 for x, y, z directions, respectively); x_i is the coordinates axes; g_i is the gravitational acceleration vector; f_i is viscous accelerations; V_F is the volume fractional filled by the fluid; ρ is the density; t is the time; and p is the hydrostatic pressure.

2.3.2 Turbulence models (TMs)

Both FLOW-3D and ANSYS-Fluent models offer a comprehensive set of turbulence models (TMs) that cater to different types of flow simulations, from simple boundary layer flows to complex, high Reynolds number turbulent flows. TMs in CFD range in complexity from simple zero-equation models to advanced scale-adaptive simulations. The zero-equation Prandtl Mixing Length Model is used for simple boundary layer flows, providing quick and computationally inexpensive results. The one-equation Spalart-Allmaras model, slightly more complex, is favored in aerospace applications. Two-equation models like standard k-ε, re-normalization group (RNG) k-ε, and k-ω are widely used for general-purpose simulations, balancing accuracy, and computational cost. The k-ε models are suitable for a broad range of flows, while k-ω models are better for flows with strong adverse pressure gradients and separation. Moving to higher complexity, Large Eddy Simulation (LES) provides a detailed capture of large-scale turbulence structures, suitable for transient and high-resolution simulations. Finally, Detached Eddy Simulation (DES), a hybrid of RANS and LES, is used for high Reynolds number flows and complex geometries.

2.4 Study scenario

To analyze and assess the flow dynamics and energy dissipation of both PKW and LW using various CFD models, it is imperative to review the constraints on geometric and hydraulic parameters established by previous researchers until now. It is crucial to provide designers with valuable instructions for the future. **Tables 1** and **2** present a comparative overview of different PKW and LW designs from previous CFD studies. These tables specifically address two aspects: discharge performance and energy dissipation. The studies employed different CFD models, such as ANSYS and

Discharge performance							
Ref.	**CFD model**	**H_t/P**	**Type**	**Shape**	**TMs**	**RE of validation**	**Notes**
[23]	Ansys-Fluent	0.11-0.26	A, B	Trap., Tri., Rec.	Standard k-ε	< 5.0%	Limited head range
[27]	Ansys-Fluent	0.035 – 0.10	A	Rec.	RNG k-ε	< 10.0%	Extremely low heads
[28]	Ansys-Fluent	0.09 – 1.13	A	Rec.	RNG k-ε	< 7.0%	—
[29]	Ansys-Fluent	0.10 – 0.70	A	Trap., Rec.	RNG k-ε	*	Curved PKW
[20]	FLOW-3D	0.05–1.05	A	Trap.	RNG k-ε	~ 5.4%	—
[30]	FLOW-3D	0.12–1.00	A	Rec.	Standard k-ε	~ 2.2%	—
[22]	FLOW-3D	0.09–1.00	A	Rec.	RNG k-ε	< 16.6%	Featured with Tri. and Rec. notches
[31]	FLOW-3D	0.10–0.90	A	Rec.	RNG k-ε	< 5.0%	Lateral crest with zigzag indentations
[32]	FLOW-3D	0.30–1.40	B	Rec.	Standard k-ε	< 9.5%	Used as a side weir
[33]	FLOW-3D	0.06–0.32	B	Rec.	Standard k-ε & RNG, k-ω	< 3.4%	Curved PKW and Limited head range
Energy dissipation							
[4]	FLOW-3D	0.10–0.85	A, B	Rec.	RNG k-ε	< 5.7%	—

*Represents not reported.

Table 1.
An overview of various PKW designs from previous CFD studies with regard to discharge performance and energy dissipation.

Discharge performance						
Ref.	**CFD model**	**H_t/P**	**Shape**	**TMs**	**RE of validation**	**Notes**
[17]	Ansys-Fluent	Up to 2.0	Trap.	RNG k-ε, LES	< 3.0%	Extremely high heads
[34]	FLOW-3D	0.15 – 0.92	Tri., Trap.		~ 10.2%	Flow splitters installed on the crest
[18]	Ansys-CFX	0.15 – 1.00	Trap.	RNG k-ε	< 3.6%	—
[19]	FLOW-3D	0.05 – 0.70	Trap.	RNG k-ε	< 4.8%	Featured with inclining crest edge & cutting walls
[35]	Ansys-Fluent	0.05 – 0.80	Trap.	Standard k-ε	< 4.9%	—
[36]	Ansys-CFX	*	Tri., Rec.	RNG k-ε	~ 3.0%	Used as a side weir
Energy dissipation						
[24]	FLOW-3D	0.10 – 0.70	Trap., Tri.	RNG k-ε	< 3.0%	—
[4]	FLOW-3D	0.10 – 0.85	Trap.	RNG k-ε	< 5.7%	—

Represents not reported.

Table 2.
An overview of various LW designs from previous CFD studies with regard to discharge performance and energy dissipation.

FLOW-3D, to investigate different PKW and LW shapes such as trapezoidal, triangular, and rectangular shapes. These tables include details regarding the range of headwater ratios (H_t/P), the turbulence models (TMs) employed, and the relative error (RE) in validating experimental data. However, this chapter will primarily focus on the findings of these studies and will discuss them to draw conclusions and develop recommendations regarding the geometric and hydraulic variables.

3. Results and discussion

3.1 Optimization of the PKW geometry

PKWs are a highly intricate geometrical structure, and their efficient design is dependent upon both hydraulic and geometrical parameters. H_t/P ratio is the parameter that has the greatest impact on hydraulics. At the same time, the primary geometrical parameters that influence the design are the magnification ratio (L/W), PKW height ratio (P/W), inlet-to-outlet key width ratio (W_i/W_o), down-to-upstream overhang lengths ratio (B_i/B_o), and PKW shape.

3.1.1 Effect of headwater ratio (Ht/P)

The flow patterns over PKWs are greatly affected by the H_t/P. As H_t/P increases, the flow transitions from a state of free-flow to a more submerged condition. **Figure 5** illustrates the 3D perspective of the flow pattern over PKW at low (H_t/P = 0.146), intermediate (H_t/P = 0.373), and high (H_t/P = 1.129) head conditions. Under low head conditions, the nappes from adjacent sidewalls enter the outlet key without interference,

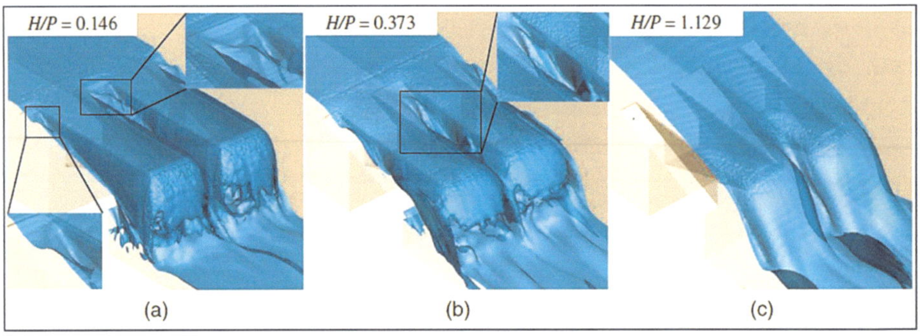

Figure 5.
3D view of flow pattern above the PKW under different H_t/P ratios [28].

converging in the middle and forming a small spray. As the head increases, the nappes from both sidewalls collide with the central nappe in the outlet key, leading to an elevated water surface profile at the end of the outlet key and resulting in a more turbulent flow. Consequently, the PKW demonstrates superior discharge performance at lower heads in comparison with higher heads. This can be explained by the occurrence of the local submergence phenomenon in the outlet keys, which results in an increase in water levels above the crests. Additionally, the energy dissipation of PKWs is highest at low flow conditions and decreases in a logarithmic-like manner as the flow increases.

3.1.2 Effect of magnification ratio (L/W)

An increased L/W ratio resulted in enhanced discharge capacity, particularly when the H_t/P is low. Nevertheless, the submergence of the crest and interference between falling lateral jets and oblique jets along the incline of the outlet key ramps are among the adverse consequences of increasing L/W. However, the overall turbulence intensity and energy dissipation downstream of the PKW tend to decrease as the L/W increases. This is because higher L/W ratios result in improved discharge efficiency and the PKW, which in turn leads to a reduction in residual energy downstream.

3.1.3 Effect of height ratio (P/W)

From a discharge standpoint, a high PKW with a P/W_u ratio approximately equal to 1.3 is optimal. However, this height ratio might not be practical for all construction scenarios, particularly for dam rehabilitation projects where a lower PKW height ratio ($P/W_u \approx 0.5$) may be more economical and feasible. The following are the reasons for the recommendation of these ratios:

1. At high PKW heights ($P/W_u > 1.3$), the flow characteristics are such that the outlet keys can effectively evacuate the flow from the inlet keys across all upstream heads. This results in a quasi-horizontal free surface along the inlet key, suggesting that flow velocities are relatively constant.

2. At very PKW heights ($P/W_u < 0.5$), the outlet key slope is insufficient to achieve supercritical flow conditions, and the tailwater level frequently remains below the outlet key toe. This leads to a reduction in the side crest discharge as a result of the outlet flow's submersion.

Conversely, the overall energy dissipation downstream tends to increase as the P/W ratio of PKW increases, as it increases the turbulence and flow contraction within the structure.

3.1.4 Effect of inlet-to-outlet key width ratio (W_i/W_o)

All studies agree that discharge performance is enhanced when the W_i/W_o ratio exceeds 1.0, as opposed to when it is less than 1.0. The flow velocity is reduced as the Wi/Wo ratio increases, which is followed by an increase in the cross section of the inlet. However, the outlet key becomes excessively narrow as the W_i/W_o ratio increases, as the $W_i + W_o$ remains constant. This results in substantial interferences between the descending lateral nappes of adjacent side crests in the outlet key, particularly at the higher heads (H_t/P). This leads to a significant decrease in discharge performance and a substantial increase in the submergence of the PKW crest.

The energy dissipation downstream of the PKW is also reduced when the W_i/W_o is increased. This was due to the (W_o) being too narrow, which resulted in the highest W_i/W_o ratio values. Subsequently, the PKW becomes more susceptible to submersion as a result of the early interaction of falling nappes from adjacent side walls in the outlet keys, particularly at the highest heads. Consequently, it was found that the energy dissipation was relatively constant, regardless of W_i/W_o, within the H_t/P range of 0.80–0.95.

3.1.5 Effect of down-to-upstream overhang lengths ratio (B_i/B_o)

The primary impact of the overhangs' position is to alter the cross section and bottom slope of the keys. Considering the inlet, for instance, the flow depth and width are less limited along the upstream overhang compared to the area above the base length of the PKW. Moving the overhangs upstream increases, thus, the inlet key cross section, subsequently causing a reduction in the flow velocities along the inlet key. However, this also decreases the outlet key slope and cross-section, causing a more rapid filling of the outlet as well as the submergence of the lateral crest by the outlet flow.

According to the discussion above, type-B PKWs provide the best discharge performance at low heads ($H_t/P \leq 0.40$). The type-A PKW exhibited the highest at high heads (i.e., $H_t/P > 0.40$). The higher discharge performance of type-B PKW is likely related to a reduction in inlet flow contraction, approach velocity, and head loss entering the inlet key attributable to the increase in wetted perimeter associated with the apex overhang configuration, which produced a nearly horizontal water surface profile. The improved discharge performance of type-B PKW can be attributed to several factors, including a decrease in inlet flow contraction, approach velocity, and head loss at the inlet key. These improvements are a result of the increased wetted perimeter caused by the apex overhang configuration, which leads to a nearly horizontal water surface profile.

To examine the effect of increasing the up and downstream overhangs on the energy dissipation by PKWs, different types of PKW were compared, as shown in **Figure 6**. To standardize the process of identifying nappe interference regions, a parameter known as interference length, caused by the side nappe in the outlet keys (B_{int}), is introduced. The type-C PKW exhibited the highest energy dissipation rates as a result of its longer B_{int}, while the type-B PKW had the lowest energy dissipation rates due to its shorter B_{int}.

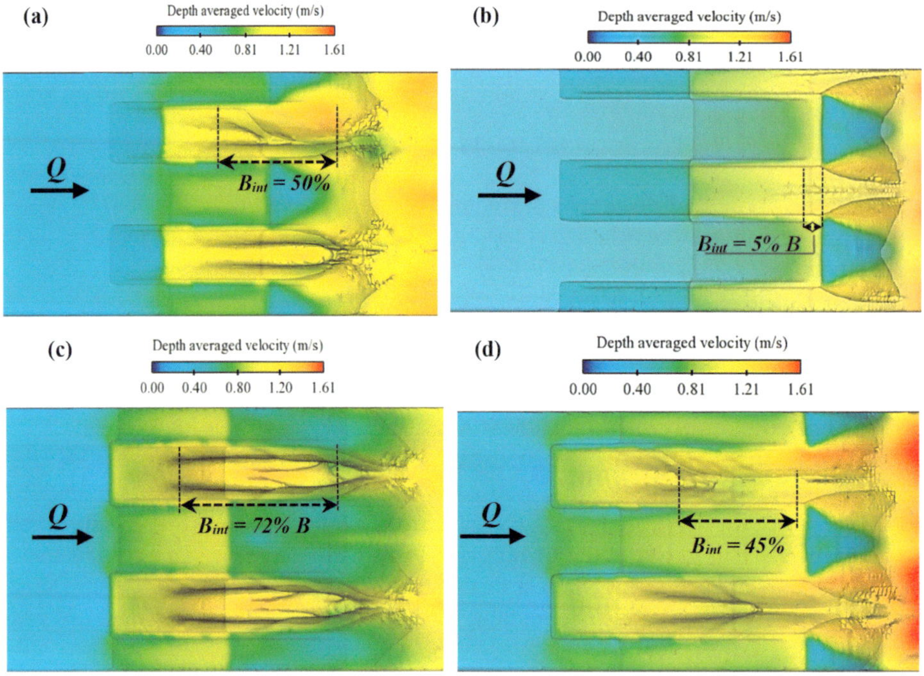

Figure 6.
Depth-averaged velocity distribution over (a) type A; (b) type B; (c) type C; and (d) type D PKWs.

3.1.6 Effect of PKW shape

The trapezoidal PKW exhibited superior discharge efficiency compared to the traditional rectangular PKW. This is partially associated with the flow conditions at the inlet key. The trapezoidal shape of the geometry enlarges the area through which water enters, leading to a decrease in the speed along the key axis. Consequently, flow is distributed more uniformly over the side crests. It explains in part why the streamlines of the water surface of the trapezoidal PKW are flatter compared to the rectangular PKW, as depicted in **Figure 7**. This discrepancy can be partially explained.

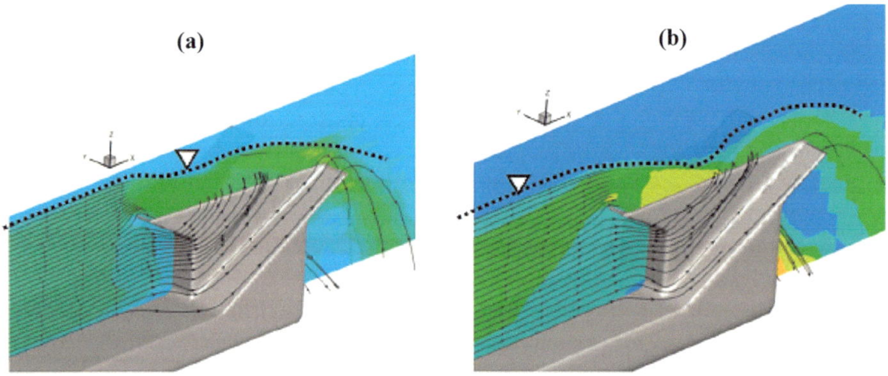

Figure 7.
Flow streamline comparison between (a) trapezoidal PKW and (b) rectangular PKW [20].

The gradual change in the shape of the inlet key limits the development of recirculation zones along the side walls and also prevents the formation of a critical section along the inlet key. These phenomena lead to an increase in the effective PKW crest length and thus discharge capacity increases in trapezoidal PKW compared to a rectangular one. It is important to note that increasing the number of cycles (N) for a given PKW width (W) also results in more flow interference and contraction, which enhances the energy dissipation process.

3.2 Optimization of the LW geometry

Optimizing the geometry of LW is crucial for enhancing their hydraulic performance. Two key parameters significantly impact the efficiency of these structures: the sidewall angle (α) and the cycle width ratio (w/P). A well-optimized ratio can enhance flow control and reduce the energy dissipation of the flows.

3.2.1 Effect of sidewall angle (α)

As the sidewall angle (α) decreases, the crest length (L) of the LW increases in a specific width. This longer crest length provides more area for water to flow over, thereby increasing the discharge performance. Despite the fact that a smaller α increases the L, it also tends to reduce the discharge coefficient (C_d). The reason behind this is the increased potential for nappe interference and local submergence as the angle decreases.

In contrast, the dissipation energy over the TL weirs increased as the sidewall angles increased. This can be explained by the fact that the spacing between the TL cycles was narrower at smaller angles than at larger angles, resulting in the submergence condition occurring earlier at smaller angles than at larger angles for the same discharge. Consequently, the energy dissipation rate caused by the nappes collision near the upstream apexes was reduced. Designers must therefore meticulously evaluate these consequences in order to achieve an optimal balance for the most efficient LW performance.

3.2.2 Effect of cycle width ratio (w/P)

The optimal w/P ratio also depends on other design factors like α, crest shape, and approach flow conditions. Wider LWs (higher w/P) allow for a longer crest length within a given footprint and experience less interference between the flows over adjacent cycles, which increases the overall discharge performance. So, to achieve a balance between the efficiency of the weir and the practical constraints of construction, the recommended ratio of w/P was between 2 and 4. Additionally, it was discovered that the half-round crest shape generally results in a higher discharge efficiency, particularly at lower heads ($H_t/P \leq 0.40$). This is due to the half-round shape, which enables the flow to remain affixed to the downstream face of the LW, thereby reducing flow separation and enhancing efficiency.

However, very high ratios might not be practical due to structural constraints and the potential for increased nappe interference. Designers must ensure that the chosen w/P ratio is feasible for the specific site conditions and that it provides the necessary hydraulic performance without compromising structural integrity.

3.3 Comparative analysis between PKWs and LWs

From the discharge standpoint, PKWs are recommended at lower heads (i.e., $H_t/P \leq 0.20$) than any shapes of LW weirs. At higher heads ($H_t/P > 0.20$), trapezoidal LWs with a smaller sidewall angle ($\alpha \leq 12^o$) are recommended, regardless of construction cost.

From the energy dissipation standpoint, trapezoidal LWs with larger angles ($\alpha > 12^o$) are superior to PKWs. Three primary reasons exist for the high energy dissipation over trapezoidal LWs as illustrated in **Figure 8a**:

1. Nappes collision occurs near the upstream apexes.

2. Hydraulic jumps occur at the downstream of LWs.

3. Turbulent recirculation behind the nappes followed oblique jet that strikes the channel bed.

PKWs were also capable of dissipating the flow energy over their crests in a manner similar to LWs (**Figure 8b**):

1. collisions and interaction between opposing falling nappes from the sidewalls at outlet key cycles.

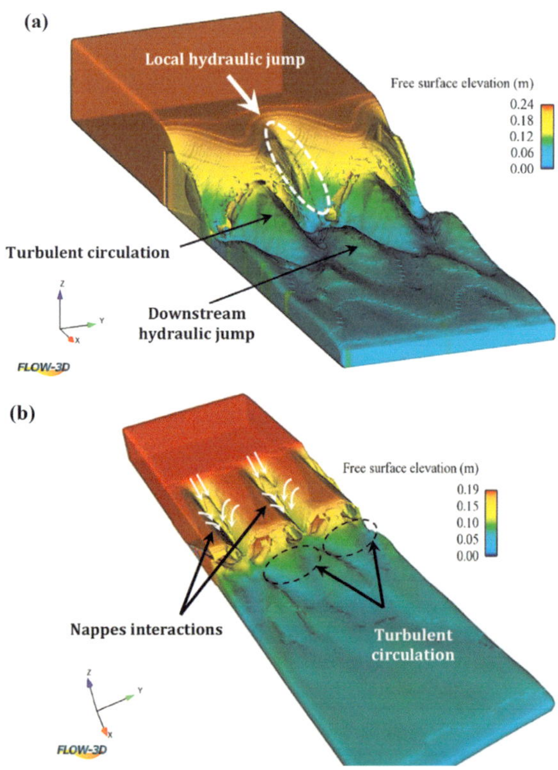

Figure 8.
3D views of energy dissipation behavior over (a) trapezoidal PKW and (b) rectangular PKW [4].

2. Turbulent recirculation is caused by a continuous collision of high-velocity free-flow jets from the downstream of inlet key crests into the downstream pool.

Figure 9 demonstrates a comparison between the energy dissipation rate ($\Delta E/E_o$) between two heights PKWs and trapezoidal LWs to the maximum energy dissipation rate (E_{max}/E_o) calculated from Eq. (9). The relative critical depth (y_c/E_o) was plotted against the $\Delta E/E_o$ to illustrate this comparison. The figure showed that the PK and TL weirs (both heights) exhibited energy dissipation rates that were exceedingly close to the (E_{max}/E_o) rate. Consequently, the $\Delta E/E_o$ by the PKWs and trapezoidal LWs was found to be approximately close to the maximum limit of energy dissipation.

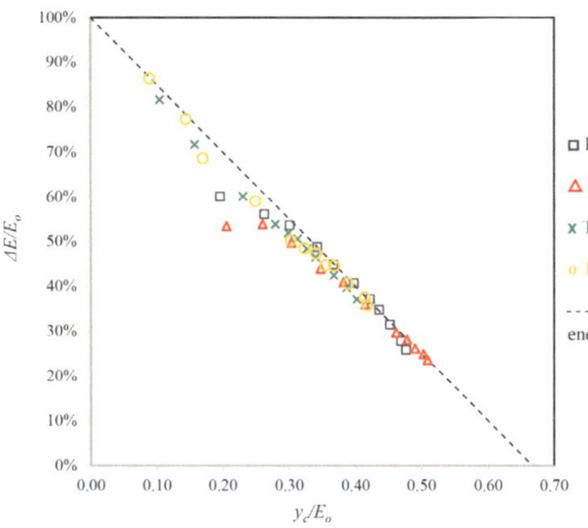

Figure 9.
Energy dissipation rates ($\Delta E/E_o$) versus relative critical depth (y_c/E_o) for both heights of PKW and LW.

4. Conclusions and recommendations

Most studies focus on adopting PKW and LW for higher discharging capacity, while others only consider downstream energy dissipation. Optimizing the geometric parameters of PKW and LW can lead to improved discharge performance while at the same time also increasing downstream energy dissipation rate, and vice versa. In order to assist with the development of the optimal and workable configuration for PKWs and LWs in both respects, the following recommendations for designing PKWs.

1. By limiting the range of H_t/P ratios, designers can fine-tune PKW geometry to achieve desired hydraulic characteristics while balancing discharge capacity, energy dissipation, and flow stability. Typically, the range of H_t/P used in PKW designs is between 0.10 and 1.20.

2. The PKW with a trapezoidal plan shape and an L/W ratio between 4 and 6 are the most effective in achieving improved discharge efficiency and cost savings.

3. The most optimal overall performance is achieved when the W_i/W_o ratio is 1.25, with a range of 1.29–1.57.

4. A higher PKW ($P/W_u \approx 1.3$) is recommended for new dam projects with high hydraulic demands. A lower PKW height ($P/W_u \approx 0.5$) is recommended for dam

rehabilitation or economic design considerations. Nevertheless, it is advisable to maintain a *P/W* ratio between 1.0 and 1.3 to achieve a balance between energy dissipation and discharge efficiency.

5. Type-B PKW is the preferred option from a discharge performance perspective, as it has a longer upstream overhang at lower heads ($H_t/P \leq 0.40$). Type-A PKW exhibits a significantly superior discharge at higher heads ($H_t/P > 0.40$) and it is suggested to rely on a self-balanced structure.

6. Type-C PKW has the highest energy from an energy dissipation perspective due to its longer downstream overhangs, followed by type-A, type-D, and finally type-B.

7. Smaller sidewall angles increase crest length, enhancing discharge capacity. Conversely, smaller angles lead to early submergence and reduced nappe interaction, decreasing the energy dissipation rate.

8. The optimal *w/P* ratio for LWs is generally between 2 and 4. Within this range, the weir achieves a balance between structural feasibility and hydraulic efficiency.

9. Half-round crest shapes of LW improve discharge efficiency at lower heads by maintaining clinging flow, while quarter-round shapes may be more practical for construction.

10. Regardless of the cost of construction, trapezoidal LWs with smaller sidewall angles (i.e., $\alpha \leq 12^{\circ}$) are the optimal choice for applications where the base length is unrestricted (e.g., rivers and canals) in terms of discharge performance.

11. Since the PKWs and trapezoidal LWs dissipate energy close to the maximum limit, they gained a new function as structures for dissipating energy, in addition to their primary role of controlling flow. Hence, it can be argued that using energy dissipation mechanisms, such as baffle blocks and sill-controlled stilling basins, is needless in the downstream zone of PKWs and LWs.

12. A good validation result was found between the experimental and numerical models, indicated by lower values of relative error, suggesting that the selected turbulence models had very little effect on CFD analysis results. Also confirms the ability of either FLOW-3D or ANSYS models to accurately simulate and predict the flow characteristics over the PKWs and LWs.

Acknowledgements

The authors acknowledge Qujing Normal University, Qujing, Yunnan, China, for facilitating the Structural Concrete Lab.

Conflict of interest

The authors declare no conflict of interest.

Notes/thanks/other declarations

The authors thank the editor and the reviewers for their valuable comments.

Author details

Haytham F. Isleem[1]*, Mohamed K. Elshaarawy[2] and Abdelrahman K. Hamed[2]

1 School of Applied Technologies, Qujing Normal University, Qujing, Yunnan, China

2 Faculty of Engineering, Civil Engineering Department, Horus University-Egypt, New Damietta, Egypt

*Address all correspondence to: haythamisleem@mail.qjnu.edu.cn

References

[1] Said MB, Ouamane A. Discharge capacity of an improved form of Labyrinth Weir. In: Kalinowska MB, Mrokowska MM, Rowiński PM, editors. Advances in Hydraulic Research. ISH 2023. GeoPlanet: Earth and Planetary Sciences. Cham: Springer; 2024. DOI: 10.1007/978-3-031-56093-4_2

[2] Laugier F. Design and construction of the first piano key weir spillway at Goulours dam. International Journal on Hydropower and Dams. 2007;**14**(5): 94-100

[3] Lempérière F, Ouamane A. The piano keys weir: A new cost-effective solution for spillways. International Journal on Hydropower and Dams. 2003;**10**(5): 144-149

[4] Selim T, Hamed AK, Elkiki M, Eltarabily MG. Numerical investigation of flow characteristics and energy dissipation over piano key and trapezoidal labyrinth weirs under free-flow conditions. Modeling Earth Systems and Environment. 2023;**10**:1253-1272. DOI: 10.1007/s40808-023-01844-w

[5] Machiels O, Pirotton M, Pierre A, Dewals B, Erpicum S. Experimental parametric study and design of piano key weirs. Journal of Hydraulic Research. 2014;**52**(3):326-335. DOI: 10.1080/00221686.2013.875070

[6] Leite Ribeiro M, Bieri M, Boillat J-L, Schleiss AJ, Singhal G, Sharma N. Discharge capacity of piano key weirs. Journal of Hydraulic Engineering. 2012; **138**(2):199-203. DOI: 10.1061/(asce) hy.1943-7900.0000490

[7] Anderson RM, Tullis BP. Piano key weir hydraulics and labyrinth weir comparison. Journal of Irrigation and Drainage Engineering. 2013;**139**(3): 246-253. DOI: 10.1061/(asce) ir.1943-4774.0000530

[8] Khode BV, Tembhurkar AR, Porey PD, Ingle RN. Experimental studies on flow over labyrinth weir. Journal of Irrigation and Drainage Engineering. 2012;**138**(6):548-552

[9] Crookston BM, Tullis BP. Hydraulic design and analysis of labyrinth weirs. I: Discharge relationships. Journal of Irrigation and Drainage Engineering. 2013;**139**(5):363-370

[10] Kabiri-Samani A, Javaheri A. Discharge coefficients for free and submerged flow over piano key weirs. Journal of Hydraulic Research. 2012; **50**(1):114-120

[11] Mohammadzadeh-Habili J, Heidarpour M, Samiee S. Study of energy dissipation and downstream flow regime of labyrinth weirs. Iranian Journal of Science and Technology, Transactions of Civil Engineering. 2018;**42**(2):111-119. DOI: 10.1007/s40996-017-0088-6

[12] Lopes R, Matos J, Melo J. Discharge capacity and residual energy of labyrinth weirs. In: Proceedings of the International Junior Researcher and Engineer Workshop on Hydraulic Structures (IJREWHS '06), Montemor-o-Novo, Portugal. Brisbane, QLD, Australia: Division of Civil Engineering, The University of Queensland. 2-4 Sept, 2006. pp. 47-55

[13] Haghiabi AH, Ghaleh Nou MR, Parsaie A. The energy dissipation of flow over the labyrinth weirs. Alexandria Engineering Journal. 2022;**61**(5): 3729-3733. DOI: 10.1016/j.aej.2021.08.075

[14] Idrees AK, Al-Ameri R. Investigation of flow characteristics and energy

dissipation over new shape of the trapezoidal labyrinth weirs. Flow Measurement and Instrumentation. 2023;**89**:102276. DOI: 10.1016/j.flowmeasinst.2022.102276

[15] Farhadi H, Moghadam MK, Sabzevari T, Noroozpour S. Study of relative energy dissipation of trapezoidal and arced piano key weirs equipped with baffles. Water Supply. 2023;**23**(1):80-93

[16] Eltarabily MG, Elshaarawy MK, Elkiki M, Selim T. Computational fluid dynamics and artificial neural networks for modelling lined irrigation canals with low-density polyethylene and cement concrete liners. Irrigation and Drainage. Jul 2024;**73**(3):910-927. DOI: 10.1002/ird.2911.

[17] Savage BM, Crookston BM, Paxson GS. Physical and numerical modeling of large headwater ratios for a 15 labyrinth spillway. Journal of Hydraulic Engineering. 2016;**142**:1-7. DOI: 10.1061/(ASCE)HY.1943-7900.0001186

[18] Carrillo JM, Matos J, Lopes R. Numerical modeling of free and submerged labyrinth weir flow for a large sidewall angle. Environmental Fluid Mechanics. 2020;**20**(2):357-374. DOI: 10.1007/s10652-019-09701-0

[19] Ghaderi A, Daneshfaraz R, Abbasi S, Abraham J. Numerical analysis of the hydraulic characteristics of modified labyrinth weirs. International Journal of Energy and Water Resources. 2020;**4**:425-436

[20] Safarzadeh A, Noroozi B. 3D hydrodynamics of trapezoidal piano key spillways. International Journal of Civil Engineering. 2017;**15**(1):89-101. DOI: 10.1007/s40999-016-0100-8

[21] Mahabadi NA, Sanayei HRZ. Performance evaluation of bilateral side slopes in piano key weirs by numerical simulation. Modeling Earth Systems and Environment. 2020;**6**(3):1477-1486. DOI: 10.1007/s40808-020-00764-3

[22] Ghanbari R, Heidarnejad M. Experimental and numerical analysis of flow hydraulics in triangular and rectangular piano key weirs. Water Science. 2020;**34**(1):32-38. DOI: 10.1080/11104929.2020.1724649

[23] Bekheet AA, Aboul Atta NM, Saad NY, El-Molla DA. Effect of the shape and type of piano key weirs on the flow efficiency. Ain Shams Engineering Journal. 2022;**13**(3):101620. DOI: 10.1016/j.asej.2021.10.015

[24] Ghaderi A, Daneshfaraz R, Dasineh M, Di Francesco S. Energy dissipation and hydraulics of flow over trapezoidal-triangular labyrinth weirs. Water (Switzerland). 2020;**12**(7):3729-3733. DOI: 10.3390/w12071992

[25] Henderson FM. Open Channel Flow. N.Y: Macmillan Publishing Company; 1966

[26] Dabling MR, Tullis BP. Piano key weir submergence in channel applications. Journal of Hydraulic Engineering. 2012;**138**:661-666. DOI: 10.1061/(ASCE)HY.1943-7900.0000563

[27] Abhash A, Pandey KK. Numerical study of discharge-head relationship of piano key weirs for low heads. Water Resources. 2021;**48**(2):235-244. DOI: 10.1134/S0097807821020020

[28] Li S, Li G, Jiang D. Physical and numerical modeling of the hydraulic characteristics of type-A piano key weirs. Journal of Hydraulic Engineering. 2020;**146**(5):1-11. DOI: 10.1061/(ASCE)HY.1943-7900.0001716

[29] Karimi Chahartaghi M, Nazari S, Mahmoodian Shooshtari M. Experimental and numerical simulation of arced trapezoidal piano key weirs. Flow Measurement and Instrumentation. 2019;**68**:101576. DOI: 10.1016/j.flowmeasinst.2019.101576

[30] Hu H, Qian Z, Yang W, Hou D, Du L. Numerical study of characteristics and discharge capacity of piano key weirs. Flow Measurement and Instrumentation. 2018;**62**(May):27-32. DOI: 10.1016/j.flowmeasinst.2018.05.004

[31] Behroozi AM, Vaghefi M. Experimental and numerical study of the effect of zigzag crests with various geometries on the performance of A-type piano key weirs. Water Resources Management. 2022;**36**(12):4517-4533. DOI: 10.1007/s11269-022-03261-7

[32] Alabedi HJ, Khassaf SI. Numerical study on discharge capacity of piano key side weir with various ratios of the crest length to the width. Open Engineering. 2024;**14**(1). DOI: 10.1515/eng-2022-0536

[33] Khanahmadi E, Dehghani AA, Alenabi SN, Dehghani N, Barry E. Hydraulic of curved type-B piano key weirs characteristics under free flow conditions. Modeling Earth Systems and Environment. 2023;**10**:1-18. DOI: 10.1007/s40808-023-01790-7

[34] Aghashirmohammadi G, Heidarnejad M, Purmohammadi MH, Masjedi A. Experimental and numerical study the effect of flow splitters on trapezoidal and triangular labyrinth weirs. Water Science. 2023;**37**(1):28-39. DOI: 10.1080/23570008.2023.2210391

[35] Mattos-Villarroel ED, Ojeda-Bustamante W, Díaz-Delgado C, Salinas-Tapia H, Flores-Velázquez J, Bautista Capetillo C. Methodological proposal for the hydraulic design of labyrinth weirs. Water. 2023;**15**(4):722

[36] Hussein BS, Jalil SA. Hydrodynamic behavior simulation of flow performance over labyrinth side weir. Polish Journal of Environmental Studies. 2024;**33**(2):1159-1171. DOI: 10.15244/pjoes/173442

Chapter 9

Computational Fluid Dynamics Simulations to Deepen Understanding of the Hemodynamic Underlying Atrial Fibrillation and Improve Therapeutic Approaches

Camilla Cortesi, Matteo Falanga, Sachal Hussain, Nadia D'Alessandro, Corrado Tomasi, Alessandro Masci and Cristiana Corsi

Abstract

This chapter provides an overview of some possible applications of patient-specific computational fluid dynamics (CFD) simulations to gain insight into the fluid dynamic mechanism underlying the onset and maintenance of atrial fibrillation (AF), improve treatment planning, and control disease progression. CFD represents a valuable non-invasive approach to determine and evaluate significant parameters in a very complex fluid dynamic system such as the left atrium in AF. Different applications of our CFD model to derive parameters such as left atrial appendage ostium velocity, wall shear stress, and endothelial cell activation potential, which are shown to be promising indicators for a quantitative prognostic evaluation of AF, are presented and critically evaluated from the perspective of defining a quantitative approach for stroke risk prediction in the context of precision medicine.

Keywords: atrial fibrillation, left atrial appendage, left atrium digital twin, stroke risk assessment, patient-specific analysis

1. Introduction

The left atrium (LA) is a structurally complex organ with a vital mechanical role in regulating left ventricular (LV) filling. It acts as a reservoir during systole, a conduit during early diastole, and a pressure-boosting pump in late diastole [1]. Any mechanical dysfunction in the left atrial appendage (LAA) can contribute to clot formation, primarily due to reduced contraction, impaired blood flow, and incomplete blood washout in the LAA [2].

 IntechOpen

Atrial fibrillation (AF) is considered one of the epidemics of our century [3]. Basic epidemiological studies of AF conducted between the late twentieth and early twenty-first centuries in developed countries estimated that the prevalence of AF in the general population is between 0.5 and 1% [4].

The estimated prevalence of AF in the United States is approximately 2.2 million individuals, which includes both paroxysmal and persistent forms of the condition [5]. Similarly, the ATRIA study [6] reported a prevalence of about 0.95% in the North American general population, with consistent findings observed in the UK. In the United States and European countries alone, there are approximately 160,000 new cases of AF every year. More than six million people in Europe are affected by AF, and its prevalence is expected to at least double in the next 50 years due to the aging population [7].

AF is a type of supraventricular arrhythmia characterized by disorganized and uncoordinated electrical activity in the atrium, leading to ineffective atrial contraction. This condition remains a predominant contributor to stroke, heart failure, sudden death, and cardiovascular morbidity worldwide.

AF is recognized as an independent risk factor for stroke. It is associated with a fivefold increased risk of cerebrovascular events, contributing to approximately 15–18% of all stroke cases [8].

Multiple features associated with AF contribute to increase this serious condition. The primary factor is anatomical remodeling, characterized by the progressive enlargement of the LA and the elongation of the LAA, which occurs as a result of reduced contractility and increased compliance caused by AF. A second factor is the loss of mechanical function, not only during the AF episodes but also in sinus rhythm (SR). This phenomenon leads to a continuous decrease in ejection fraction.

All these factors favor blood stagnation in the LA and the LAA, consequently increasing the risk of thromboembolic events. The thrombotic material linked to AF is predominantly found in the LAA: in particular, thrombi are frequently observed in AF patients who have experienced ischemic strokes compared to those who have not [9]. Additionally, it appears that LAA flow velocity is lower in AF, and it is probably related to the loss of mechanical contraction.

Unfortunately, parameters able to highlight the conditions previously described are not available in clinical practice and the metrics employed to assess stroke risk, including the widely endorsed CHA_2DS_2-VASc score, rely on broad empirical variables such as age, hypertension, and diabetes mellitus, without considering specific hemodynamic implications of AF which might improve their predictive power.

In this context, multiple clinical studies have suggested that stroke risk stratification could be enhanced by incorporating hemodynamic data from the LA and LAA [10]. These studies are founded on the idea that AF significantly alters intra-atrial blood flow dynamics, causing blood stasis and increasing the likelihood of clot formation and subsequent embolism. The precise manner these intricate mechanisms interact cannot be evaluated through experimental methods and remains predominantly uncharted.

In this respect, computational fluid dynamics (CFD) serves as a robust instrument that may establish a connection between LA structural remodeling and intracardiac thrombosis in individual patients by analyzing the patient-specific LA blood flow characteristics [11]. CFD represents a valuable non-invasive approach to determining and assessing physically meaningful parameters and indicators such as cardiac blood flow rates, vorticity, turbulent kinetic energy, etc. In particular, CFD modeling of the

LA in AF has not been thoroughly explored, despite its significant potential clinical implications. This approach offers a distinctive method for examining various boundary conditions (BCs) within a complex fluid dynamics framework, allowing for non-invasive, fully controllable, and reproducible investigations.

In this book chapter, we want to present an overview of possible applications of CFD simulations, applying a patient-specific dynamic model, which takes into account the heart wall movement and proposes the latest fully customized atrial blood flow in AF. The developed approach has been applied in different settings to better understand the pathophysiology of AF and predict stroke risk. Each of the following sections will delve into a specific application.

The following section outlines the design and implementation of the patient-specific CFD model [12]. This model has been tested on patients with AF under both SR and AF conditions. The goal is to provide a thorough evaluation of the hemodynamic implications associated with AF episodes, focusing specifically on the LA chamber and the LAA. Key aspects assessed include blood flow velocity and vortex structure within these regions.

One question that remains open concerns the BCs applied to the CFD simulation. Most of the computational models developed for the LA in the literature assume, as BCs, a uniformly distributed flow at the pulmonary veins (PVs) and mitral valve (MV) derived from literature data. Furthermore, most models used assume laminar flow in the atrium or a rigid model. This includes the possibility of improving BCs and making them realistic and patient-specific with information on the quantification of LA contraction in AF as presented in the dedicated section.

To enhance the ability to assess the risk of stroke for a given patient, the CDF simulations have been also applied to a digital twin (DT) model of the LA. The most important hemodynamic parameters from the CFD model were computed and evaluated among three groups of subjects. The comparison between the groups enabled an initial identification of possible parameters that may be used in the development of a novel index for the stratification of stroke risk.

Finally, the impact of the LAA morphology and its occlusion on stroke risk assessment in AF has been evaluated thanks to the CFD simulations. The influence of the LAA shape on the blood flow characteristic in AF condition has been evaluated through the correlation of the geometrical and fluid dynamics indexes, computed for each LAA morphology, with the blood stasis and thrombi formation. The CFD model was also applied to simulate the fluid dynamics effects of LAA occlusion (LAAO) in AF patients to predict hemodynamic changes due to the occlusion. Blood velocity, particle washout, and endothelial damage were calculated to quantify changes in flow pattern after occlusion concerning the thrombogenic risk.

The CFD models should be held as permanent work in progress. There are always additional phenomena that can be included and, by increasing the grade of patient-specificity, a better understanding of the hemodynamics of AF will be achieved to improve therapeutic approaches.

2. Development of the patient-specific CFD model and initial evaluation

This section sets out the main steps of the procedure for implementing the CFD model and it is applied in all the different applications presented in this chapter.

The model personalization involved the extraction and reconstruction of the anatomical model of the LA atrium for each analyzed patient.

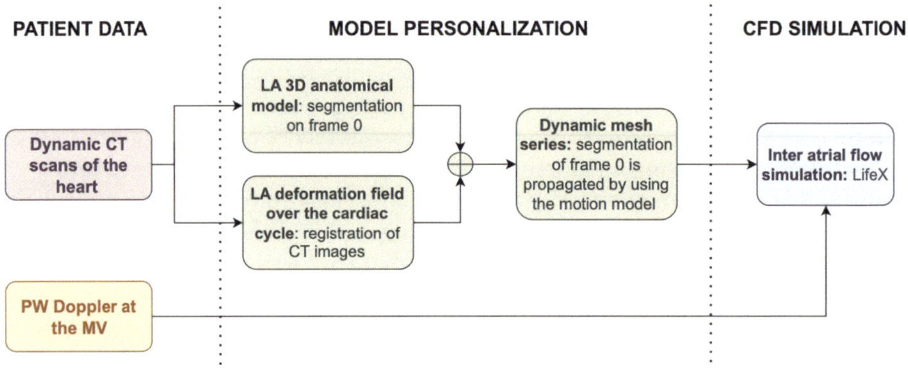

Figure 1.
Flowchart illustrating the sequential steps in constructing the CFD model. The blue box represents patient-specific data, while the purple boxes indicate the processing of this data to derive the LA anatomical structure. The associated deformation field is shown in the green boxes, serving as inputs for the personalized CFD model. The orange box highlights the BCs applied to the model, which are sourced from the literature and are not patient-specific. Source: Author's original work.

In all the presented applications, the workflow follows the same steps as shown in **Figure 1**. The differences lie in the BCs applied to the simulations and the type of image acquisition used from which to extract the anatomical model.

2.1 Definition of the LA anatomical model

The LA was identified through the analysis of the first volume of the Computed Tomography (CT) and Magnetic Resonance Imaging (MRI) dynamic acquisition, contingent upon the type of data accessible for the particular application. A volume of interest (VOI) encompassing the LA was manually delineated. The LA was subsequently detected in each image within the defined VOI, leading to the reconstruction of a 3D representation of the LA. By detecting the LA contour in 2D space, LA segmentation was performed slice by slice in a MATLAB working environment [12]. A rough segmentation was computed for each slice using intensity thresholds: the peak associated with the LA in the image histogram was identified to establish two thresholds, which were subsequently utilized for hysteresis segmentation, providing greater flexibility compared to using a single threshold. Further processing was designed to improve LA segmentation accuracy: morphological operators and a curvature motion were, respectively, applied to disconnect regions and to regularize the LA contour in each slice. Once all the steps described previously have been carried out, the 3D anatomy was then obtained by stacking the two-dimensional segmentation.

Furthermore, to ensure a seamless geometric representation of the computational domain for CFD simulation, our 3D anatomical model underwent two key processes (i) it was refined through Laplace smoothing with MeshLab software [13]; and (ii) five cutting planes were utilized to analyze the four PVs and the MV, delineating specific subsets of the inflow and outflow boundaries of the anatomical model for the CFD simulation. This refined anatomical model provided the necessary input for the subsequent labeling and volume mesh generation processes. For this purpose, the Vascular Modeling Toolkit (VMTK) library was employed, ensuring accurate representation and effective mesh creation for the CFD analysis [14]. An example of the final result of the LA mesh in a patient is presented in **Figure 2**.

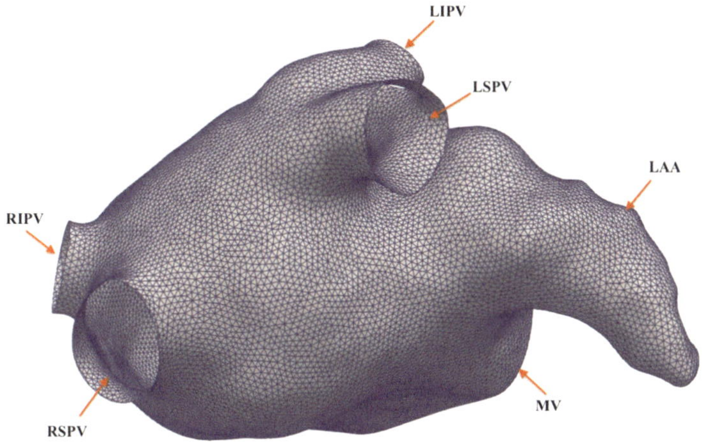

Figure 2.
Illustration of the LA mesh in a patient, highlighting key anatomical structures including the right inferior pulmonary vein (RIPV), right superior pulmonary vein (RSPV), left superior pulmonary vein (LSPV), left inferior pulmonary vein (LIPV), left atrial appendage (LAA), and mitral valve (MV). Source: Author's original work.

2.2 Computation of LA deformation

The deformation of the LA during various phases of the cardiac cycle was determined by applying a 3D nonrigid image registration step of the acquired volumes, resulting in the displacement $\mathbf{s}_{i\to i+1}(\mathbf{x})$ between two successive time volumes $\text{Im}_i(\mathbf{x})$ and $\text{Im}_{i+1}(\mathbf{x})$. A first step based on an affine transformation was applied and, afterwords the result obtained was combined with the nonrigid transformation based on B-spline [15] model defined as:

$$\mathbf{T}(\mathbf{x}) = \mathbf{x} + \sum_{\mathbf{x}_k \in \mathcal{N}_x} \mathbf{p}_k \boldsymbol{\beta}^3\left(\frac{\mathbf{x} - \mathbf{x}_k}{\sigma}\right) \tag{1}$$

with $\mathbf{x}_k$ the control points, $\boldsymbol{\beta}^3(\mathbf{x})$ represents the cubic multidimensional B-spline polynomial, $\mathbf{p}_k$ represents the vectors of B-spline coefficients, which correspond to the displacements of the control points. The symbol σ denotes the spacing between these B-spline control points, while $\boldsymbol{\mathcal{N}}_x$ refers to the set of all control points that lie within the compact support of the B-spline at a specific point x. The control points $\mathbf{x}_k$ were defined on a uniform grid that is overlaid on the fixed image, ensuring a structured approach to the deformation and transformation processes applied to the image data. The configuration of this control point grid is determined by the spacing between the control points $\sigma = (\sigma_1, \ldots, \sigma_{\text{dim}})$, which may vary for each direction.

The parameters outlined in Eq. (1) were carefully chosen to enhance the quality of tracking, as evaluated through visual assessment. Furthermore, we employed the mean square difference as a metric for measuring the similarity of image registration. Following this step, the overall displacement of a given volume at time i in relation to the reference volume at time 0 was calculated. This was accomplished by incrementally summing the inter-frame displacements that occurred between successive time points: $s_{i\to 0}(x) = s_{i\to n-1}(x)\, s_{i-1\to 0}(x)$, with $s_{0\to 0}(x) = 0$. Consequently, let x_0 represent the position of a mesh vertex at time 0. The updated position x_i at time i was determined using the formula $x_i = x_0 + s_{n\to 0}(x_0)$. In this equation, s_n denotes the incremental displacement applied to the vertex, which effectively accounts for the changes in position over time.

The motion field was then employed to extend the LA computational domain across the entire cardiac cycle, thereby enabling the simulation of LA motion in the SR state. LA motion in AF is characterized by irregularities, disorganization, rapid fluctuations, and a significant reduction in contraction strength. Previous studies [16] have approached this phenomenon by simulating the walls of the LA as rigid structures. This method aimed to replicate the severe conditions associated with chronic AF, where effective atrial contraction becomes impossible. To simulate conditions that precede chronic AF (paroxysmal and persistent AF conditions), our study employed a modeling approach for atrial contraction. This was achieved by applying a random displacement to each vertex of the LA anatomical model. The displacement was characterized by a sine function with a frequency of 4 Hz, which was further modulated by a random factor derived from a uniform probability density function ranging from zero to one. This frequency selection was deliberately aligned with the typical rates observed during episodes of atrial fibrillation. Additionally, the sine wave's amplitude was constrained to a small value of 0.1 mm, effectively simulating the diminished contraction often seen in AF patients. This careful modulation not only reflected the pathological state of reduced atrial contractility but also helped mitigate numerical challenges associated with significant degradation of the mesh quality during computational simulations.

Considering the contraction model implemented in this study, the analysis of the two patients with persistent AF revealed only minimal variations in LA volume during the cardiac cycle, ranging from 1 to 2%. This slight fluctuation indicates that the anatomical model of the LA could be maintained in a relatively stable state throughout the cardiac cycle, supporting the underlying assumption of anatomical stability in patients experiencing persistent AF. However, the implemented model could benefit from a more realistic approach rather than maintaining a static anatomical framework.

To enhance the temporal resolution, we utilized Fourier interpolation on the displacement field. This methodology facilitated the reconstruction of a continuous and periodic function from the available discrete data points. The ability to create a continuous displacement function was crucial for ensuring the stability of the numerical CFD model. By transforming the discrete data into a smooth, continuous representation, the model was able to more accurately simulate the dynamic behaviors of the LA during the cardiac cycle.

Taking into account the assumed periodic nature of the heartbeat, we simulated a series of cardiac cycles to analyze the required conditions for ensuring physiological blood flow velocity at the onset of each cycle. By simulating multiple cycles, we aimed to establish a steady-state condition that would prevent the presence of unphysiological initial states in blood flow velocity.

2.3 The computational fluid dynamic model

The blood flow within the model was represented as an incompressible fluid, governed by the Navier-Stokes equations. These equations were formulated in the Arbitrary Lagrangian-Eulerian (ALE) framework, which allows for the flexible handling of moving boundaries and deforming geometries. This approach is particularly advantageous in fluid dynamics simulations involving the cardiovascular system, as it effectively captures the complexities of blood flow dynamics and the interactions between the fluid and the surrounding anatomical structures [17]. This assumption facilitated the division of the overall problem into two interrelated subproblems: the fluid dynamics problem, which focuses on the behavior and motion of the fluid, and the geometric problem, which concerns the movement of the computational domain relevant to the fluid dynamics. The geometric aspect is crucial, as it determines the displacement of the fluid, denoted as $\hat{\mathbf{d}}_f$ which in turn defines the ALE mapping. To accurately capture the fluid's movement within the domain $\Omega \subset \mathbb{R}^3$, we consider $\hat{\mathbf{d}}_f$ as a harmonic extension to the reference domain $\hat{\Omega}$ of the displacement $\hat{\mathbf{d}}_{wall}$, registered at the boundary of the LA Ω (lateral wall, PVs, MV), that is, for all t (0,T):

$$\begin{cases} -\Delta\, \hat{\mathbf{d}}_f = 0 \text{ in } \hat{\Omega} \\ \hat{\mathbf{d}}_f = \hat{\mathbf{d}}_{wall} \text{ on } \partial\hat{\Omega} \end{cases} \tag{2}$$

The solution to the geometric problem defines the ALE map $\mathcal{A}(\hat{\boldsymbol{x}}) = \hat{\boldsymbol{x}} + \hat{\mathbf{d}}_f(\hat{\boldsymbol{x}}, t)$ for each $\hat{\boldsymbol{x}} \in \Omega$ and the current configuration of the fluid domain. The Navier-Stokes equations for an incompressible fluid written in ALE coordinates are given below:

$$\begin{cases} \rho_f \left.\dfrac{\partial \mathbf{u_f}}{\partial t}\right|_{\hat{x}} + \rho_f((\mathbf{u_f} - \mathbf{w}) \cdot \nabla)\mathbf{u_f} - \nabla \cdot \boldsymbol{\sigma}_\mathbf{f}(\mathbf{u_f}, p_f) = 0 \text{ in } \Omega_t \\ \nabla \cdot \mathbf{u_f} = 0 \text{ in } \Omega_t \\ \mathbf{u_f} = \mathbf{z_f} \text{ on } \Gamma_D \\ \boldsymbol{\sigma}_\mathbf{f}\mathbf{n_f} = \mathbf{g_f} \text{ on } \Gamma_N \end{cases} \tag{3}$$

where $\left.\frac{\partial}{\partial t}\right|_{\hat{x}} = \frac{\partial}{\partial t} + \mathbf{w} \bullet \nabla$ is the derivative ALE, $\mathbf{w}(\mathbf{x}) = \frac{\partial \mathcal{A}_t(x)}{\partial t}$ is the velocity of the fluid domain, u_f and p_f are the velocity and pressure of the fluid, respectively. The p_f denoted the density of the fluid and σ_f the Cauchy stress tensor

$$\boldsymbol{\sigma}_f(\mathbf{u}_f, p_f) = \mu\left(\nabla \mathbf{u}_f + (\nabla \mathbf{u}_f)^T\right) - p_f \mathbf{I} \tag{4}$$

with $\boldsymbol{I}$ identity tensor, μ dynamic viscosity of the fluid and $\boldsymbol{n}_f$ the outgoing $\partial\Omega_t$ versor. The functions $\boldsymbol{z}_f$ and $\boldsymbol{g}_f$ denote the Dirichlet and Neumann conditions applied to the Γ_D and Γ_N contours of Ω_t, respectively.

Regarding the spatial discretization of the model, the Galerkin finite element approximation was considered using linear Lagrange polynomials of degree one for both velocity and pressure (P1-P1), with a Variational Multiscale-Streamline Upwind Petrov-Galerkin (VMS-SUPG) stabilization scheme [18]. As for temporal discretization, a semi-implicit second-order backward differential scheme was applied, as reported in [19].

2.4 Boundary conditions imposed on the CFD model

The inflow sections, which change in response to the heartbeat and have non-circular cross-sectional shapes, were modeled using a pseudo-parabolic velocity profile. To derive this profile, we solved a Laplacian problem for each pulmonary vein

$$\begin{cases} -\Delta \mathbf{u}_{BC} = 1 \text{ in } \Gamma_{PV} \\ \mathbf{u}_{BC} = 0 \text{ on } \partial\Gamma_P \end{cases} \tag{5}$$

where $\mathbf{u}_{BC}$ was the solution of Eq. (5). Notice that Γ_{PV} was either Γ_1, Γ_2, Γ_3, or Γ_4 (one for each PV).

In this process, a pseudo-parabolic flow profile was applied to each PV to be suitably rescaled according to the flow rate for t (0,T). The pseudo-parabolic velocity profile for each PV was obtained by weighting the solution of the Laplacian problem, $\mathbf{u}_{BC}$, according to the current flow rate at the corresponding PV.

A realistic MV flow rate in AF patients during SR, denoted Q^0, was obtained from reference [20]. The flow rate at each PV was then determined by applying the mass balance equation for all t (0,T)

$$Q_1^{pv} + Q_2^{pv} + Q_3^{pv} + Q_4^{pv} + Q^0 + \frac{dV}{dt} = 0 \tag{6}$$

where Q_i^{pv}, (i = 1, 2, 3, 4) were the flow rates of each pulmonary vein, Q^0 was the flow rate at the MV section, and (dV/dt) was the LA volume variation. From Eq. (6), we defined Q_{tot}^{pv}, the total flux at the PVs

$$Q_{tot}^{pv} = Q_1^{pv} + Q_2^{pv} + Q_3^{pv} + Q_4^{pv} \tag{7}$$

To account for the number and position of the PVs, which vary significantly among patients [21] and can affect LA fluid dynamics, the total pulmonary vein flow Q_{tot}^{pv} was distributed among the four PVs based on a criterion related to their respective cross-sectional areas

$$Q_l^{pv} = \frac{A_l}{A_{tot}} Q_{tot}^{pv}; l = 1, 2, 3, 4 \tag{8}$$

where A_l represents the sectional area of each pulmonary vein, and A_{tot} is the sum of all the PVs' sectional areas. Thus, for each t ∈ (0,T), the flow rate for each pulmonary vein could be determined and applied to the computational model.

Eq. (8) was then modified by incorporating the flow resulting from the mesh displacement Q_l^w for each pulmonary vein throughout the cardiac cycle

$$Q_l^{pv} = \frac{A_l}{A_{tot}} Q_{tot}^{pv} - Q_l^w; l = 1, 2, 3, 4 \tag{9}$$

In the context of AF, the inflow BCs were adjusted by excluding the A-wave from the representative MV flow rate signal Q^0. Following this adjustment, each PV flow rate was recalculated according to the previously described method.

To reduce the occurrence of backflows, which can lead to numerical instabilities at the outflow boundary Γ_{out}, we applied a natural-type boundary condition [22] with backflow penalization ($\mathbf{g}_f = \mathbf{0}$):

$$-p_f\mathbf{n} + \mu\left(\nabla\mathbf{u}_f + (\nabla\mathbf{u}_f)^T\right)\cdot\mathbf{n} - \rho_f\left(\{\mathbf{u}_f\cdot\mathbf{n}\}_-\right)\mathbf{u}_f = \mathbf{0} \text{ on } \Gamma_{out} \tag{10}$$

where $\mathbf{n}$ was the outward directed unit vector normal to Γ_{out}, and $\{\mathbf{u}_f\cdot\mathbf{n}\}_-$ denoted the negative part of $\mathbf{u}_f\cdot\mathbf{n}$:

$$\{\mathbf{u}_f\cdot\mathbf{n}\}_- = \begin{cases} \mathbf{u}_f\cdot\mathbf{n} \text{ if } \mathbf{u}_f\cdot\mathbf{n} < 0, \\ 0 \text{ if } \mathbf{u}_f\cdot\mathbf{n} \geq 0 \end{cases} \tag{11}$$

The previously mentioned boundary condition was implemented to impose a slight penalty on the reverse flow that may arise from the backflows at Γ_{out}, which could potentially destabilize the discrete formulation of the problem. It was observed that the last term on the left-hand side of Eq. (11) became active only when $\mathbf{u}_f \cdot \mathbf{n} < \mathbf{0}$, on Γ_{out}; otherwise, when $\mathbf{u}_f \cdot \mathbf{n} \geq \mathbf{0}$, the outflow boundary condition simplified to the conventional stress-free condition. The LA mesh along with its corresponding $\hat{\mathbf{d}}_{wall}$ was supplied as inputs to the finite element library LifeX [23], which was utilized to simulate interatrial blood flow with the computational model.

2.5 Numerical simulations and preliminary results

The first application of the developed model was conducted on two AF patients, representing a preliminary step toward enhancing stroke risk stratification and therapy delivery.

For each patient, two numerical simulations were performed: one corresponding to SR and one corresponding to AF condition. The motion models utilized in SR and AF exhibit distinct characteristics: SR utilizes a patient-specific motion model derived from the previously described CT volumes, while AF is simulated by independently applying a random displacement to each node.

The simulations for the SR and AF conditions were conducted throughout seven cardiac cycles. To mitigate the effects of the non-physiological initial conditions on blood flow, the outcomes from the first three cycles were excluded from the analysis. Then, the results from the remaining cycles were phase averaged. An average heartbeat was derived in this way reflecting the inherent variability present in the solution as a result of the nonlinearity associated with the Navier-Stokes equations. Although steady states were not achieved, which is inherent to the nature of the problem, a phase-average solution was derived, contingent upon the inclusion of a sufficiently large number of heartbeats in the averaging procedure. Moreover, this kind of phase-averaging allowed a statistical study of the blood flow, in terms of cycle-to-cycle variations and assessment of turbulent effects (if any).

The simulated LA blood velocity for the first patient is displayed in **Figure 3**. Zero velocity was measured at MV and PVs, at the end of diastole (first row). During LV systole (second row), the velocity increased to 20 cm/s at the four PVs in SR due to atrial diastole In contrast, under AF, the velocities at the PVs predominantly remained negligible due to the random motion model. During the filling phase of LV (third row), the E-wave in the MV flow reaches a peak velocity of about 65/70 cm/s in both SR and AF simulations. The simulation results also confirmed a corresponding rise in PV velocity (mean value 40 cm/s). In atrial systole (fourth row), the expected A-wave was clearly observed at the MV with peak velocity ranging between 35 and 40 cm/s in the SR simulation, while it was absent in the AF scenario. Beyond peak and mean values, the model provides insights into the complex dynamics of LA blood flow (see **Figure 3**). Notably, the model captured intricate flow characteristics, including the presence of small vortices.

Blood stasis within the LAA may lead to the formation of blood clots [25]. To assess blood stasis, we introduced 500 point particles into the LAA and monitored the number that remained within the LAA after each cardiac cycle. Following three cycles, 26% of the particles persisted in the LAA in SR, while 45.6% remained in AF for patient 1. In the case of patient 2, 39% remained in SR and 50% in AF. These findings suggest that AF, particularly evident in patient 1, is associated with a diminished washout of the LAA, which could potentially facilitate thrombus formation over time.

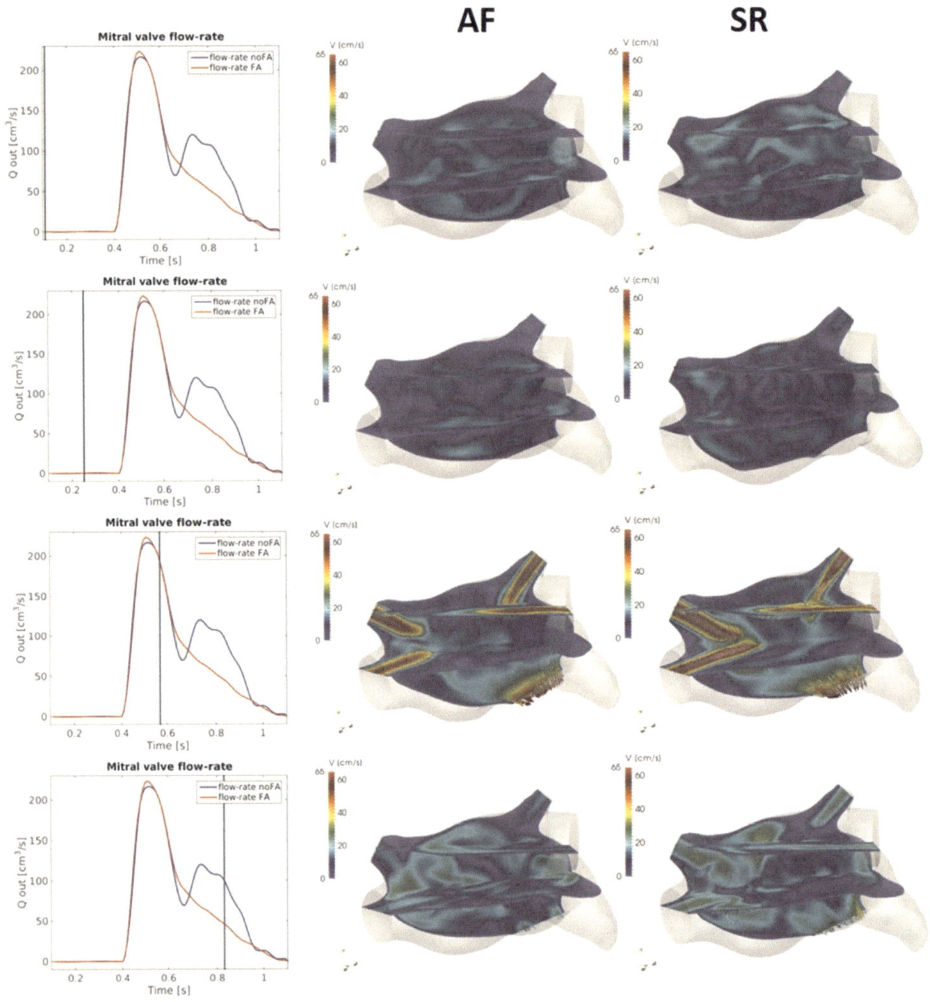

Figure 3.
Example of the computed LA blood flow in one patient. Flow rates at the MV in SR (blue) and AF (red) are reported in the first column. The black vertical line corresponds to a specific phase of the cardiac cycle. Flow velocity in AF (second column) and SR (third column) are shown at the end of diastole (first row), LV systole (second row), LV filling (E-wave) (third row), and atrial systole (A-wave) (fourth row). Source: Ref. [24]. CC-BY-NC. Used with permission.

In this preliminary analysis, realistic blood flow patterns for both SR and AF conditions were obtained. The proposed model confirmed that AF episodes contribute to a decreased LAA washout, thereby increasing the risk of thrombi formation. However, a significant drawback is that the model has not been evaluated against empirical flow data. Additionally, a comparative analysis with a healthy atrium is essential for a more comprehensive understanding of the hemodynamic implications of AF. Finally, further evaluations involving a larger cohort of patients with varied anatomical and hemodynamic profiles are warranted. To overcome some of the limitations encountered in this preliminary study, and better characterize LA contraction in AF, a regionalization approach is presented in the next sessions of the chapter.

3. LA contraction quantification for realistic BCs

The quantification of AF contractions tailored to individual patients may yield accurate BCs for CFD simulations in research contexts.

An automated method for segmenting the LA into distinct regions to facilitate a thorough 3D assessment of LA contraction has been proposed [26]. Currently, the assessment of this nature is exclusively accessible via two-dimensional echo imaging. In this context, we conducted an analysis of left atrial global and regional contraction in both control subjects and patients with AF, detailing the mechanical abnormalities linked to AF.

The complete surface regionalization of LA yielded five regions: anterior, posterior, lateral, septal, and inferior.

Parameters for evaluating the global LA function were calculated, encompassing the LA volume-time curve and the left atrial ejection fraction (EF).

To investigate regional alterations, parameters for region-specific contraction were established. Regional volumes were assessed across the cardiac cycle, alongside the calculation of regional ejection fraction (REF) for each designated area. REF was derived from the regional volumetric curves by analyzing the end-diastolic and end-systolic phases of the left atrial volumes. Additionally, for each region, the standard deviation (SD) of the time intervals at which regional volumes attained their maximum value was calculated, serving as a measure of left atrial asynchrony (IA) [27]. Strain analysis was computed along the longitudinal, radial, and circumferential directions. Longitudinal strain (LS) was determined by computing the changes in the length of the long axis with respect to the end-diastolic ventricular frame across the cardiac cycle:

$$\mathbf{LS(x)} = \frac{\mathbf{LA_{len}(x)} - \mathbf{LA_{len\_ED}}}{\mathbf{LA_{len\_ED}}} \mathbf{\cdot 100} \tag{12}$$

where $LA_{len(x)}$ represents the length of long axis at frame x, LA_{len_ED} denotes the length of long axis at the end-diastolic ventricular frame.

To assess radial strain (RS), the regional radial dimension (RRD) was determined for each region by averaging the Euclidean distances from each vertex within that region to the long axis. Subsequently, the regional radial strain (RRS) was computed as follows:

$$\mathbf{RRS(x)} = \frac{\mathbf{RRD(x)} - \mathbf{RRD_{ED}}}{\mathbf{RRD_{ED}}} \mathbf{\cdot 100} \tag{13}$$

Six circumferential evenly spaced curves were analyzed, positioned between the midpoint of the long axis and the MV barycenter to compute the circumferential strain (CS). For each curve, the percentage strain was calculated [28] utilizing the following formula:

$$\mathbf{CS(x)} = \frac{\mathbf{L(x)} - \mathbf{L_{ED}}}{\mathbf{L_{ED}}} \mathbf{\cdot 100} \tag{14}$$

where L(x) represents the length of the curve at frame x and L_{ED} denotes the length of the curve at end-diastolic ventricular frame. Then CS was determined by averaging the strain values from the six curves.

Following the computation of global strains, the contraction of each specific region was evaluated through regional wall motion (RWM), which was derived from the difference between the current regional radial displacement (RRD) and the RRD measured at the end of diastole for each respective region:

$$\mathbf{RMW}(\mathbf{x}) = \mathbf{RRD}(\mathbf{x}) - \mathbf{RRD_{ED}} \tag{15}$$

In the AF group, LA exhibited larger volumes in comparison to CTRL group. Despite this dilation, no significant differences in regional contraction timing were observed between the two groups, as indicated by the IA analysis.

Both LA EF and REFs, assessed across the entire LA and within individual regions, were significantly lower in the AF group relative to the CTRL group.

Furthermore, global and regional strain values were found to be greater in CTRL group than in AF group, reflecting the anticipated normal functioning of the LA throughout the cardiac cycle in the CTRL cohort. Notably, variations in global LA strain were more pronounced in the circumferential and radial orientations ($p < 0.005$), while regional strain disparities were particularly significant in the radial strain (RS) at the inferior, lateral, and septal regions ($p < 0.005$).

A similar pattern was observed concerning the mean peak RWM, which exhibited elevated values in the CTRL group relative to the AF group across all regions, except for the posterior region. The most significant disparity was noted in the septal region ($p < 0.005$). In the CTRL group, peak RWM was synchronized across all regions, except for the posterior region; conversely, in patients with AF, a notable lack of synchrony in contraction was apparent throughout all regions.

In this preliminary study, the contraction parameters were made available also in controls and by 3D CT imaging. Several contraction parameters, alongside strains, have been suggested for patients with AF. To the best of our knowledge, this chapter represents the first attempt to quantitatively characterize the alterations in LA contraction associated with a history of AF, including an analysis at the regional level.

The limitation arises from the fact that CT imaging data were obtained in an SR for all patients, which precludes the ability to establish a definitive cause-and-effect relationship between alterations in LA function and the onset of AF. However, a notable association has been reported.

The findings indicate that: (1) the overall dysfunction of LA contraction linked to AF is notably more significant in both the radial and circumferential directions; (2) the marked regional decline in radial strain observed in the septal, inferior, and lateral areas suggests a greater decrease in mechanical efficiency in these regions relative to the posterior and anterior areas.

4. A digital twin approach for stroke risk assessment in AF patient

AF has been previously noted to present a fivefold increased risk of cerebrovascular events, contributing to 15–18% of all stroke occurrences. In clinical practice, the prevention of strokes is guided by the CHA_2DS_2-VASc score, which is founded on general clinical risk factors; however, it does not provide sufficient predictive accuracy at the patient-specific level.

The association between the AF pattern and the stroke risk, independent of CHA_2DS_2-VASc score, is currently a topic of considerable discussion. Investigating the correlation between AF pattern and clinical outcomes is challenging due to the distinct patient characteristics associated with paroxysmal AF, which typically include a younger demographic and a reduced incidence of structural heart disease and other comorbid conditions (HF, chronic kidney disease, chronic obstructive pulmonary disease, peripheral vascular disease). Consequently, this group tends to exhibit lower estimated risks for both thromboembolic events and bleeding [29].

Therefore, a patient-specific assessment of stroke risk on a patient-specific basis would have a great impact on clinical practice.

The analysis of blood flow patterns within the LA and LAA derived using CFD simulations has been proposed as a new approach to mechanistic determinants of atrial thrombogenesis at an individual level. Previous studies have demonstrated the ability to replicate realistic 3D blood flow patterns in the LA by employing patient-specific dynamic anatomical models, displacement fields derived from registered imaging data, and boundary conditions determined through Doppler measurements at the PVs and the MV. In this context, we propose a DT model of the LA, coupled with CFD simulations, to improve the assessment of stroke risk for individual patients.

4.1 Patient data and data analysis

Thirty non-consecutive patients were analyzed in this study. According to the clinical evidence of AF, patients were subdivided into three groups: paroxysmal AF (PAR group, 10 subj.), persistent AF (PER group, 10 subj.), and controls (CTRL group, 10 subj.) who had to be with no or minimal heart disease and without any evidence or suspect of AF or other atrial arrhythmia.

Contrast Enhanced Computed Tomography (CECT) data were acquired in SR. A total of 10 volumes (170 axial slices, 0.4 mm pixel size, 1 mm slice thickness) were reconstructed to encompass a complete cardiac cycle starting from the end of ventricular diastole. These reconstructions were utilized for subsequent analysis employing retrospective ECG gating. Additionally, Doppler measurements were obtained at the V and PVs.

The data processing workflow to extract the LA anatomical model, obtain the patient-specific displacement model, and mesh propagation, follow the same steps explained in Section 2 of this chapter. To run fluid dynamic simulations in SR conditions, the dynamic model was provided as input to the customized CFD model. CFD simulations were executed utilizing realistic boundary conditions that considered the MV flow rate and the dimensions of each PV, as previously described.

The dynamic viscosity was established at 0.035 poise, the density at 1.06 g/cm^3, and the time step was set to 0.001 seconds. Simulations were performed using the LifeX package. To remove non-physiological initial conditions on fluid velocity and lessen the impact of the initial conditions on output, it was necessary to simulate an arbitrary number of cardiac cycles. Five cardiac cycles were run via each simulation.

The blood velocity field was analyzed for both LA and LAA across all models. Furthermore, time-average wall shear stress (TAWSS), oscillatory shear index (OSI), relative residence time (RRT), and endothelial cell activation potential (ECAP) indexes were evaluated. These metrics can be utilized to assess endothelial shear stress, the development of new tissues and plaques, and the promotion of neointimal hyperplasia [30].

First of all, the viscous stress vector $\vec{\tau}$ exerted by the wall on the fluid was defined as:

$$\vec{\tau} = \mu \left[\nabla \vec{u} + \left(\nabla \vec{u} \right)^T \right] \bullet \vec{n} \tag{16}$$

where μ represents the viscosity of the blood, $\nabla\vec{u}$ denotes the velocity gradient, and $\vec{n}$ is the unit normal vector extending from the fluid to the wall.

The transition from the overall viscous stress vector to the wall shear stress (WSS) vector encompasses both the magnitude and direction:

$$\overrightarrow{WSS} = \vec{\tau} - \left(\vec{\tau} \bullet \vec{n}\right) \vec{n} \tag{17}$$

The force exerted by the wall on the fluid per unit area is represented by the WSS, oriented along the local tangent plane. Because of the periodic nature of the heartbeat, it is possible to calculate an average measure known as the TAWSS, which is defined as follows:

$$TAWSS = \frac{1}{T}\int_0^T \left\|\overrightarrow{WSS}\right\|_2 dt \tag{18}$$

where T represents the overall length of the simulated cardiac cycles and $\|\bullet\|_2$ is the Euclidean norm of a vector.

Now let us consider the OSI, defined as follows:

$$OSI = \frac{1}{2}\left(1 - \frac{\left\|\int_0^T \overrightarrow{WSS}\, dt\right\|_2}{\int_0^T \left\|\overrightarrow{WSS}\right\|_2 dt}\right) \tag{19}$$

This dimensionless metric yields values between 0 and 0.5, with elevated readings observed in regions where the WSS exhibits considerable variation throughout the cardiac cycle. Notably, it is the directional changes in flow that predominantly influence the attainment of high OSI values.

Another index is the RRT, defined as follows:

$$RRT = \frac{1}{(1 - 2OSI) * TAWSS} \tag{20}$$

This indicator highlights areas along the wall where shear stress is low and the flow displays oscillatory behavior. As a result, the duration of blood residence near the wall is prolonged due to this particular hemodynamic condition. This circumstance may permit the deposition of the particles and result in an inflammatory reaction in the endothelial cells.

To identify areas of the vessel wall that are subjected to elevated OSI and diminished TAWSS, Di Achille et al. [31] proposed a new index that employs the ratio of these two parameters to evaluate the "thrombogenic susceptibility" of the vessel wall. This index is referred to as ECAP:

$$ECAP = \frac{OSI}{TAWSS} \tag{21}$$

Elevated ECAP index values are indicative of conditions characterized by high OSI and low TAWSS, which suggest a heightened susceptibility of endothelium. This metric is commonly employed in the assessment of thrombogenic risk associated with the LAA [32].

Furthermore, to evaluate blood stasis, the residence time T_r was characterized as the duration that blood particles spend within the LA chamber. This parameter is represented by the subsequent equation:

$$\frac{\partial T_r}{\partial t} + \vec{v} * \nabla T_r = 1 \tag{22}$$

Rather than solving the equation, the simulation-generated 3D flow velocity field was employed to monitor the Lagrangian flow tracers, allowing for the estimation of the time taken along their path. Then, the spatio-temporal evolution of blood residence time inside the LAA was evaluated. We observed that, despite possible numerical errors, solving the equation is identical to this approach [33].

These parameters were calculated for all subjects involved in the study and the comparison between the groups was conducted.

4.2 Results and discussion

In all study subjects, data processing and simulations were feasible. In the LA, control subjects showed higher average velocity compared to AF subjects (0.5 ± 0.12, 0.2 ± 0.05, 0.07 ± 0.01 m/s for controls, PAR, and PER respectively).

In the LAA, the lowest average velocity was observed in a PER subject at 0.02 m/s, while the highest mean velocity was observed in a CTRL subject at 0.18 m/s. Overall, the CTRL group demonstrated a significantly higher mean velocity of 0.11 ± 0.03 m/s compared to PAR (0.05 ± 0.02 m/s) and PER (0.04 ± 0.02 m/s) groups ($p < 0.05$).

Similar results were obtained at the LAA ostium, where the CTRL group exhibited a higher mean velocity of 0.28 ± 0.05 m/s compared to PAR (0.14 ± 0.03 m/s) and PER (0.11 ± 0.04 m/s) groups ($p < 0.05$). In summary, CTRL subjects demonstrated a more effective washout effect due to the elevated blood velocities.

The next step was the evaluation of CFD parameters for the LAA. These are shown in **Figure 4** as a representative subject for each group. The TAWSS map illustrates an average value that depends on the velocity gradient, as it is determined by integrating the WSS throughout the complete cardiac cycle.

Our analysis indicated that CTRL subjects displayed higher TAWSS values at the ostium of the LAA compared to those with AF. Furthermore, the TAWSS was evaluated throughout the entire LAA region. Overall, the velocity gradient was greater in CTRL subjects.

Also, higher OSI values both at the ostium and in the entire appendage in the AF subjects concerning the CTRL subjects were noticed. One probable explanation for this fluctuation is the characteristics of its movement, which first enables an emptying phase where blood flows out, followed by a filling phase that reverses the flow direction within a single heartbeat. The alteration in the direction of blood flow led to elevated values of OSI, with pronounced variations observed in the AF subject, particularly at the LAA tip. Nevertheless, this parameter did not exhibit statistically significant differences across the groups ($p = 0.2$).

Higher values of RRT were found in the PAR and PER groups as compared to the CTRL group ($p < 0.05$). This means that there is a combination of higher oscillations and lower wall shear stress in the AF subjects as compared to the CTRL subjects, this results in an extended duration of blood flow near the vessel wall, facilitating the deposition of particles and possibly triggering an inflammatory response in the endothelial cells.

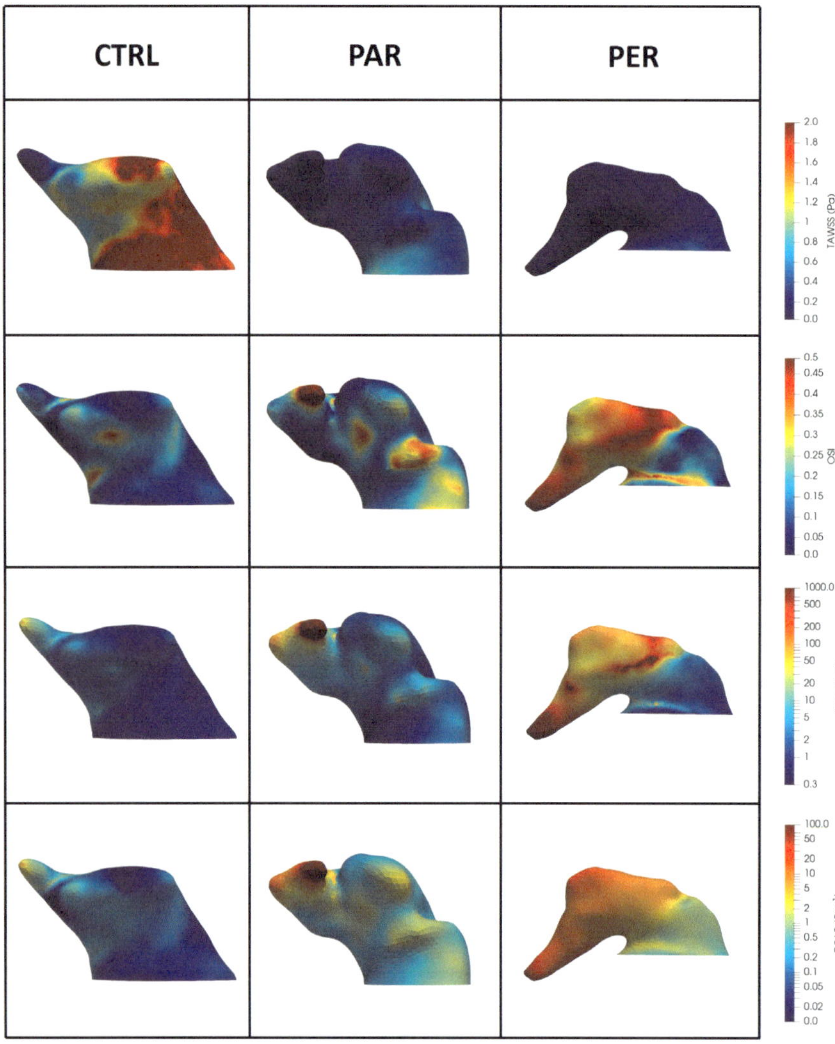

Figure 4.
LAA analysis in one representative subject for each group: TAWSS (1st row), OSI (2nd row), RRT (3rd row), ECAP (4th row). Source: Ref. [34]. CC-BY-NC 4.0. Used with permission.

Finally, the ECAP maps showed higher values in the PAR and PER groups as compared to the CTRL group ($p < 0.05$), suggesting the presence of potentially thrombogenic areas in AF subjects.

The residence time T_r was then evaluated considering the velocity field computed from the CFD simulations. In general, the lowest T_r values were observed around the PV inlets, where blood enters the chamber, and the velocity reaches moderate values (70–80 cm/s). According to Durán et al. [35], secondary flow patterns (parallel to the MV ostium) are not as restricted by global chamber mass conservation as main flow patterns (perpendicular to the MV ostium), allowing them to vary greatly with anatomical characteristics, boundary conditions, and other factors. Specifically, modifications to the secondary flow patterns close to the LAA ostium may have an impact on the flow within the LAA and, consequently, the LAA residence time, while the washout process of the LA body mainly depends on the main flow patterns.

The suggested methodology facilitated the development of individualized models for each patient utilizing CECT data. This process is readily adaptable to accommodate different data inputs. For instance, it can be applied to 3D from various imaging techniques, such as MRI or real-time 3D echocardiography, as long as the segmentation phase is adapted to meet the particular requirements of the data.

The use of B-spline registration assisted in generating an accurate representation of the left atrium's contraction behavior. Even though it was used in a different condition than AF, previous research has shown that including patient-specific information in the model leads to more accurate simulation results [10]. Enhanced accuracy and reliability of our CFD results are achieved by incorporating realistic boundary conditions at the MV and PVs in our model.

The primary conclusion of this study is that, on average, WSS was significantly lower, while OSI, ECAP, and RRT were markedly higher in subjects with AF compared to the control group within the LAA. Statistically significant differences were observed for WSS, ECAP, and RRT. These findings indicate that the LAA in AF patients is more prone to slower and more variable blood flow compared to controls.

In vitro investigations have shown that exposure to low and oscillating flow can activate endothelial cells, leading to a proinflammatory state [36]. Consequently, the interplay of low WSS and high OSI flow may activate the endothelial cell layer of the LAA wall, promoting the recruitment of local platelets and leukocytes [37]. This activation then triggers the coagulation cascade near the endothelial cells through mechanisms associated with venous thrombus formation [38], ultimately leading to the development of a local clot that could potentially result in an ischemic event.

The computation of these parameters is inherently complex and necessitates a comprehensive understanding of their physical significance. Furthermore, it is important to note that not all CFD solvers automatically provide these parameters during the simulation of hemodynamics. Therefore, comparing the values of CFD parameters between different studies and subjects is challenging.

Having available all these indices, assessment of stroke risk could be ideally improved compared to the mere use of CHA_2DS_2-VASc score on a quantitative and patient-specific basis, enabling personalized blood clot prevention treatment for individual patients and facilitating physicians to schedule examinations based on patients' likelihood of specific outcomes.

5. The role of LAA on stroke risk assessment and thrombogenicity

LAA represents the vestige of the developing left atrium, which arises from the primordial pulmonary vein and its associated branches. Due to its distinctive hooked shape, the LAA is identified as the region within the LA that is most susceptible to blood stasis, thereby elevating the likelihood of thrombus development and subsequent stroke. Notably, approximately 90% of intracardiac thrombi found in patients with cardioembolic stroke/transient ischemic attack are believed to originate from the LAA.

Until recently, the sole available option for anticoagulation therapy was oral medication.

However, it increases the risk of bleeding and interferes with other medications and multiple organ functions, and its risk may outweigh the otherwise notable benefits in thromboembolic events [39]. For these reasons, other approaches have been developed in recent years, including percutaneous LAA closure, which seems to be more effective in reducing thromboembolic risk than warfarin anticoagulation therapy [40]. However, LAA occlusion (LAAO) is now confined to a limited number of patient subgroups, as the risks and costs associated with the procedure may exceed the benefits of its preventive anti-embolic effects [41].

Percutaneous closure of the LAA is a complex intervention that necessitates a significant degree of skill and specialized training to ensure the procedure is carried out effectively and without complications.

Furthermore, the anatomical variability of the LAA is crucial in selecting the appropriate occlusion device and ensuring its accurate placement relative to the LAA ostium during the implantation process.

In recent years, there has been a growing interest in the morphological and quantitative features of the LAA. Di Biase et al. [42] introduced a classification system for LAA morphology based on its morphological complexity and examined the correlation between this feature and patient-specific stroke history.

AF determines changes in the LAA wall and its adjacent structures, which may elucidate the observed increase in LAA volumes among patients who have experienced a stroke. However, despite these attempts, the relationship between the above anatomical characteristics of the LAA and stroke risk appears to be inconclusive, and conflicting results have been published.

In this scenario, hemodynamic information about the LA and especially the LAA is crucial. CFD can significantly enhance silico simulations of blood flow patterns following occlusion, tailored to individual patients. This approach allows for the evaluation of flow stagnation, thereby offering valuable biophysical insights within a complex fluid dynamics framework.

This section aims to present two applications of the CFD model in the LAA domain: the primary objective is elucidating novel findings regarding the impact of AF on LAA hemodynamics, thereby enhancing the comprehension of the associated risk for thrombus development, and the second aim is to simulate the effects of LAA occlusion on fluid dynamics in AF patients to predict LAAO-induced hemodynamic changes concerning thrombosis risk.

5.1 The impact of LAA morphology on stroke risk assessment in AF

Five 3D LA anatomical models (including LAA), obtained from real clinical data (CT and MRI), were selected, all of which differed in terms of LAA. In all cases, the LAA was separated from the LA body. Several geometric parameters were calculated for each LAA. A framework was then created to link the different LAAs of four patients to an LA chamber model that was selected among the five patients. These anatomical models provided the computational domain for the previously described CFD model.

Simulations of hemodynamics within the LA and LAA under AF conditions were performed. The relationship between geometric and fluid dynamic indices was then calculated to evaluate the correlation between blood stasis risk and thrombus formation for each LAA [43].

5.2 Patient data and LAA extraction

The dataset consisted of five 3D anatomical models of the LA. One of the models was derived from CT data and the other four models were extracted from MRI data, all coming from patients with AF.

The data underwent processing through custom-developed image segmentation algorithms, as detailed in Section 2.1. Surface meshes were created from the 3D binary masks of the LA utilizing the iso2mesh toolbox in MATLAB.

The subsequent phase concentrated on the implementation of an algorithm designed to autonomously identify and separate the LAA from the LA body in the anatomical models. For this objective, the shape diameter function (SDF) proposed by Shapira et al. [44] was employed. This approach facilitated the semantic clustering of our 3D models in the form of a label vector for each node of the mesh. By determining the mode (statistics) of the label vector, we were able to identify the atrial chamber, the PVs, and the LAA that was detached from the LA surface mesh.

Applying this approach, five LAA surface meshes were obtained and their specific shape was geometrically characterized (see Section 5.1.3).

5.3 LAA alignment and definition of the new models

The alignment of the LA template with the LAA meshes was achieved by the identification of the vertices associated with the LAA ostia and the LA ostium template. The local correspondence of vertices between the LAA and LA template ostia was attained through the application of the iterative closest point (ICP) algorithm [45], which corresponded to the closest points between two point datasets: one dataset designed as a fixed dataset and the other is used as a floating. In this instance, the fixed dataset comprised the vertices of the LA ostium, whereas the floating dataset included the vertices of the LAA ostium. After this step, a collection of five LA anatomical models was created, each depicting the same atrial chamber but distinct left atrial appendage geometries. The models underwent Laplacian smoothing filtering and Poisson surface reconstruction techniques to refine the anatomical models.

5.4 LAA geometrical and fluid dynamic parameters

The morphology and geometric characteristics of the LAAs exhibited variability and to assess the variations among the LAAs and their impact on the hemodynamic properties, a detailed analysis was conducted, that included measurements of LAA volume, LAA surface area, length, tortuosity, and LAA opening. Length was assessed by calculating LA centerlines. Tortuosity was defined as [46]:

$$\mathcal{X} = \frac{L}{D} - 1; \tag{23}$$

where L is the length of the center line and D is the Euclidean distance between the endpoints of the center line. Tortuosity values well above 0 indicate a high degree of complexity of the LAA shapes, while values close to 0 suggest a more ordered LAA geometry. To calculate LAA volumes and orifice areas (i.e., LAA ostium), LAA volume meshes were created.

The CFD model and the BCs employed for simulating the hemodynamics of each anatomical model were those previously detailed. For each LA anatomical model, a simulation under AF condition was run for five cardiac cycles to mitigate the occurrence of unphysiological fluid velocity conditions. The findings presented pertain to the final cardiac cycle of the simulation.

The key parameters, which describe LAA fluid dynamics, including velocity and vortex structures, LAA opening velocity, and mainly residence time, were also calculated. The Q criterion was used to calculate the vortex structures. It is defined as follows:

$$Q = \frac{1}{2}\left(\mathbf{W}_{ij}\mathbf{W}_{ij} - \mathbf{S}_{ij}\mathbf{S}_{ij}\right); \qquad (24)$$

being $\mathbf{S}_{ij} = \left(\frac{\partial \mathbf{u}_{fi}}{\partial \mathbf{x}_j} + \frac{\partial \mathbf{u}_{fj}}{\partial \mathbf{x}_i}\right)$ and $\mathbf{W}_{ij} = \left(\frac{\partial \mathbf{u}_{fi}}{\partial \mathbf{x}_j} - \frac{\partial \mathbf{u}_{fj}}{\partial \mathbf{x}_i}\right)$ the symmetric and antisymmetric parts of the velocity gradient tensor $\frac{\partial \mathbf{u}_{fi}}{\partial \mathbf{x}_j}$. The analysis revealed and illustrated the interconnected areas where $Q > 0$ and the pressure is below the ambient level at the vortex core.

5.5 Results

5.5.1 LAA geometric descriptors

Table 1 displays the geometric parameters computed for each LAA. Significant variability in the LAA characteristics, with volume, length, and tortuosity values varying between 2.04 and 2.60 cm^3, was found. The LAA with the simplest geometry and linear morphology had the lowest tortuosity values, indicating high variability in LAA geometric characteristics and their impact on appendage fluid dynamics.

5.5.2 Velocity and vortex structure analysis

LAA1 had a low entering velocity during the ventricular systole (7–9 cm/s). After the end of this phase, the blood flow exited from the LAA with very low velocity (2–3 cm/s) and these values could be associated with the morphology of the LAA1. The morphology of the LAA2 exhibited distinct characteristics when compared to the LAA1, and these variations were evident in the velocity analysis. LAA2 had a higher velocity (15–20 cm/s) and a greater quantity of blood flow within it. LAA2, having a greater volume than LAA1 while exhibiting reduced complexity and length,

	LAA_1	LAA_2	LAA_3	LAA_4	LAA_5
V (cm^3)	2.04	2.57	2.60	2.21	2.12
A_s (cm^2)	8.94	9.25	10.10	9.11	8.31
A_o (cm^2)	1.86	2.13	1.87	1.62	1.15
P_o (cm)	4.98	5.65	5.06	4.90	3.88
L (cm)	3.54	2.26	2.80	2.52	2.36
χ	0.46	0.05	0.08	0.03	0.26

Table 1.
LAA geometrical parameters: V, volume; A_s, surface area; A_o, orifice area; P_o, orifice perimeter; L, length; χ, tortuosity.

accommodated a significant volume of blood flow that extended to its distal region. The velocity profile of LAA3 was also observed, with changes in velocity profile related to different geometric parameters: LAA3 had a higher length but lower than LAA1 exhibiting a tortuosity that was an order of magnitude less than LAA1 and twice that of LAA2. Therefore, blood flow entered and exited from this appendage with intermediate values between LAA1 and LAA2 (6–7 cm/s). LAA4 had the most "linear" geometry, with the smallest tortuosity compared to other auricles. However, blood flow did not reach the distal parts of this LAA. LAA4 had a high orifice perimeter and a smaller volume, allowing blood to enter but not infiltrate the more constricted part of the auricle and it was compelled to exit at elevated velocities (10–12 cm/s). LAA5 velocity values were much lower than the other appendages and did not exceed 7 cm/s. The low values of A_o, V, A_s, and P_o, coupled with high tortuosity, likely explain the difficulty of blood flow entering this LAA due to its reduced orifice length.

The presence of vortex structures in the LAA may indicate better blood flow washout, preventing blood stasis. The vorticity analysis confirmed that vortices were mostly localized near the LAA orifice, with the highest velocity observed in LAA1. After the MV opening, vortices had low velocities and did not reach the distal part of the LAA, indicating scarce blood washout in AF. LAA2 and LAA3 had higher velocity values and reached all anatomical parts of the LAA. LAA4 and LAA5 had low velocities throughout the cardiac cycle, with vortex structures being infrequent and exhibiting minimal velocities.

These results underline that LAA geometric characteristics impact the hemodynamic pattern within the LAA, highlighting the need to consider not only appendage morphology type for stroke risk assessment in patients with AF since complex LAA morphologies can have high thrombogenic risk, while simple morphologies can have equal or higher risk.

6. LAAO simulations applying different endocardial devices

LAAO is a transcatheter procedure aimed at reducing the risk of stroke in AF patients by mechanically excluding the LAA from the bloodstream. Using CFD, it is also possible to model the fluid dynamics effects associated with LAA occlusion in AF patients, thereby enabling the prediction of hemodynamic alterations resulting from the device's implantation [47]. LAAO was modeled with two distinct categories of occluders that are presently offered in the market, which are based on the principles of plugs and pacifiers [48]. 3D anatomical models of the LA were created using real clinical data collected from five patients with AF. CFD simulations were carried out on the LA model both before and after the LAAO procedure, with all simulations conducted under AF conditions. Various fluid dynamics indices, such as blood velocity, particle washout, endothelial damage, and device-related thrombosis, were calculated to measure changes in flow pattern following occlusion concerning the risk of thrombosis.

6.1 Patient data and creation of LA models

The simulations were conducted within a computational domain comprising 3D anatomical models of the LA extracted from dynamic CT images acquired in five AF patients using the same workflow described in Section 2. To generate the 3D models using LAAO, the LAA has been removed from the entire anatomical models that had been acquired earlier. This process involved the application of the SDF, as detailed in Section 5.1.1. The anatomical location of the LAA was exploited to detect and remove the LAA. If the incision was too distant from the pulmonary ridge, a manual adjustment of the LAA incision was performed.

After the removal of the LAA, the occlusion was simulated by generating a surface at the location of the opening. The final surfaces were created using MeshLab, resulting in tetrahedral meshes consisting of 15 to 17 million finite elements. Key geometric features of the devices and their compatibility with the shape of the patient-specific LAA ostium were considered during this process. It was assumed that the simulated LAAO procedure was performed correctly and thus a complete occlusion was simulated without peri-device blood leaks and involving the pulmonary ridge.

The occluders designed with a pacifier mechanism feature a flap along with an extra disc that effectively seals the LAA ostium from the left side of the LA, as they are designed with a slightly concave shape. In contrast, the Poisson surface reconstruction algorithm was utilized to accurately model the occluders based on the plug principle, which is defined by a convex shape.

In summary, a final set of three anatomical LA models was generated for each patient: (1) the initial model with the LAA; (2) the LAAO model utilizing the pacifier principle; (3) the LAAO model employing the plug principle.

6.2 Fluid dynamics parameter computation

Blood velocity was directly derived from the CFD simulations. To assess particle washout, the central region of the LA chamber was initially filled with 20,000 massless particles at the start of the last cardiac cycle simulated. The number of particles still present in the LA at the conclusion of the cycle was then counted.

ECAP values were normalized by the mean TAWSS value calculated inside a VOI around the LAA opening, which was determined from the pre-occlusion simulation.

As already seen, regions prone to thrombogenesis exhibit elevated ECAP values, which signify significant oscillatory shear flows coupled with reduced wall shear stresses.

Additionally, a localized analysis was carried out considering the identical volume utilized for ECAP normalization, which involved 2000 massless particles. The integrated evaluation of hemodynamic indices such as blood flow velocity, particle washout, and ECAP can more easily quantify the thrombogenic risk in AF patients [49] and enable the assessment of device-related thrombus (DRT) within the designated volume close to the occlusion [50].

To conduct an initial evaluation of potential leaks peripheral to the device, a partial dislocation of the device following the closure procedure was simulated. The dislocation led to the identification of a leak measuring 14.2 mm along the long axis and 5.05 mm along the short axis.

6.3 Hemodynamic analysis

The alterations in blood flow velocity, particle washout, and ECAP resulting from LAA occlusion exhibited comparable trends across all patients.

At the beginning of ventricular diastole, the model with the "pacifier" device showed higher velocities at the PVs and MV, and a more organized flow toward the MV compared to the one with "plug". During atrial systole, the LAA simulation showed lower velocities and a high risk of blood stagnation in the LAA. The LAAO simulations with the "pacifier" device showed a more organized flow compared to the "plug" device, where higher rooftop velocities were noted. At the end of atrial systole, lower velocities were observed in correspondence with the MV in the simulation of the pre-occlusion LA. In the LAAO simulations, we observed a higher velocity at the PVs and the MV to the pre-LAAO case, with higher values using the "pacifier" device.

Table 2 displays the average blood velocity readings for each patient in the area close to the surface of the implanted LAAO device. Overall, the LAAO simulations resulted in higher speeds with the "pacifier" device.

Residual particle count analysis was conducted for all patients, showing that simulations involving the LAAO devices led to improved washout of the atrial chambers, as indicated by residual particle counts: 392 $\pm$ 79 with LAA compared to 254 $\pm$ 102 with "pacifier" device and 294 $\pm$ 95 with "plug" device. These results are also consistent with studies in the literature [51] where LA fluid dynamics were examined before and after LAAO and the authors found higher velocities near the device and the MV. These findings were verified for the local analysis, wherein 26 $\pm$ 5 particles were counted during occlusion using the "plug" device, while only 20 $\pm$ 5 particles remained in the chosen volume when occlusion was carried out with the "pacifier" device.

When compared to the "plug" configuration, the "pacifier" configuration demonstrated a marginally higher degree of atrial washout in all patients, as evidenced by the consistently lowest amount of residual fluid particles.

In all the patients, the distal region of the LAA exhibited significantly reduced or negligible blood flow velocity. The ECAP map adjacent to the device delineates areas where the ECAP exceeds 0.5 Pa^{-1}, indicating the presence of thrombogenic regions.

A similar investigation was conducted on the post-occlusion models using the two devices. In every patient, the spatial distribution of ECAP revealed that regions of elevated values were reduced in size when contrasted with the pre-occlusion simulations.

Consistent with previous studies [26, 27, 40], the findings presented indicate that velocities below 0.2 m/s and ECAP values above 0.5 Pa−1 are both significant risk factors for predicting DRT. The simulations suggest that occlusion mitigates this risk by enhancing blood velocity and decreasing ECAP peak.

When comparing the "pacifier" and "plug" devices, the former appears to be more effective at reducing the risk of DRT than the latter. Higher ECAP values are

Average velocities in a volume near the implanted device (m/s)			
	Model with the LAA	**LAAO with the "pacifier" device**	**LAAO with the "plug" device**
P1	0.17	0.22	0.19
P2	0.22	0.23	0.25
P3	0.18	0.42	0.34
P4	0.19	0.23	0.20
P5	0.15	0.17	0.14

Source: Author's original work.

Table 2.
Mean velocities throughout the cardiac cycle, within a volume surrounding the implanted device in the simulations performed in the five patients enrolled in the study.

measured at the edge of the plug device, suggesting that there may be blood clots between the LA wall and the curvature of the plug.

The simulation of the peri-device leak demonstrated that the existence of the leak facilitated the entry of blood into the LAA, where it subsequently became stagnant. In this configuration, the ECAP map reveals extensive, high-quality areas of damage located in the distal segment of the LAA. Comparing the results from the peri-device leak simulation with those from the pre-occlusion simulation indicates a heightened risk of DRT. This increase in risk is attributed to a decrease in maximum velocity (0.26 m/s during pre-occlusion compared to 0.15 m/s during the leak simulation) and a rise in ECAP value which worsens (2.23 Pa−1 before closure vs. 3.84 Pa−1 leak simulation). The analysis of particle washout in conjunction with the dislocation revealed a total of 540 residual particles, which accounts for 2.7%.

7. Conclusion

All the CFD model applications presented so far emphasize the potential of CFD simulation in deepening our understanding of the fluid dynamic mechanisms underlying the onset and maintenance of AF, to contribute to progress toward stroke risk stratification and the delivery of optimized therapies on a patient-specific basis.

The preliminary results suggested that there are clear differences between patients diagnosed with AF and individuals with no or minimal heart disease and without any evidence of AF. Using CFD simulations, it is also possible to evaluate quantitatively the contribution of the LAA for stroke risk assessment as well as the efficacy of several devices nowadays used in clinical practice for LAA occlusion. These results suggest that applying the CFD model has the potential to create a DT model for assessing the risk of stroke in AF patients.

To attain this objective, it is essential to address various challenges associated with both imaging and modeling. Cardiac imaging serves as a crucial technique for assessing the risk of thrombus formation in AF patients. Transesophageal echocardiography (TEE) is recognized as the definitive imaging modality for the detection of LAA thrombus, demonstrating a sensitivity and specificity ranging from 95 to 100% [52]. However, the analysis of blood clotting within vessels and the assessment of coagulative status cannot be conducted. By incorporating more metrics like the residence time in a specific area, vorticity, and kinetic energy, we can mitigate certain limitations.

Research has indicated that a blood hypercoagulability index can be established through a simplified framework of reaction-diffusion-convection equations that elucidate the dynamics of thrombus formation. This model primarily examines critical proteins, including thrombin, fibrinogen, and fibrin [53], which play vital roles in the coagulation process. Incorporating these additional indicators into our methodology would significantly enhance our comprehension of the influence each factor has on the risk of stroke. More indices will help explain the phenomenon of clot formation from different perspectives. An additional critical element not to neglect concerns the comprehensive validation of our methodology, which necessitates the use of 4D flow magnetic resonance imaging acquisition [54]. This technique is infrequently employed in clinical settings, primarily due to the specialized expertise required for its proper implementation.

This means that by weighting each component of the model and parameter computable from the CFD simulations, a new clinical score could be developed to measure stroke risk on a patient-specific basis.

Author details

Camilla Cortesi[1,2*], Matteo Falanga[1], Sachal Hussain[1], Nadia D'Alessandro[3], Corrado Tomasi[4], Alessandro Masci[1] and Cristiana Corsi[1]

1 DEI, University of Bologna, Cesena Campus, Italy

2 Precision Medicine Department, University of Palermo, Italy

3 Sant'Anna School of Advanced Studies, The BioRobotics Institute, Pisa, Italy

4 Santa Maria delle Croci Hospital, AUSL della Romagna, Ravenna, Italy

*Address all correspondence to: camilla.cortesi3@unibo.it

References

[1] Hoit BD. Left atrial size and function. Journal of the American College of Cardiology. 2014;**63**(6):493-505

[2] Al-Saady NM, Obel OA, Camm AJ. Left atrial appendage: Structure, function, and role in thromboembolism. Heart. 1999;**82**(5):547-554

[3] Lippi G, Sanchis-Gomar F, Cervellin G. Global epidemiology of atrial fibrillation: An increasing epidemic and public health challenge. International Journal of Stroke. 2021;**16**(2):217-221

[4] Murphy NF, Simpson CR, Jhund PS, Stewart S, Kirkpatrick M, Chalmers J, et al. A national survey of the prevalence, incidence, primary care burden and treatment of atrial fibrillation in Scotland. Heart. 2007;**93**(5):606-612

[5] Chugh SS, Blackshear JL, Shen WK, Hammill SC, Gersh BJ. Epidemiology and natural history of atrial fibrillation: Clinical implications. Journal of the American College of Cardiology. 2001;**37** (2):371-378

[6] Go AS, Hylek EM, Phillips KA, Chang Y, Henault LE, Selby JV, et al. Prevalence of diagnosed atrial fibrillation in adults: National implications for rhythm management and stroke prevention: The anticoagulation and risk factors In atrial fibrillation (ATRIA) study. Journal of the American Medical Association. 2001; **285**(18):2370

[7] January CT, Wann LS, Alpert JS, Calkins H, Cigarroa JE, Cleveland JC Jr, et al. 2014 AHA/ACC/HRS guideline for the management of patients with atrial fibrillation: Executive summary: A report of the American College of Cardiology/American Heart Association Task Force on practice guidelines and the Heart Rhythm Society. Circulation. 2014;**130**(23):2071-2104. DOI: 10.1161/CIR.0000000000000040. Epub 2014 Mar 28. Erratum in: Circulation. 2014 Dec 2;130(23):e270-1

[8] Singleton MJ, Imtiaz-Ahmad M, Kamel H, O'Neal WT, Judd SE, Howard VJ, et al. Association of atrial fibrillation without cardiovascular comorbidities and stroke risk: From the REGARDS study. Journal of the American Heart Association. 2020;**9**(12):e016380

[9] Chimowitz MI, DeGeorgia MA, Poole RM, Hepner A, Armstrong WM. Left atrial spontaneous echo contrast is highly associated with previous stroke in patients with atrial fibrillation or mitral stenosis. Stroke. 1993;**24**(7):1015

[10] Fluckiger JU, Goldberger JJ, Lee DC, Ng J, Lee R, Goyal A, et al. Left atrial flow velocity distribution and flow coherence using four-dimensional flow MRI: A pilot study investigating the impact of age and pre- and postintervention atrial fibrillation on atrial hemodynamics. Journal of Magnetic Resonance Imaging. 2013;**38**(3): 580-587

[11] Otani T, Al-Issa A, Pourmorteza A, McVeigh ER, Wada S, Ashikaga H. A computational framework for personalized blood flow analysis in the human left atrium. Annals of Biomedical Engineering. 2016;**44**(11):3284-3294

[12] Masci A, Alessandrini M, Forti D, Menghini F, Dedé L, Tomasi C, et al. A proof of concept for computational fluid dynamic analysis of the left atrium in atrial fibrillation on a patient-specific basis. Journal of Biomechanical Engineering. 2020;**142**(1):011002

[13] Cignoni P, Callieri M, Corsini M, Dellepiane M, Ganovelli F, Ranzuglia G.

MeshLab: An Open-Source Mesh Processing Tool. Eurographics Italian Chapter Conference. The Eurographics Association; 2008. Available from: http://diglib.eg.org/handle/10.2312/LocalChapterEvents.ItalChap.ItalianChapConf2008.129-136

[14] Izzo R, Steinman D, Manini S, Antiga L. The vascular modeling toolkit: A python library for the analysis of tubular structures in medical images. Journal of Open Source Software. 2018;**3**(25):745

[15] Unser M. Splines: A perfect fit for signal and image processing. IEEE Signal Processing Magazine. 1999;**16**(6):22-38

[16] Sanatkhani S, Menon PG. Relating atrial appendage flow stasis risk from computational fluid dynamics to imaging based appearance paradigms for cardioembolic risk. In: Cardoso MJ, Arbel T, JMRS T, Aylward S, Li S, Boctor E, et al., editors. Imaging for Patient-Customized Simulations and Systems for Point-of-Care Ultrasound [Internet]. Cham: Springer International Publishing; 2017. pp. 86-93. (Lecture Notes in Computer Science; vol. 10549). Available from: https://link.springer.com/10.1007/978-3-319-67552-7_11 [Accessed: agosto 16, 2024]

[17] Khurram Rooh A, Masud A. A Multiscale/stabilized formulation of the incompressible Navier–Stokes equations for moving boundary flows and fluid–structure interaction. Computational Mechanics. 2006;**38**(4–5):403-416

[18] Scovazzi G. A discourse on Galilean invariance, SUPG stabilization, and the variational multiscale framework q. Computer Methods in Applied Mechanics and Engineering. 2007;**196**(4):1108-1132

[19] Forti D, Dedè L. Semi-implicit BDF time discretization of the Navier–Stokes equations with VMS-LES modeling in a high performance computing framework. Computers & Fluids. 2015;**117**:168-182

[20] Koizumi R, Funamoto K, Hayase T, Kanke Y, Shibata M, Shiraishi Y, et al. Numerical analysis of hemodynamic changes in the left atrium due to atrial fibrillation. Journal of Biomechanics. 2015;**48**(3):472-478

[21] Mansour M, Holmvang G, Sosnovik D, Migrino R, Abbara S, Ruskin J, et al. Assessment of pulmonary vein anatomic variability by magnetic resonance imaging: Implications for catheter ablation techniques for atrial fibrillation. Journal of Cardiovascular Electrophysiology. 2004;**15**(4):387-393

[22] Bazilevs Y, Gohean JR, Hughes TJR, Moser RD, Zhang Y. Patient-specific isogeometric fluid–structure interaction analysis of thoracic aortic blood flow due to implantation of the Jarvik 2000 left ventricular assist device. Computer Methods in Applied Mechanics and Engineering. 2009;**198**(45–46):3534-3550

[23] Pasquale Claudio Africa. lifex: A flexible, high performance library for the numerical solution of complex finite element problems. SoftwareX. 2022;**20**:101252. ISSN 2352-7110. DOI: 10.1016/j.softx.2022.101252

[24] Masci A et al. A patient-specific computational fluid dynamics model of the left atrium in atrial fibrillation: Development and initial evaluation. In: Pop M, Wright G, editors. Functional Imaging and Modelling of the Heart. FIMH 2017, Lecture Notes in Computer Science. Vol. 10263. Cham: Springer; 2017. DOI: 10.1007/978-3-319-59448-4_37

[25] Gupta DK, Shah AM, Giugliano RP, Ruff CT, Antman EM, Grip LT, et al.

Left atrial structure and function in atrial fibrillation: ENGAGE AF-TIMI 48. European Heart Journal. 2014;**35**(22): 1457-1465

[26] Hussain S, Falanga M, Chiaravalloti A, Tomasi C, Corsi C. Patient-specific left atrium contraction quantification associated with atrial fibrillation: A region-based approach. Computer Methods and Programs in Biomedicine. 2024;**249**:108138

[27] Corsi C, Lang RM, Veronesi F, Weinert L, Caiani EG, MacEneaney P, et al. Volumetric quantification of global and regional left ventricular function from real-time three-dimensional echocardiographic images. Circulation. 2005;**112**(8):1161-1170

[28] Chen L, Zhang C, Wang J, Guo L, Wang X, Liu F, et al. Left atrial strain measured by 4D auto LAQ echocardiography is significantly correlated with high risk of thromboembolism in patients with non-valvular atrial fibrillation. Quantitative Imaging in Medicine and Surgery. 2021; **11**(9):3920-3931

[29] Botto GL, Tortora G, Casale MC, Canevese FL, Brasca FAM. Impact of the pattern of atrial fibrillation on stroke risk and mortality. Arrhythmia & Electrophysiology Review. 2021;**10**(2): 68-76

[30] Ku DN, Giddens DP, Zarins CK, Glagov S. Pulsatile flow and atherosclerosis in the human carotid bifurcation. Positive correlation between plaque location and low oscillating shear stress. Arteriosclerosis. 1985;**5**(3):293-302

[31] Di Achille P, Tellides G, Figueroa CA, Humphrey JD. A haemodynamic predictor of intraluminal thrombus formation in abdominal aortic aneurysms. Proceedings of the Royal Society A Mathematical, Physical and Engineering and Science. 2014;**470** (2172):20140163

[32] Dueñas-Pamplona J, García JG, Sierra-Pallares J, Ferrera C, Agujetas R, López-Mínguez JR. A comprehensive comparison of various patient-specific CFD models of the left atrium for atrial fibrillation patients. Computers in Biology and Medicine. 2021;**133**:104423

[33] Rossini L, Martinez-Legazpi P, Vu V, Fernández-Friera L, Pérez Del Villar C, Rodríguez-López S, et al. A clinical method for mapping and quantifying blood stasis in the left ventricle. Journal of Biomechanics. 2016;**49**(11):2152-2161

[34] Falanga M et al. A digital twin approach for stroke risk assessment in atrial fibrillation patients. Heliyon. 2024; **10**:20. DOI: 10.1016/j.heliyon.2024.e39527

[35] Durán E, García-Villalba M, Martínez-Legazpi P, Gonzalo A, McVeigh E, Kahn AM, et al. Pulmonary vein flow split effects in patient-specific simulations of left atrial flow. Computers in Biology and Medicine. 2023;**163**: 107128

[36] Passerini AG, Polacek DC, Shi C, Francesco NM, Manduchi E, Grant GR, et al. Coexisting proinflammatory and antioxidative endothelial transcription profiles in a disturbed flow region of the adult porcine aorta. Proceedings of the National Academy of Sciences of the United States of America. 2004;**101**(8): 2482-2487

[37] Torisu T, Torisu K, Lee IH, Liu J, Malide D, Combs CA, et al. Autophagy regulates endothelial cell processing, maturation and secretion of von Willebrand factor. Nature Medicine. 2013;**19**(10):1281-1287

[38] Mackman N. New insights into the mechanisms of venous thrombosis. The Journal of Clinical Investigation. 2012; **122**(7):2331-2336

[39] Hankey GJ, Eikelboom JW. Dabigatran etexilate: A new oral thrombin inhibitor. Circulation. 2011; **123**(13):1436-1450

[40] Reddy VY, Möbius-Winkler S, Miller MA, Neuzil P, Schuler G, Wiebe J, et al. Left atrial appendage closure with the watchman device in patients with a contraindication for oral anticoagulation. Journal of the American College of Cardiology. 2013;**61**(25):2551-2556

[41] Fountain RB, Holmes DR, Chandrasekaran K, Packer D, Asirvatham S, Van Tassel R, et al. The Protect AF (watchman left atrial appendage system for embolic protection in patients with atrial fibrillation) trial. American Heart Journal. 2006;**151**(5):956-961

[42] Di Biase L, Santangeli P, Anselmino M, Mohanty P, Salvetti I, Gili S, et al. Does the left atrial appendage morphology correlate with the risk of stroke in patients with atrial fibrillation? Journal of the American College of Cardiology. 2012;**60**(6):531-538

[43] Masci A, Barone L, Dedè L, Fedele M, Tomasi C, Quarteroni A, et al. The impact of left atrium appendage morphology on stroke risk assessment in atrial fibrillation: A computational fluid dynamics study. Frontiers in Physiology. 1938;**201, 9**:9

[44] Shapira L, Shamir A, Cohen-Or D. Consistent mesh partitioning and skeletonisation using the shape diameter function. The Visual Computer. 2008;**24**(4):249-259

[45] Besl PJ, McKay ND. A method for registration of 3-D shapes. IEEE Transactions on Pattern Analysis and Machine Intelligence. 1992;**14**(2):239-256

[46] Piccinelli M, Veneziani A, Steinman DA, Remuzzi A, Antiga L. A framework for geometric analysis of vascular structures: Application to cerebral aneurysms. IEEE Transactions on Medical Imaging. 2009;**28**(8):1141-1155

[47] D'Alessandro N, Falanga M, Masci A, Severi S, Corsi C. Preliminary findings on left atrial appendage occlusion simulations applying different endocardial devices. Frontiers in Cardiovascular Medicine. 2023;**10**: 1067964

[48] Glikson M, Wolff R, Hindricks G, Mandrola J, Camm AJ, Lip GYH, et al. EHRA/EAPCI expert consensus statement on catheter-based left atrial appendage occlusion – An update. Europace. 2020;**22**(2):184-184

[49] Qureshi A, Balmus M, GYH L, Williams S, Nordsletten DA, Aslanidi O, et al. Mechanistic modelling of Virchows triad to assess thrombogenicity and stroke risk in atrial fibrillation patients. European Heart Journal – Digital Health. 2022;**3**(4):ztac076.2788. DOI: 10.1093/ehjdh/ztac076.2788

[50] Mill J, Agudelo V, Olivares AL, Pons MI, Silva E, Nuñez-Garcia M, et al. Sensitivity analysis of in silico fluid simulations to predict thrombus formation after left atrial appendage occlusion. Mathematics. 2021;**9**(18):2304

[51] Aguado AM, Olivares AL, Yagüe C, Silva E, Nuñez-García M, Fernandez-Quilez Á, et al. In silico optimization of left atrial appendage occluder implantation using interactive and modeling tools. Frontiers in Physiology. 2019;**10**:237

[52] Melillo E, Palmiero G, Ferro A, Mocavero PE, Monda V, Ascione L. Diagnosis and management of left atrium appendage thrombosis in atrial fibrillation patients undergoing cardioversion. Medicina (Mex). 2019;**55**(9):511

[53] García-Villalba M, Rossini L, Gonzalo A, Vigneault D, Martinez-Legazpi P, Durán E, et al. Demonstration of patient-specific simulations to assess left atrial appendage thrombogenesis risk. Frontiers in Physiology. 2021;**12**: 596596

[54] Markl M, Lee DC, Furiasse N, Carr M, Foucar C, Ng J, et al. Left atrial and left atrial appendage 4D blood flow dynamics in atrial fibrillation. Circulation: Cardiovascular Imaging [Internet]. 2016;**9**(9):e004984. Disponibile su: https://www.ahajournals.org/doi/10.1161/CIRCIMAGING.116.004984 [Accessed: agosto 18, 2024]

Chapter 10

Applications of Particle Image Velocimetry in Internal Combustion Engines Research

Gengxin Zhang, Penghua Shi and Yu Jin

Abstract

Particle image velocimetry (PIV) has become a key tool in the internal combustion engines (ICEs) research. This chapter comprehensively reviews the current applications of PIV in ICEs research, particularly in the fuel injection, the combustion processes, and the in-cylinder flow of engines. Detailed insights obtained from PIV measurements have significantly contributed to optimizing engine design, improving combustion efficiency, and reducing emissions. Despite its advantages, PIV technology still faces several challenges, especially in the context of ammonia–hydrogen-fueled engines. Selecting appropriate tracer particles and organizing the complexity in-cylinder flow field pose significant obstacles. Moreover, this chapter explores corresponding measures to address these challenges, including the development of specialized tracer particles, multimodal measurement approaches, and AI-enhanced data processing. These measures have the potential to further improve the accuracy and applicability of PIV measurements in carbon-neutral engines research. This chapter aims to provide researchers and engineers with a comprehensive understanding of the most advanced PIV applications, encountered challenges, and corresponding measures in the field of ICEs research.

Keywords: particle image velocimetry (PIV), internal combustion engines (ICEs), fuel spray, combustion, flow field, artificial intelligence (AI)

1. Introduction

To continuously optimize the design of fluid dynamic machinery products, there has been a longstanding interest in measuring fluid motion. In the early days, intrusive hot-wire anemometers were invented and widely used. Later, to avoid interference with the flow field, non-intrusive laser Doppler velocimeters (LDV) were developed and applied, though they were still limited to single-point measurements of the flow field. Building on LDV technology, phase Doppler particle analyzers (PDPA) were developed to measure both particle velocity and size at single points simultaneously. Particle image velocimetry (PIV) emerged after the 1980s, evolving from flow visualization techniques and incorporating modern computer technology, optical technology, and image analysis technology to become a robust method for flow measurement [1]. Compared to traditional single-point velocity measurement methods, PIV enables full-field, non-contact, instantaneous measurements of complex, unsteady flows. This advanced optical measurement technique is now widely used for quantitative research in fluid dynamics within both scientific research and industrial applications.

In the context of internal combustion engines (ICEs), understanding the complex interactions between fuel injection, combustion processes, and in-cylinder airflow is crucial for optimizing engine performance and reducing emissions. PIV has become an indispensable tool in ICEs research [2]. Specifically, PIV allows for detailed visualization and quantitative analysis of fuel spray dynamics, including in-nozzle flow, the velocity distribution of atomized droplets, spray penetration, and air entrainment rate, all of which are essential for achieving optimal fuel-air mixing. PIV also aids in understanding flame propagation, turbulence, and combustion stability during the combustion process, which are vital for improving combustion efficiency and reducing emission formation. Additionally, PIV can provide comprehensive data on in-cylinder air motion, such as swirl and tumble flows, which are critical for enhancing the in-cylinder mixing process. The data provided by PIV help in designing intake and exhaust systems, piston shapes, and injectors, thereby improving overall engine performance.

As described above, PIV provides high-resolution, non-intrusive flow field measurements, making it particularly valuable in ICEs research. The primary aim of this chapter is to provide a comprehensive review of the progress, challenges, and prospects of PIV technology in ICEs research. The chapter is structured as follows: Firstly, it briefly outlines the basic measurement principles of PIV, along with typical technical advancements and optical setups used in ICEs applications. Then, based on specific case studies and experimental results, it discusses the effectiveness of PIV in ICEs research, focusing on fuel spray characterization, combustion analysis, and in-cylinder flow field studies. Additionally, the chapter addresses practical challenges and countermeasures related to the future application of PIV technology in ICEs research. This includes the selection of tracer particles for new carbon-neutral engine research and solutions to the complexities of flow fields in combustion environments. Specifically, the application of artificial intelligence (AI) in the post-processing and prediction of PIV flow field data is highlighted as a promising innovation that could further enhance the accuracy and applicability of PIV measurement in future ICEs research. By covering these topics, this chapter aims to provide researchers and engineers with a thorough understanding of the current state of PIV applications in ICEs research, the challenges that need to be addressed in the future, and the corresponding countermeasures.

2. Principles of particle image velocimetry

2.1 Basic principles and methodology for PIV

PIV technology originates from flow visualization techniques, primarily using image analysis to track the movement of tracer particles dispersed within the flow field to measure fluid velocity [3]. As illustrated in **Figure 1**, the specific principle involves measuring the displacement vector Δx of a tracer particle moving from position A [x(t), y(t), z(t)] to A' [x(t + Δt), y(t + Δt), z(t + Δt)] after a time

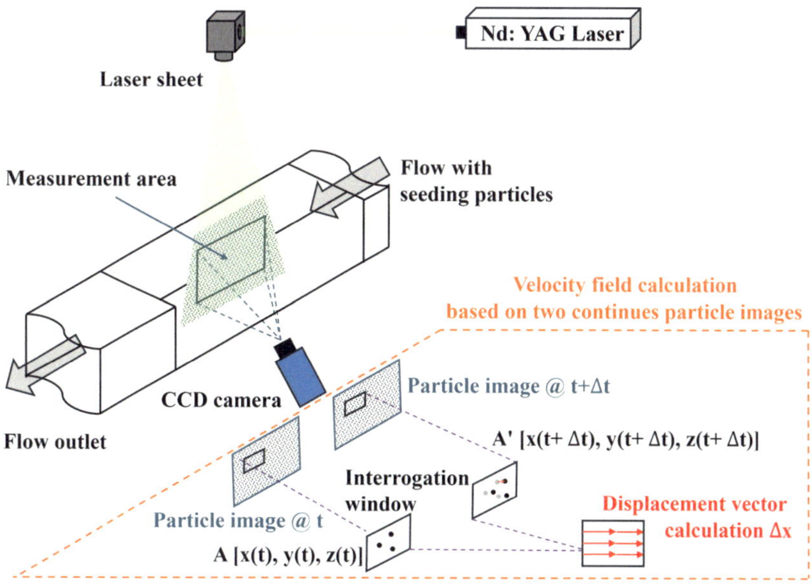

Figure 1.
Basic principles of PIV.

interval Δt. When the time interval Δt is sufficiently small, comparable to the Taylor microscale of the Lagrangian velocity field constrained by accuracy, the particle's velocity can be approximated by the ratio of the displacement Δx to the time interval Δt, as shown in Eq. (1).

$$\begin{cases} v_x = \lim\limits_{dt \to 0} \dfrac{(dx(t))}{dt} \approx \dfrac{x(t+\Delta t)-x(t)}{\Delta t} \\ v_y = \lim\limits_{dt \to 0} \dfrac{(dy(t))}{dt} \approx \dfrac{y(t+\Delta t)-y(t)}{\Delta t} \\ v_z = \lim\limits_{dt \to 0} \dfrac{(dz(t))}{dt} \approx \dfrac{z(t+\Delta t)-z(t)}{\Delta t} \end{cases} \tag{1}$$

According to Eq. (1), once the displacement information of the particles is obtained through image analysis, and combined with the camera's calibration parameters, the spatial displacement of the particles can be determined. Additionally, since Δt is fixed during a single experiment, the velocity vector of the particle in the Cartesian coordinate system can be calculated. When the displacement information of a sufficient number of particles is known, the velocity distribution of the entire flow field can be reconstructed.

The key steps and principles involved in PIV operations are as follows: (1) Seeding the Fluid: Tracer particles are seeded into the fluid. These particles are small enough to accurately follow the flow but large enough to scatter sufficient light for imaging. The selection of tracer particles depends on the fluid properties and flow conditions. (2) Illumination: A thin laser sheet is used to illuminate the tracer particles in a specific plane of the fluid. The time delay between laser pulses is chosen based on the expected velocity in the flow. (3) Image Capture: Cameras capture images of the illuminated particles at continuous time intervals. This interval must ensure that the particles move an appropriate distance between images. (4) Image Processing: The captured images are processed using cross-correlation algorithms to determine the displacement of groups of particles between image pairs. The displacement data are then used to calculate the velocity vectors of the fluid. (5) Post-Processing: The velocity field data may undergo post-processing to eliminate spurious vectors, smooth data, and interpolate missing vectors. Techniques such as median filtering and Gaussian smoothing are commonly used to enhance the accuracy and readability. (6) Data Visualization and Analysis: The final velocity field is visualized using vector plots, contour maps, and other graphical representations. Further analysis can include calculating derived quantities such as vorticity and turbulence intensity to gain deeper insights into flow characteristics. By following these operational steps and principles, PIV can provide a comprehensive and detailed understanding of the fluid dynamics characteristics involved in ICEs research.

2.2 Typical optical arrangement of PIV system for ICEs research

Over the past decade, PIV technology for ICEs research has been continuously refined and developed beyond classical PIV. In particular, extended three-dimensional three-component (3D3C) velocity field analysis has become crucial for understanding in-cylinder spray, combustion, and flow fields. This includes scanning PIV, holographic PIV (Holo-PIV), stereo PIV (S-PIV), and tomographic PIV (T-PIV) [4]. Holo-PIV measures instantaneous 3D velocity fields by constructing holographic recordings of the luminous particles. However, digital planar holographic PIV (DS-PIV) has not been widely accepted due to its complexity. Therefore, the following will primarily outline the general optical setup of classical PIV, high-speed PIV (HS-PIV), and the rapidly developing tomographic-stereo PIV (TS-PIV) technology [5].

2.2.1 Optical layout for classic PIV system

Figure 2 shows the cross-flow wind tunnel at Hiroshima University, designed to provide a uniform cross-flow field. The downstream section connects to a chamber,

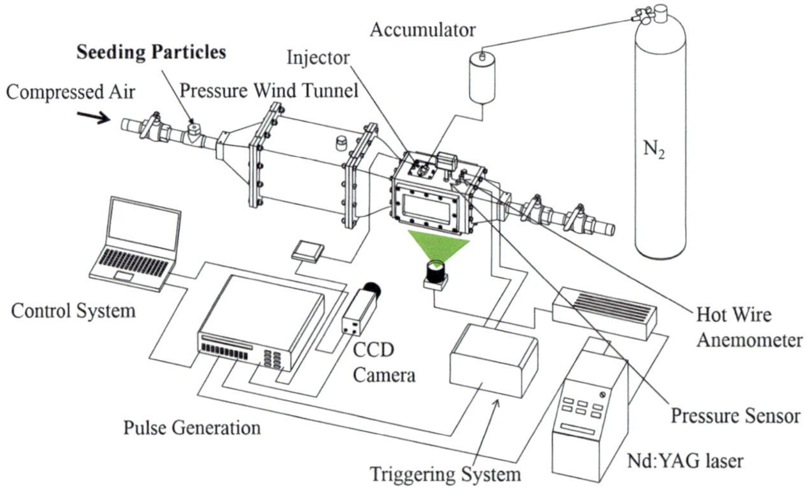

Figure 2.
PIV measurement in cross-flow flow field [6].

where a fuel injector is embedded in the upper surface. Three optical windows, each measuring 100 mm × 200 mm, are installed on the front, rear, and bottom sides for observing the behavior of free and wall-impinging spray in both vertical and horizontal planes. A metal cap and a throttle valve connected downstream of the chamber are employed to adjust the cross-flow speed and ambient pressure, simulating in-cylinder conditions of an engine. When the metal cap is closed, the desired cross-flow speed and ambient pressure greater than atmospheric pressure can be achieved by adjusting the downstream valve. A hot-wire anemometer probe mounted to measure the cross-flow speed and a pressure sensor continuously monitor the internal ambient pressure inside the chamber. The acquired cross-flow speed and ambient pressure signals are transmitted in real-time to a programmable logic controller (PLC), which synchronizes the triggering of the injector, laser equipment, and cameras.

Figure 2 shows the detailed setup of the classic PIV optical system for capturing spray images and velocity distributions, representing the arrangement for vertical plane imaging. An Nd: YAG laser (DPIV-N50) generates a laser sheet that vertically illuminates the tracer particles inside the observation chamber from below. The laser wavelength is 532 nm, and the sheet thickness is 1 mm. A charge-coupled device (CCD) camera (PCO1600) is positioned at a 90° angle to the laser sheet. When measuring the cross-flow field velocity, the double-pulse interval of the laser is set to 40 μs. In contrast, when measuring the velocity field of spray, the pulse interval is set to 15 μs. To observe the velocity distribution on a horizontal plane, the positions of laser sheet and the camera can be swapped. In the experiment, atomized olive oil droplets (approximately 5 μm) are employed to track the velocity within the flow field, while the Mie scattering of the atomized spray droplets themselves is used to track spray movement.

2.2.2 Optical layout for high-speed PIV (HS-PIV) system

Figure 3 illustrates the HS-PIV optical setup used by the University of Leeds for measuring flame development. The spherical constant volume vessel (CVV) has an

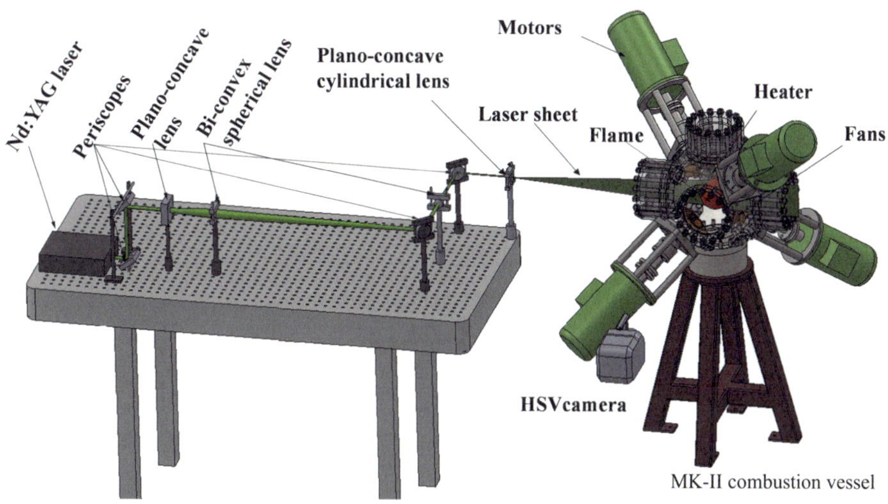

Figure 3.
High-speed flame PIV measurement in combustion vessel [7].

internal diameter of 190 mm, with two pairs of vertically arranged optical windows each having a diameter of 150 mm. Additionally, electric fans are installed at the four corner positions to promote the mixing of the premixed fuel. In the experiment, olive oil is used as the tracer particles, generated by an atomizer (approximately 1 μm) and carried by air into the CVV where it is premixed with the fuel. The laser source is a double-pulsed Nd: YAG laser (DM60–532-DH) with a wavelength of 532 nm. The laser beam is expanded into a sheet approximately 0.5 mm thick using a series of optical lenses, with the thickness adjustable by the spacing of three spherical lenses shown in this figure. Unlike the CCD cameras used in classical PIV systems, a high-speed video (HSV) camera is used to record the continuous development of the flame.

2.2.3 Optical layout for tomographic: Stereo PIV (TS-PIV) system

As shown in **Figure 4**, to investigate the string cavitation regime at the condition of low needle lift, Hiroshima University and Kobe University conducted a study on the internal flow of an enlarged model of a sac-type transparent diesel injector using TS-PIV. In this study, fluorescent particles with a diameter of 15 μm and a specific gravity of 1.1 (Kanomax, FLUOSTAR F1015) were used as tracer particles. The TS-PIV used a 5 mm thick double-pulsed Nd: YAG laser sheet (pulse interval is 60 μs), four connected CCD cameras (JAI AM-200GE), and a HSV camera (Photron SA-Z). All were synchronized by a delay generator (DG). The introduction of the HSV aimed to further elucidate the relationship between the sac's internal flow structures and the transient behavior of string cavitation. This is because string cavitation occurs within a short time (approximately 10 ms), and the CCD cameras with a relatively slow capture speed of 10 fps are not enough for use. Therefore, high-speed visualization

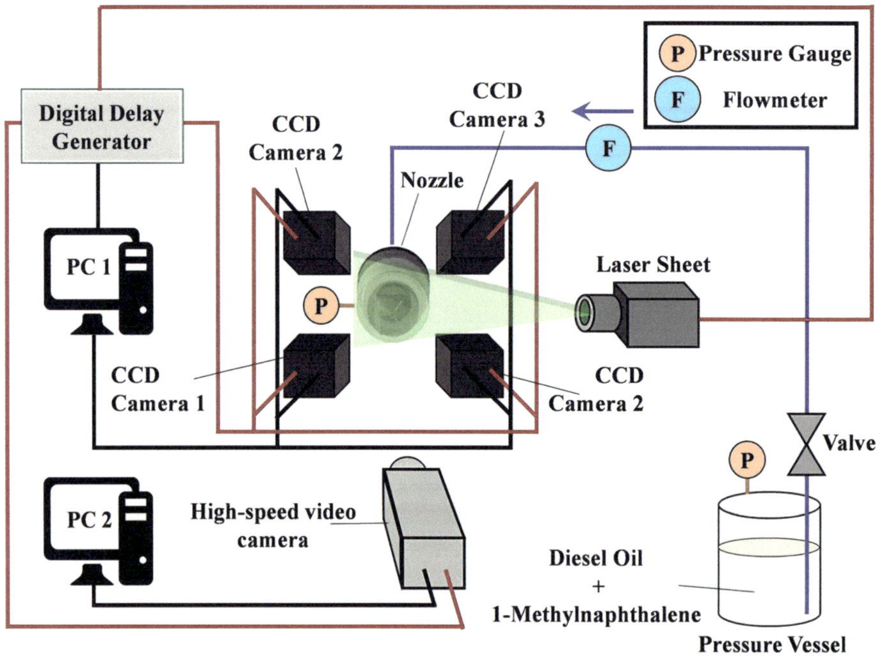

Figure 4.
Tomographic–Stereo PIV measurement for flow field in-nozzle [8].

was simultaneously performed at a frame rate of 3000 fps. It is noteworthy that the positions and intensities of the particles captured by each camera in the system must be calibrated before the formal experiment using a calibration plate. The calibration plate, with an outer diameter of 8 mm, consists of 49 dots, each with a diameter of 0.423 mm. Specifically, the camera position is calibrated by gradually moving the calibration plate at intervals of 0.25 mm within the thickness range of the laser sheet.

Regardless of the form of PIV technology used, accurate PIV measurements heavily depend on the tracer particles dispersed within the flow field. If these particles exhibit a high flow-following behavior, their movement could accurately reflect the flow field's motion. Hence, high-quality particles used in PIV system must meet the following specifications: (1) Density Matching: The density of the tracer particles should match the experimental fluid as feasible. (2) Small Scale: The particles should be sufficiently small in size. (3) Shape and Distribution: The particles should be as spherical as possible, with a uniform size distribution. (4) High Light Scattering Efficiency: The particles must have a high light scattering efficiency. Additionally, it is necessary to consider the potential for tracer particles to contaminate or damage optical windows and their ability to withstand high-temperature environments inside the engine cylinder.

3. Progress in the applications of PIV in ICEs research

3.1 Flow inside the nozzle and spray motion

Fuel spray dynamics are important to the combustion process of ICEs, and accurately characterizing these dynamics is essential for optimizing fuel–air mixture. PIV can provide detailed insights into various aspects of fuel spray, including in-nozzle cavitation, the velocity distribution of atomized droplets at nozzle exit, spray penetration, and the interaction between the spray and flow field.

3.1.1 Flow inside the nozzle

Due to the occurrence of in-nozzle cavitation and its impact on the spray atomization at nozzle exit, the internal flow within the nozzle is a critical factor to consider in injector design. Conducting flow measurements within the sac to study the conditions that results in string cavitation is of great significance [8]. **Figure 5** presents the averaged velocity distribution at center plane of nozzle capsule under different flow conditions using TS-PIV experiments. As shown in **Figure 5a** and **b**, at an extremely low needle lift (Z/D = 0.5), the downward velocity near the sac center is lower than that near the wall. This velocity gradient can induce vortex circulation, leading to string cavitation. This conclusion is further validated by the observation of instantaneous flow field around the string cavitation. However, further research indicates that string cavitation emerges only when the vortices are especially strong, serving as an indicator of intense vortical flow. In such cases, a hollow cone-shaped spray jet is produced at the exit, increasing the jet angle.

3.1.2 Free spray

Currently, increasing the ambient gas density to improve thermal efficiency results in different spray behaviors. PIV technology has also been used to measure the velocity field of free sprays, thereby assessing the location of the maximum velocity

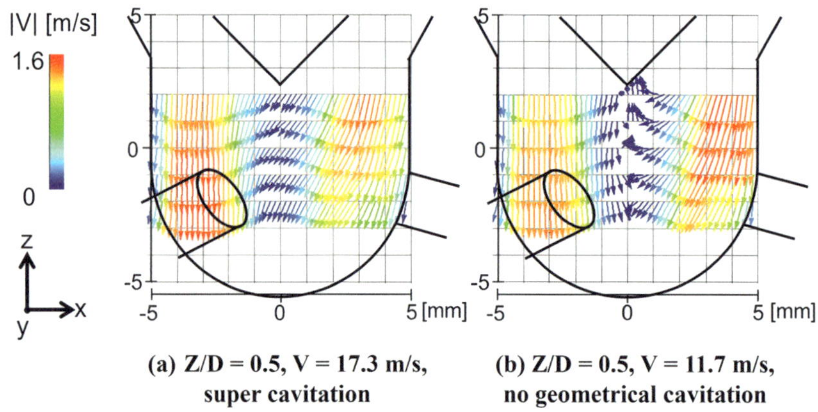

Figure 5.
Average velocity distribution at center plane of nozzle capsule [8].

gradient within the spray and the local equivalence ratio distribution. The findings demonstrate that under high ambient density conditions, the position of the maximum velocity gradient shifts from the periphery of the spray to the center, with the region of sufficient air moving upstream [9]. PIV has also been used to characterize the vector field and air entrainment rate of gasoline sprays under ultra-high injection pressures. The results indicate that the air motion induced by the spray increases with higher injection pressure and mass, although this is a transient phenomenon. Moreover, compared to convergent nozzles, divergent nozzles further enhance the interaction between the spray and the surrounding air [10].

Laser absorption scattering (LAS) technology has been used to analyze the spray tip penetration and the equivalence ratio distribution of mixtures. It was noted that split fuel injection greatly influences the characteristics of the mixture, helping to improve combustion efficiency and reduce emissions. To elucidate the underlying mechanisms, the combination of PIV technology and laser induced fluorescence (LIF) (LIF-PIV) was employed to study the air motion induced by the spray. The surrounding air motion induced by the first injection promotes the evaporation of the second spray in the case of split injection, reducing the fuel replenishment to the first spray and preventing high-density droplets from accumulating at the front edge [11]. LIF-PIV technology was further used to quantitatively analyze the interaction between non-evaporating diesel spray and surrounding gas. The spray motion in the tip region is important during the air entrainment process. Higher injection pressure enhances the gas entrainment effect, although the increase gradually tends to saturate [12]. Superheated sprays, due to the flash boiling effect, have the potential to improve spray atomization and evaporation processes. HS-PIV technology was used in a high-pressure CCV to study the interactions between multi-hole spray plumes under different superheat conditions in both cross-sectional and axial planes. The results showed that a mixing zone is founded in the cross-sectional direction, and the velocity vectors in this region increase with superheat [13]. Additionally, the axial spray structure collapses from a single narrow jet to a solid cone-shaped spray as superheat increases, with stronger vorticity forming at the boundary near the spray tip [14]. LIF-PIV technology has also been used to investigate and demonstrate the potential of split injection strategies in enhancing the fuel-air mixing of flash-boiling sprays [15].

Equivalency ratio, velocity magnitude, vorticity values, and shear strain rate all grow during the injection process but reduce by an order of magnitude at the time of ignition, according to research using HS-PIV and planar laser-induced fluorescence (PLIF) technology [16]. The turbulence induced by the late-stage spray can enhance fuel-air mixing. The combination of TPIV and HS-PIV helps to resolve the 3D3C turbulent structures induced by the spray. Results indicate that fuel injection increases the amplitude of turbulence kinetic energy (TKE), shear rate (S), and vorticity (Ω) by 400%, enhancing fuel-air mixing [17]. A novel high-speed two-color PIV (HS-TC-PIV) technique was proposed to measure the velocity fields of fuel spray and the surrounding gas field at the same time, revealing their interaction mechanisms. In the experiment, a tracer system was developed to differentiate between the phases of fuel droplets and the surrounding gas. Fluorescent seed particles (4.8 μm) were adopted to trace the surrounding gas, while Mie scattering signals reflected the spray phase. The motion characteristics of the surrounding gas in varying regions during the injection process varied significantly, indicating notable differences in momentum transfer processes [18]. This method was further employed to study the interaction between sprays and cross-flow flow field in our research [6]. As shown in **Figure 6a**, PIV measured the velocity vector distribution in the chamber downstream of the cross-flow wind tunnel at a speed U_x of 5 m/s and an ambient pressure P_a of 0.4 MPa. The results showed that the cross-flow flow field was generally uniform and stable, except near the nozzle and close to the upper and lower wall positions. To further observe detailed quantitative velocity information, **Figure 6b** extracted the velocity values at the nozzle position (x = 0 mm) and downstream (x = 50 mm), averaging the values from 10 times measurements, with error bars indicating the maximum and minimum values. It can be seen that velocity distribution near the nozzle was the most unstable, while other positions remained relatively stable.

As shown in **Figure 7**, the spray structure and velocity distribution on the vertical surface at the end of injection (EOI) are depicted. The injection pressure is set to 10 MPa, the ambient pressure to 0.4 MPa, the cross-flow speed up to 10 m/s, and the injection duration to 7 ms. The higher velocities are concentrated in the dense spray

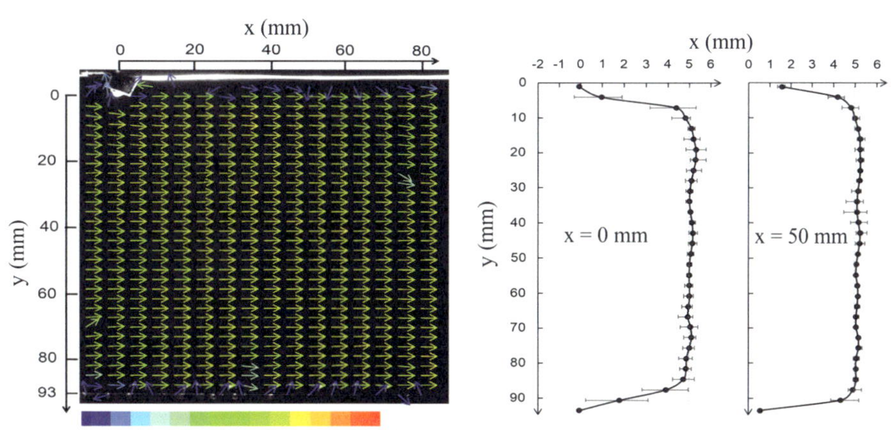

(a) Velocity distribution in observation section (Ux = 5 m/s and Pa = 0.4 MPa)

(b) Velocity distribution in lines of x = 0 and 50 mm.

Figure 6.
Velocity distribution in the observation chamber downstream of cross-flow flow field.

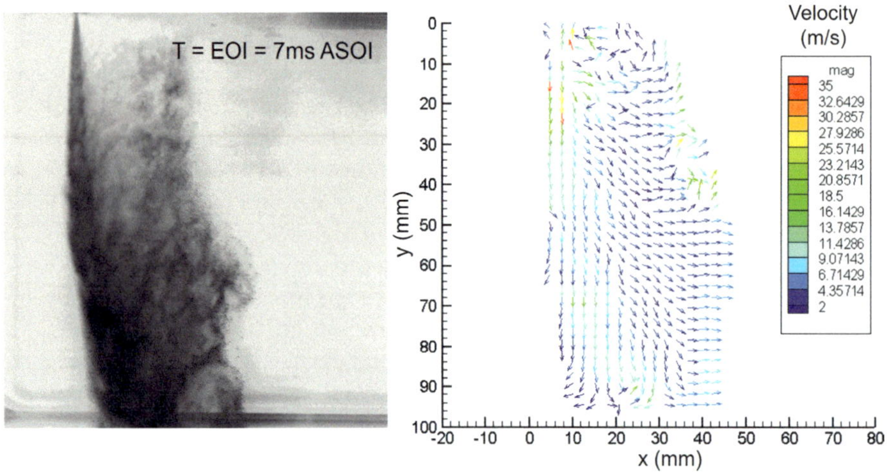

Figure 7.
Spray velocity distribution on the vertical surface inside observation chamber downstream of the cross-flow flow field.

region on the windward side, with the direction predominantly vertical downward. On the leeward side, the velocity vectors are mainly horizontal and lower than those on the windward side, which means the smaller atomized droplets gain horizontal momentum and move horizontally downstream with the cross-flow. It also can be seen that some larger velocity vectors appear in the upper right boundary region, considered as measurement errors due to the sparse distribution of droplets in that region.

As shown in **Figure 8**, the spray structure and its velocity distribution on the horizontal surface at y = 25 mm in the observation chamber at the EOI are depicted. The experimental conditions are set the same as **Figure 7** above. It is evident that a Karman vortex street structure forms in the downstream region, which can also be seen in the velocity vector distribution map. The velocity within vortex cores of the Karman vortex street is significantly lower compared to the surrounding

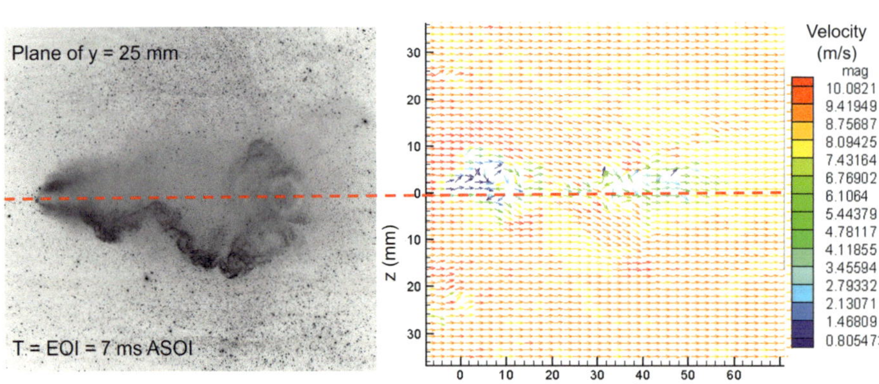

Figure 8.
Spray velocity distribution on the horizontal surface inside observation chamber downstream of the cross-flow flow field.

environment. This indicates that the smaller atomized droplets are carried to the surrounding areas as the Karman vortex street forms and the vortices shed.

3.1.3 Wall-impingement spray

The combination of S-PIV and HS-PIV was used to comparatively study the effects of two biofuels, ethanol and methyl ethyl ketone (MEK), on in-cylinder flow structures. During late injection in the compression stroke, tumble vortices significantly influence the wetting effects and spray morphology. Compared to ethanol, MEK exhibited lower wetting effects [19]. Understanding the gas flow field and droplet behavior under free spray and wall-impinging spray conditions is crucial for optimizing injection strategies and improving combustion efficiency. LIF-PIV technology was employed to assess the flow field around wall-impingement sprays under non-evaporative conditions, thoroughly analyzing spray droplet velocity and vorticity under different wall-impingement distances and injection pressures. The results showed that shorter wall impingement distances limit the increase in surrounding gas mass flow, but the effect of injection pressure on surrounding gas flow improvement is limited [20]. LIF-PIV technology was also employed to comparatively study the gas entrainment characteristics of multi-hole diesel-free and wall-impingement sprays. Due to the momentum interaction of multiple jets, stronger wall vortices and more gas entrainment were observed under wall impingement conditions [21]. Understanding the concentration and velocity distribution of sprays under low-temperature conditions can provide insights for improving diesel engine cold start performance. LIF-PIV technology was also used to analyze wall-impinging sprays under low-temperature conditions. The results indicated that lower ambient temperatures lead to a notable decrease in fuel evaporation rates and the uniformity of gas-phase distribution [22].

Currently, there is limited information about the velocity field of post-impingement sprays. PIV technology has been used to further analyze the velocity distribution of wall-impingement diesel sprays at different impingement angles. The results revealed that post-impingement sprays form high-speed and low-turbulence sub-layers near the wall and the actual impingement mass decrease with increasing angle [23]. Meanwhile, it was found that an increase in ambient gas density reduces the radial peak velocity of the post-impingement spray [24]. Furthermore, higher ambient gas density and lower injection pressure both lead to reduced turbulence kinetic energy near the wall for impinging sprays, affecting the spray flow and mixing characteristics [25]. We further analyzed the near-wall flow field structure of wall-impinging sprays within a cross-flow effect. As shown in **Figure 9**, PIV technology was employed to measure the cross-flow flow field equipped with a wall-impinging plate firstly (the impingement distance set to 50 mm and cross-flow speeds set to 10/20/30 m/s). Through data analysis, the velocity vectors at different positions were extracted and characterized in this figure below [26]. A comparative analysis of the velocity distribution at different positions revealed no significant differences in upstream and downstream velocity distributions in the cross-flow flow field, confirming the uniformity of flow field where the wall-impingement spray occurs.

Figure 10a displays the velocity distribution PIV measurements of wall-impingement sprays at 3 and 4 ms after the start of injection (ASOI) without cross-flow. The wall-jet vortex can be noticed where it is enlarged in this figure. To have a better grasp of the velocity distribution in this region, velocity values extracted along the x-axis and y-axis are shown in **Figure 10b**. The black and red lines represent the

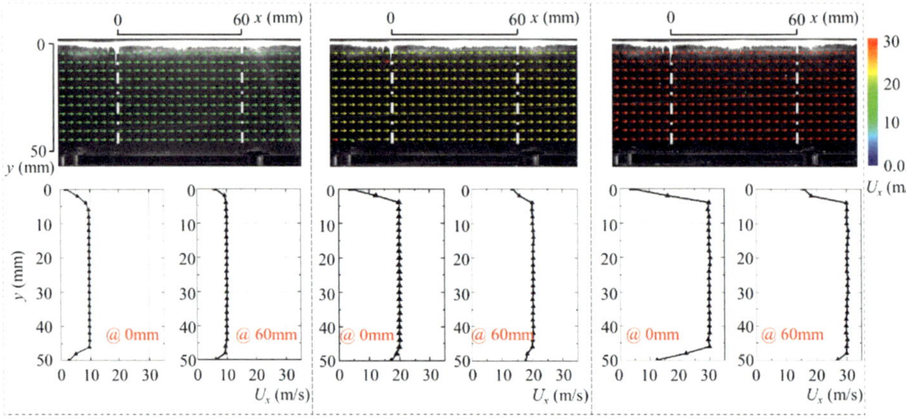

Figure 9.
Velocity distribution in the observation chamber with flat-wall downstream of cross-flow flow field.

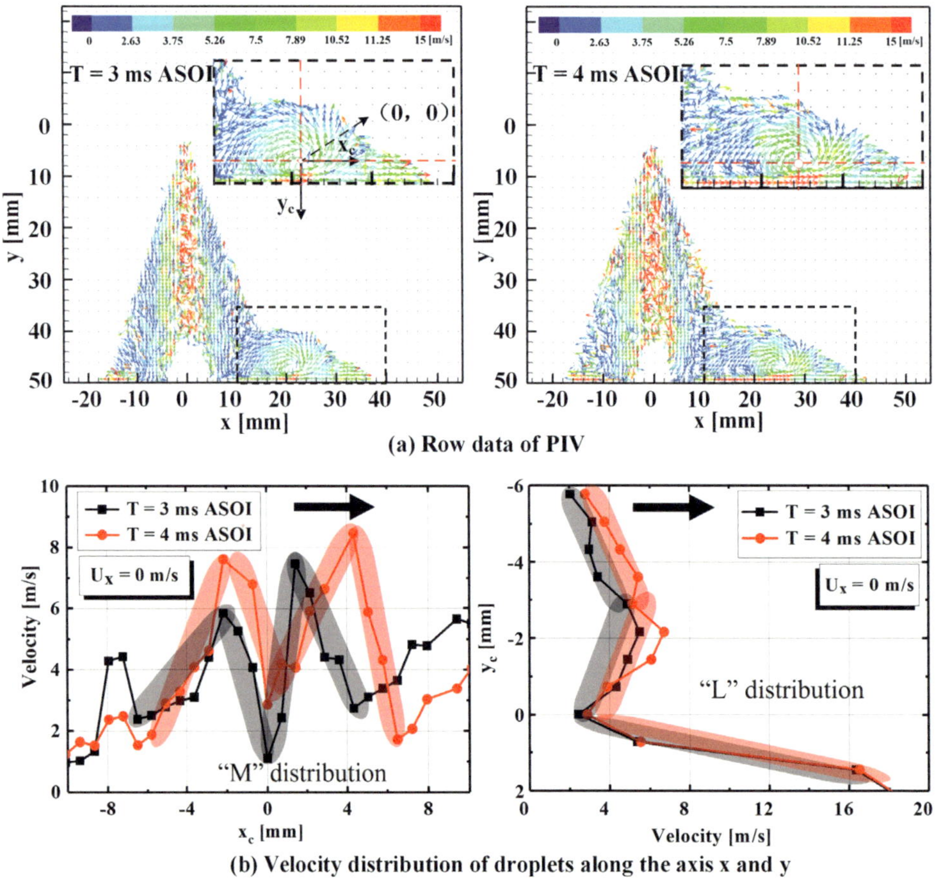

Figure 10.
Velocity distribution around wall-jet vortex core without cross-flow.

velocity distributions at 3 and 4 ms ASOI, respectively. It can be seen that the x-axis velocity distribution exhibits an "M" shape, with the velocity at the center of the vortex core almost being 0 m/s, indicating that there are almost no moving droplets within this region. The two peaks suggest that droplet velocities are higher on the sides and gradually decrease toward the edges. Notably, the vortex region shifts to the right over time, and the peak velocity increases, which can be attributed to the replenishment of kinetic energy to the wall-jet vortex during the injection process. Additionally, the y-axis velocity distribution generally exhibits an "L" shape, with the velocity on the far-wall side lower than that on the near-wall side. The air entrainment on the far-wall side, which provides resistance to droplet movement, reduces their momentum. Additionally, it was found that the droplet velocity in the far-wall region increases significantly over time, while it remains relatively unchanged in the near-wall region [27].

Figure 11 displays the velocity distribution PIV measurements of wall-impingement sprays at 3 and 4 ms ASOI within a cross-flow flow field. Compared to the no-wind static flow field in **Figure 10**, the spray density at the impingement point is reduced due to the influence of the cross-flow, and the previously unmeasurable white areas have largely disappeared. Additionally, the wall-jet vortex structures are dispersed by the cross-flow and are no longer clear. Furthermore, a large number of small droplets are carried downstream by cross-flow, with the velocity direction on the leeward side aligning horizontally to the right, consistent with the cross-flow direction, and the values are approximately close to the cross-flow speed. Notably, the spray area within the cross-flow field, including the area of the wall-jet vortex region, increases significantly over time.

3.2 Combustion analysis

Understanding flame propagation and combustion stability is crucial for optimizing ICE performance. By capturing the flame propagation during the combustion process, PIV enables researchers to visualize the flame front, measure its speed, and analyze the interactions between flame and surrounding flow field.

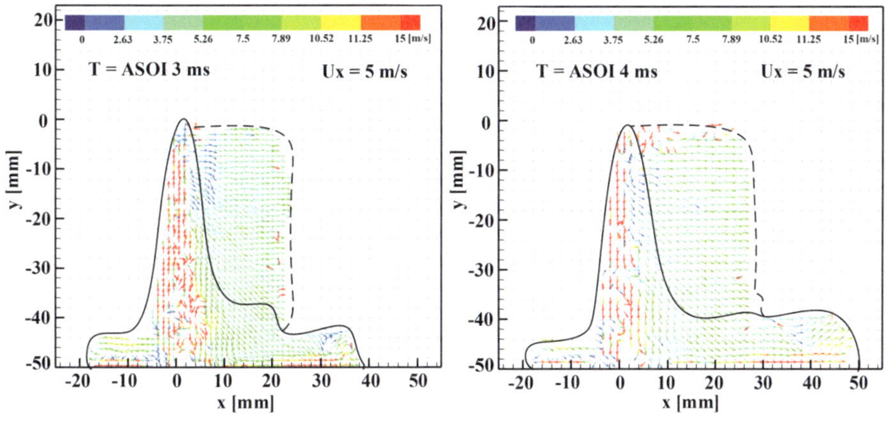

Figure 11.
Velocity distribution around wall-jet vortex core with cross-flow.

HS-PIV experiments have been conducted to characterize the impact of spark energy on misfires. The results revealed that low velocity near the spark plug leads to low spark energy, causing misfire cycles [28]. Further comparison through PLIF-PIV of partial combustion and misfire cycles with fully burned cycles in an optical engine showed that the latter had identical speed and equivalence ratio near the spark plug at the start of spark. This demonstrated that failure during flame propagation is the main cause of partial combustion and misfires [29]. Direct flame visualization using endoscopes and flow visualization using PIV measurements were employed to comprehensively examine combustion characteristics related to late intake valve closure (LIVC). The study results revealed that the LIVC strategy reduces particulate matter (PM) formation and promotes particulate oxidation by increasing in-cylinder flow velocity and turbulence kinetic energy, lowering the effective compression ratio, and decreasing initial in-cylinder temperature [30]. PIV technology can also be combined with temperature measurements to study the effect of pressure and turbulence intensity on flame-wall interactions (FWI) by comparing the heat flux under laminar and turbulent combustion conditions. The results indicated that increases in pressure and turbulence intensity lead to increased heat flux, signifying stronger heat losses [31].

Quantitative PIV studies of local flame speed have revealed the interactions between turbulence and obstacles during flame propagation. The results showed that vortex shedding and local wake recirculation lead to turbulence formation, which in turn causes wrinkling of the flame front in the combustible mixture, increasing the flame surface area and thereby improving combustion speed [32]. HS-PIV has also been used to characterize instantaneous flame fronts, phase-averaged vorticity, and streamlines, describing the dynamics of unsteady flames and flows. It was found that the characteristic stretch rate determines the critical flame speed at which flame structure modes transition, thereby affecting the stability characteristics of the combustion chamber [33]. The use of four-plane PIV (FP-PIV) technology for measuring premixed methane flames has characterized flame displacement speed and front curvature during the combustion process. The results indicated that the speed is independent of front curvature or strain rate [34].

Additionally, the instability of hydrogen-air premixed combustion flames can lead to flame self-acceleration [35, 36]. **Figure 12** uses HS-PIV to measure the velocity vector field and characterize the flame propagation during hydrogen-air combustion. In the early stage ($t = 1$ ms), the flame surface remains smooth, with velocity vectors uniformly distributed around the unburned gas ring near the front [7]. Over time (at $t = 3.0$, 3.5, and 4.3 ms), the front expansion intensifies and the stretch rate correspondingly decreases, causing the stable flame surface to become wrinkled due to flame instability. This further leads to increased fluctuations in the velocity vectors near the front, with greater anisotropy in gas flow direction, indicating enhanced flame instability.

3.3 Flow field in-cylinder investigation

Understanding in-cylinder fluid dynamics is also crucial for optimizing ICE performance. Currently, besides using PIV technology to characterize in-cylinder flow fields, it is also employed to measure complex flow patterns in intake and exhaust ports, near-wall boundary layer behavior, and coolant pipes. Research results can provide detailed analyses of in-cylinder flow patterns and structures, which can then be used to optimize the design of intake and exhaust ducts, coolant pipes, as well as cylinder and piston structures.

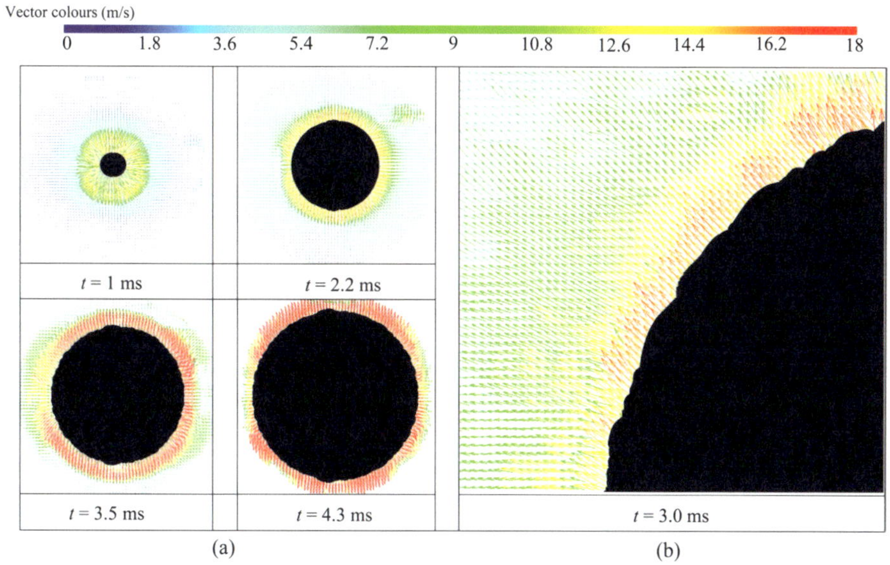

Figure 12.
Vector maps of hydrogen-air laminar flames at ϕ = 1.0, Pi = 0.1 MPa, and Ti = 300 K.

The combination of PIV and two-color methods has been used to simultaneously measure in-cylinder flow and temperature fields. The findings demonstrated that the phenomenon of temperature stratification caused by evaporative cooling shows a significant effect on the formation and evolution of cold gas regions and temperature gradients after fuel injection [37]. Differences in coolant flow velocity at different locations can lead to uneven hot and cold distribution. PIV and numerical simulation have been used in comparative validation studies to analyze coolant flow distribution, which can be used to find ways to achieve uniform cooling within the cylinder. Ultimately, by adjusting the positions of the coolant inlet and outlet, the flow rate was optimized to achieve more uniform cooling [38]. It has been suggested that two-dimensional (2D) thermoluminescent PIV could be used in optical engines to record temperature and velocity fields at the same time [39].

The combination of PIV and detached eddy simulation (DES) has been used to deeply characterize the details of intake port flow, especially the flow features near the valves. The simulation and experimental measurements showed high agreement, indicating that the high shear rates introduced by the intake port design lead to high fluctuations and turbulence [40]. Additionally, the near-wall boundary layer flow in ICEs is highly unstable and significantly impacts the wall heat transfer process. A new flying PIV (F-PIV) technique has enabled the measurement of the near-wall velocity field, characterizing the evolution of near-wall flow. The experimental results revealed the complexity and instability of near-wall flow, especially the turbulence and vortex structures generated during piston motion [41]. There is also PIV evidence showing that turbulence in the near-wall region is anisotropic, suggesting that forced convection heat transfer is the primary heat transfer mechanism within the engine cylinder [42]. Furthermore, high-resolution PIV technology has been used to detail the characterization of in-cylinder boundary layer flow. In addition to typical turbulent boundary layers, the entrainment in the outer region of the wall jet causes the appearance of small-scale vortices [43].

High cycle-to-cycle variations (CCV) are typical characteristics during ICEs operation. Wall-resolved large eddy simulation (LES) can be employed to analyze large-scale flow structures, including vortex and tumble flows, and these were validated against experimental results from PIV. The analysis revealed that the periodic changes in these flow structures were the primary source of CCV [44]. PIV measurements inside an optical engine cylinder showed that closing the tumble flaps could significantly reduce flow CCV, increase the mean and turbulent kinetic energy near the spark plug position, and improve combustion stability [45]. S-PIV experiments in an optical engine investigated the impact of pressure differentials between intake valves on in-cylinder tumble motion. Post-processing using proper orthogonal decomposition (POD) identified the most energetic flow structures. Results indicated that while higher pressure differentials significantly increased in-cylinder turbulence levels, the primary flow structures were independent of pressure differential and mainly depended on valve lift [46]. Traditional time-averaged methods for representing velocity field ensembles can reduce the size of resultant vectors, affecting the accurate characterization of intake flow velocities. Recently, a dimension-reduction method called sparse-promoted dynamic mode decomposition (SP-DMD) was proposed to extract core structures from velocity field ensembles while preserving more realistic velocity vector magnitudes [47].

Holo-PIV measurements have demonstrated the potential for capturing non-reactive 3D flow inside engine cylinders. The measurement results confirmed the existence of counter-rotating vortex pairs and tumble flows below the intake valves [48]. T-PIV technology has also been employed to reconstruct the 3D flow field inside optical engines during intake and compression strokes, providing a detailed analysis of flow characteristics. The study revealed that in-cylinder flow characteristics are highly dependent on piston and spatial location. The flow is highly turbulent during the intake stroke, while it becomes more uniform during the compression stroke [49]. As shown in **Figure 13**, endoscopic PIV demonstrated good consistency with classical PIV when testing the in-cylinder flow field. Subsequently, using high-speed endoscopic PIV, spray visualization, and combustion imaging techniques, a comprehensive analysis of the in-cylinder flow field, spray, flame propagation, and soot formation process was conducted. The results indicated that the in-cylinder flow direction and flame propagation path are critical factors influencing soot formation, particularly the flow characteristics around the injector tip [50].

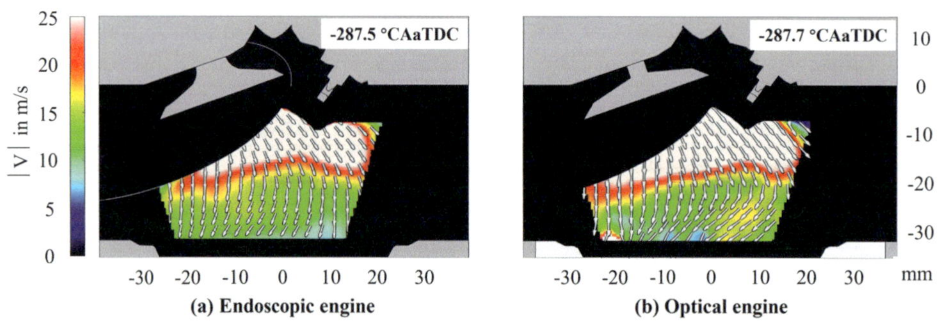

Figure 13.
Phase-averaged flow fields of at about −288°CAaTDC (top).

4. Challenges in PIV applications for ICEs research

The transition to carbon-neutral ICEs using zero-carbon fuels such as ammonia and hydrogen introduces a new set of challenges for the application of PIV. These challenges include the selection of tracer particles that can withstand high-temperature measurements and the complexity of flow structures during combustion. Addressing these issues is crucial for the accurate and reliable use of PIV in the optimized design of carbon-neutral ICEs.

4.1 Choice of tracer particles

As previously mentioned, ideal tracer particles must meet criteria for specific gravity, size, shape, and light scattering ability while also considering the potential contamination and damage to optical windows. The selection of tracer particles is equally crucial for the success of PIV measurements in carbon-neutral ICEs. Zero-carbon fuels typically have unique physical and chemical properties and combustion characteristics, which can affect tracer particles differently compared to traditional fuels. For example, hydrogen combustion produces high temperatures and water vapor, which can alter the behavior of ordinary tracer particles. Choosing tracer particles that remain stable and visible under these conditions is a significant challenge. In the future, there may be some specific challenges when selecting tracer particles as follows:

1. Particle Degradation: Many common tracer particles may degrade or evaporate at high temperatures, leading to signal loss and inaccurate measurements. For instance, organic particles may combust, and metal particles may oxidize or melt. Selecting high-temperature-resistant particles, such as ceramics or refractory metal oxides, can mitigate degradation issues. These particles can maintain their integrity and scattering properties at high temperatures.

2. Thermophoretic Effect: At high temperatures, particles may be influenced by thermophoretic forces, causing them to move toward cooler regions, which can result in flow deviations. To reduce the impact of thermophoretic forces on velocity measurements, researchers can use particles with high thermal inertia, such as larger or denser particles. Adjusting the timing and synchronization of the PIV system can also help mitigate these effects.

3. Light Scattering Efficiency: Light scattering efficiency may vary with temperature due to changes in particle size, refractive index, or surface characteristics. This can affect the quality of captured images and the accuracy of velocity calculations. Ensuring the selection of particles that do not change their light scattering characteristics in high-temperature environments is crucial.

4.2 Complex flow field in-cylinder

Currently, carbon-neutral ICEs using ammonia and hydrogen are still in the early stages of development, and the in-cylinder flow patterns are not yet well understood. The application of PIV can help elucidate the complex in-cylinder flow fields, but measuring the intricate flow structures in new combustion processes presents significant challenges.

1. Reactivity of Zero-Carbon Fuels: Fuels like hydrogen have high diffusivity and reactivity, leading to fast and highly turbulent combustion processes, whereas ammonia fuel exhibits slower dynamics. Capturing these extremely rapid or slow dynamics requires PIV systems with precise temporal resolution and advanced synchronization capabilities.

2. Combustion Byproducts: Zero-carbon fuels produce unique byproducts, such as water from hydrogen combustion or nitrogen oxides from ammonia combustion. These byproducts can interact with tracer particles, affecting their scattering properties and the accuracy of PIV measurements.

4.3 Implementation to address challenges

To overcome the challenges faced in PIV testing of carbon-neutral ICEs, the following strategies can be adopted:

1. Development of Specialized Tracer Particles: Research and develop new tracer particles specifically designed to withstand the unique conditions of zero-carbon fuel combustion. Meanwhile, these particles should be able to maintain consistent light scattering characteristics and resist chemical reactions with combustion byproducts.

2. Multimodal Measurement Approaches: Combine PIV with other diagnostic techniques to provide complementary information about the combustion process. For example, combining PIV with LIF can simultaneously map fuel concentration, temperature, and flow fields to study combustion processes. This multimodal approach allows for detailed analysis of complex interactions in multiphase and reactive flows. Future development of more integrated methods can provide a comprehensive understanding of flow dynamics and chemical reactions.

3. AI-enhanced Data Processing and Analysis: Employ advanced data processing algorithms and AI to improve the accuracy of velocity measurements in challenging environments. Machine learning techniques can identify and track flow structures and patterns that may be difficult to identify. They can also be used to correct data distortions or errors caused by high temperatures and combustion byproducts, ultimately achieving precise predictions of flow field structures.

By addressing these challenges and utilizing innovative solutions, integrated PIV technology can provide valuable data for developing efficient and clean combustion engine technologies, continuing to be a valuable tool in the development and optimization of carbon-neutral ICEs.

5. Conclusions

Particle image velocimetry (PIV) has become an indispensable tool in internal combustion engines (ICEs) research, especially in the period of transitioning to carbon-neutral fuels. This chapter explores various applications of PIV in ICEs research, detailing its role in fuel spray characterization, combustion analysis, and in-cylinder flow investigation. Additionally, this chapter discusses the specific challenges faced when applying PIV to carbon-neutral ICEs. These challenges include selecting appropriate tracer particles for high-temperature environments, ensuring accurate and reliable measurements, and addressing the complexity of flow structures during combustion, as well as the strategies and solutions to overcome these challenges. The ongoing evolution of PIV and its integration with advanced technologies will open new possibilities for optimizing engine performance, reducing emissions, and achieving a cleaner and more efficient future.

Acknowledgements

We would like to extend our sincere gratitude to Professor Hongliang Luo of Harbin Engineering University for his invaluable guidance during the writing of this chapter. This work was supported by the Natural Science Foundation of Jiangsu Province, China (No. BK20240859) and funded by the Shandong Postdoctoral Science Foundation (No. SDCX-ZG-202400247).

Conflict of interest

The authors declare no conflict of interest.

Author details

Gengxin Zhang[1], Penghua Shi[2] and Yu Jin[3*]

1 Department of Mechanical Engineering, School of Engineering, University of Birmingham, Birmingham, UK

2 Mechanical Engineering Program, Graduate School of Advanced Science and Engineering, Hiroshima University, Higashi-Hiroshima, Japan

3 Institute for Energy Research, Jiangsu University, Zhenjiang, P.R. China

*Address all correspondence to: jin_yu_1991@163.com

References

[1] Raffel M, Willert CE, Scarano F, Kähler CJ, Wereley ST, Kompenhans J. Particle Image Velocimetry: A Practical Guide. Cham: Springer; 2018

[2] Jena A, Singh AP, Agarwal AK. Challenges and opportunities of particle imaging velocimetry as a tool for internal combustion engine diagnostics. In: Singh AP, Agarwal AK, editors. Novel Internal Combustion Engine Technologies for Performance Improvement and Emission Reduction. Singapore: Springer Singapore; 2021. pp. 43-77

[3] Abdulmouti H. Particle imaging velocimetry (PIV) technique: Principles and applications, review. Yanbu Journal of Engineering and Science. 2021;**6**(1):35-65

[4] Baum E, Peterson B, Böhm B, Dreizler A. On the validation of LES applied to internal combustion engine flows: Part 1: Comprehensive experimental database. Flow Turbulence and Combustion. 2014;**92**(1-2):269-297

[5] Chen F, Liu H. Particle image velocimetry for combustion measurements: Applications and developments. Chinese Journal of Aeronautics. 2018;**31**(7):1407-1427

[6] Guo M, Shimasaki N, Nishida K, Ogata Y, Wada Y. Experimental study on fuel spray characteristics under atmospheric and pressurized cross-flow conditions. Fuel. 2016;**184**:846-855

[7] Xie Y, Yang J, Gu X. Flame wrinkling and self-disturbance in cellularly unstable hydrogen-air laminar flames. Combustion and Flame. 2024;**265**:113505

[8] Prasetya R, Sou A, Oki J, Nakashima A, Nishida K, Wada Y, et al. Three-dimensional flow structure and string cavitation in a fuel injector and their effects on discharged liquid jet. International Journal of Engine Research. 2019;**22**(1):243-256

[9] Zama Y, Ochiai W, Sugawara K, Furuhata T, Arai M. Study on mixing process of diesel spray under high ambient gas density condition. Atomization and Sprays. 2013;**23**(5):443-461

[10] Yamaguchi A, Koopmans L, Helmantel A, Dillner J, Dahlander P. Air motion induced by ultra-high injection pressure sprays for gasoline direct injection engines. SAE International Journal of Fuels and Lubricants. 2020;**13**(3):223-235

[11] Li T, Nishida K, Zhang YY, Onoe T, Hiroyau H. Enhancement of stratified charge for DISI engines through split injection (effect and its mechanism). JSME International Journal Series B-Fluids And Thermal Engineering. 2005;**48**(4):687-694

[12] Zhu JY, Nishida K, Kuti OA, Moon S. Quantitative analyses of fuel spray-ambient gas interaction by means of lif-piv technique. Atomization and Sprays. 2011;**21**(6):447-465

[13] Zhang M, Xu M, Zhang Y, Zhang G, Cleary DJ. Flow-field investigation of multi-hole superheated sprays using high-speed pint. Part i. Cross-sectional direction. Atomization and Sprays. 2012;**22**(11):983-995

[14] Zhang M, Xu M, Zhang YY, Zhang GM, Cleary DJ. Flow-field investigation of multihole superheated sprays using high-speed PIV. Part ii. Axial direction. Atomization and Sprays. 2013;**23**(2):119-140

[15] Sun Z, Wang HY, Cui ML, Nour M, Li XS, Xu M. Investigation of flash boiling injection schemes in lean-burn gasoline direct injection engines. Applications in Energy and Combustion Science. 2021;**7**:100035

[16] Peterson B, Sick V. Simultaneous flow field and fuel concentration imaging at 4.8 kHz in an operating engine. Applied Physics B-Lasers and Optics. 2009;**97**(4):887-895

[17] Peterson B, Baum E, Ding CP, Michaelis D, Dreizler A, Böhm B. Assessment and application of tomographic PIV for the spray-induced flow in an IC engine. Proceedings of the Combustion Institute. 2017;**36**(3):3467-3475

[18] Zhang M, Xu M, Hung DLS. Simultaneous two-phase flow measurement of spray mixing process by means of high-speed two-color PIV. Measurement Science and Technology. 2014;**25**(9):095204

[19] van Overbrueggen T, Braun M, Klaas M, Schroder W. Experimental analysis of the impact of injected biofuels on in-cylinder flow structures. SAE International Journal of Engines. 2016;**9**(2):1320-1348

[20] Zhu JY, Nishida K, Uemura T. Experimental study on flow fields of fuel droplets and ambient gas of diesel spray-free spray and flat-wall impinging spray. Atomization and Sprays. 2014;**24**(7):599-623

[21] Moon S, Matsumoto Y, Nishida K, Gao J. Gas entrainment characteristics of diesel spray injected by a group-hole nozzle. Fuel. 2010;**89**(11):3287-3299

[22] Wang XR, Li HM, Li GX, Gao Y, Huo HB, Zhang XQ, et al. Experimental investigation of atomization characteristics at low temperature environments during diesel engine cold start. Journal of the Energy Institute. 2024;**113**:101544

[23] Zama Y, Sugawara K, Akop MZ, Furuhata T, Arai M. Experimental study on velocity distribution of postimpingement diesel spray on a wall. Part 1: Effect of impingement angle on flow pattern. Atomization and Sprays. 2014;**24**(8):723-746

[24] Zama Y, Odawara Y, Furuhata T. Experimental investigation on velocity inside a diesel spray after impingement on a wall. Fuel. 2017;**203**:757-763

[25] Zama Y, Odawara Y, Furuhata T. Experimental study on velocity distribution of post-impingement diesel spray on a wall. Part 2: Effect of ambient gas density and injection pressure on flow pattern. Atomization and Sprays. 2016;**26**(9):921-938

[26] Shi P, Zhang G, Luo H, Ogata Y, Nishida K. Effect of wall-impingement distance on fuel adhesion characteristics of split injection spray under cross-flow condition. Fuel. 2024;**362**:130807

[27] Zhang G, Si Z, Zhai C, Luo H, Ogata Y, Nishida K. Characteristics of wall-jet vortex development during fuel spray impinging on flat-wall under cross-flow conditions. Fuel. 2022;**317**:123507

[28] Peterson B, Sick V. High-speed flow and fuel imaging study of available spark energy in a spray-guided direct-injection engine and implications on misfires. International Journal of Engine Research. 2010;**11**(5):313-329

[29] Peterson B, Reuss DL, Sick V. High-speed imaging analysis of misfires in a spray-guided direct injection engine. Proceedings of the Combustion Institute. 2011;**33**:3089-3096

[30] Shin JY, Lee S, Bae C, Jung J, Son Y, Heo H, et al. Effect of late intake valve closing on the particle number emissions of a turbocharged gasoline direct injection engine. Fuel Processing Technology. 2022;**230**:107209

[31] Padhiary A, Pilla G, Sotton J, Bellenoue M. Effect of pressure and turbulence intensity on the heat flux during flame wall interaction (FWI). Flow Turbulence and Combustion. 2023;**111**:1345-1370

[32] Jarvis S, Hargrave GK. A temporal PIV study of flame/obstacle generated vortex interactions within a semi-confined combustion chamber. Measurement Science and Technology. 2006;**17**(1):91

[33] Hong S, Speth RL, Shanbhogue SJ, Ghoniem AF. Examining flow-flame interaction and the characteristic stretch rate in vortex-driven combustion dynamics using PIV and numerical simulation. Combustion and Flame. 2013;**160**(8):1381-1397

[34] Kerl J, Lawn C, Beyrau F. Three-dimensional flame displacement speed and flame front curvature measurements using quad-plane PIV. Combustion and Flame. 2013;**160**(12):2757-2769

[35] Zhang G, Xu H, Wu D, Yang J, Morsy ME, Jangi M, et al. Deep learning-driven analysis for cellular structure characteristics of spherical premixed hydrogen-air flames. International Journal of Hydrogen Energy. 2024;**68**:63-73

[36] Zhang G, Xu H, Wu D, Yang J, Morsy ME, Jangi M, et al. Quantitative three-dimensional reconstruction of cellular flame area for spherical hydrogen-air flames. Fuel. 2024;**375**:132504

[37] Peterson B, Baum E, Böhm B, Sick V, Dreizler A. Spray-induced temperature stratification dynamics in a gasoline direct-injection engine. Proceedings of the Combustion Institute. 2015;**35**:2923-2931

[38] Qasemian A, Keshavarz A. Experimental and numerical study of an internal combustion engine coolant flow distribution. Tehnicki Vjesnik-Technical Gazette. 2016;**23**(1):257-264

[39] Kopf A, Frattina V, Bardi M, Endres T, Bruneaux G, Schulz C. In-cylinder thermographic PIV combined with phosphor thermometry using ZnO:Zn. International Journal of Engine Research. 2023;**24**(1):113-131

[40] Hartmann F, Buhl S, Gleiss F, Barth P, Schild M, Kaiser SA, et al. Spatially resolved experimental and numerical investigation of the flow through the intake port of an internal combustion engine. Oil and Gas Science and Technology-Revue D Ifp Energies Nouvelles. 2016;**71**(1):2015022

[41] Koehler M, Hess D, Brücker C. Flying PIV measurements in a 4-valve IC engine water analogue to characterize the near-wall flow evolution. Measurement Science and Technology. 2015;**26**(12):125302

[42] Fan XQ, Che ZZ, Wang TY, Lu Z. Numerical investigation of boundary layer flow and wall heat transfer in a gasoline direct-injection engine. International Journal of Heat and Mass Transfer. 2018;**120**:1189-1199

[43] Liu DM, Wang TY, Jia M, Li W, Lu Z, Zhen XD. Investigation of the boundary layer flow under engine-like conditions using particle image velocimetry. Journal of Engineering for Gas Turbines and Power-Transactions of The ASME. 2019;**141**(8):082801

[44] Wu SC, Patel SS, Ameen MM. Investigating the origins of cyclic

variability in internal combustion engines using wall-resolved large eddy simulations. Journal of Engineering for Gas Turbines and Power-Transactions of The ASME. 2023;**145**(5):1-27

[45] Tian FQ, Shi L, Che ZZ, Lu Z, Sun K, Wang TY. Experimental investigation of cyclic variation of the in-cylinder flow in a spark-ignition engine with a charge motion control valve. Flow Turbulence and Combustion. 2023;**111**(2):743-766

[46] El Adawy M, Heikal MR, Aziz ARA, Munir S, Siddiqui MI. Effect of boost pressure on the in-cylinder tumble-motion of gdi engine under steady-state conditions using stereoscopic-PIV. Journal of Applied Fluid Mechanics. 2018;**11**(3):733-742

[47] Baker SJ, Fang XH, Barbato A, Breda S, Magnani M, Fontanesi S, et al. Extracting vector magnitudes of dominant structures in a cyclic engine flow with dimensionality reduction. Physics of Fluids. 2024;**36**(2):025131

[48] van Overbrüggen T, Dannemann J, Klaas M, Schröder W. Holographic particle image velocimetry measurements in a four-valve combustion engine. Experiments in Fluids. 2014;**55**(1):1634

[49] Agarwal AK, Gadekar S, Singh AP. In-cylinder flow evolution using tomographic particle imaging velocimetry in an internal combustion engine. Journal of Energy Resources Technology-Transactions of The ASME. 2018;**140**(1):012207

[50] Fach C, Rödel N, Schorr J, Krüger C, Dreizler A, Böhm B. Multi-parameter imaging of in-cylinder processes during transient engine operation for the investigation of soot formation. International Journal of Engine Research. 2022;**23**(9):1573-1585